Methods in
RELATIVISTIC NUCLEAR PHYSICS

Methods in

RELATIVISTIC NUCLEAR PHYSICS

Michael DANOS

National Bureau of Standards, Washington, D.C.

Vincent GILLET *and* Monique CAUVIN

Commissariat à l'Energie Atomique, Centre d'Etudes Nucléaires de Saclay

1984

NORTH-HOLLAND

AMSTERDAM · OXFORD · NEW YORK · TOKYO

ISBN: 0 444 86317 6

Published by:

North-Holland Physics Publishing

a division of

Elsevier Science Publishers B.V.

P.O. Box 103
1000 AC Amsterdam
The Netherlands

Sole distributors for the U.S.A. and Canada:

Elsevier Science Publishing Company, Inc.

52 Vanderbilt Avenue
New York, N.Y. 10017
U.S.A.

Library of Congress Cataloging in Publication Data

Danos, Michael.
Methods in relativistic nuclear physics.

Bibliography: p.
Includes index.
1. Quantum field theory. 2. Many-body problems.
3. Electromagnetic interactions. I. Gillet, Vincent.
II. Cauvin, Monique, 1932– . III. Title. IV. Title:
Relativistic nuclear physics.
QC793.3.F5D35 1983 530.1′43 83-8068
ISBN 0-444-86317-6

Printed in the Netherlands

Preface

When Yukawa introduced the mesons in order to explain the short range character of the nuclear forces, the field of nuclear physics split into two parts: nuclear structure and nuclear forces. Nuclear structure developed into non-relativistic nuclear physics and led to the creation of various models to describe the emerging wealth of nuclear data. The field of nuclear forces developed into high-energy particle physics with its own immense body of phenomena and data.

For most nuclear phenomena the non-relativistic framework is fully adequate. However, this framework is too narrow in phenomena associated with exchange currents and with high-momentum transfers studied with intermediate energy nuclear accelerators. In such cases the presence of particles other than protons and neutrons in the nuclei must be explicitly accounted for. Since the concept of nuclear forces is a strictly non-relativistic construct, it must be abandoned and the forces must be replaced explicitly by their physical origin, i.e., by the interaction between baryons and mesons or, at a deeper level, between quarks and gluons.

When actually attempting to follow this program, namely the description of the nucleus as a system of interacting quantized fields, one is immediately confronted by the difficulties related to combining relativistic invariance with the fundamental characteristics of bound many-body systems such as: spatial localization, conservation and coupling of angular momenta, center-of-mass motion, etc. In order to handle these problems new tools and methods had to be devised. They are the subject of this book. Thus, in this work, the description of a bound many-body system has been formulated as a problem of relativistic quantum field theory which is solved by nuclear physics methods. Even though these tools and methods have been developed in the nuclear context, they are fully general and applicable to the description of any bound quantum systems composed both of relativistic and non-relativistic particles.

We are indebted to many colleagues for useful exchange of views on this new subject. In particular L.C. Biedenharn always has been helpful in elucidating subtle points of mathematical physics. Our special thanks go to R. Hayward for teaching us the treatment of higher spin fields and for the permission to use his material, some of which is still unpublished. We also thank T. Kohmura and T. Suzuki for useful comments and discussions, and for their contribution to the Appendix.

The authors are very grateful to Mrs. Paulette Gugenberger, head of the documentation group of the Department of Nuclear Physics of Saclay, for her dedicated assistance in preparing the manuscript. We owe a special gratitude to Mrs. Eliane Thureau, a member of this group, who performed miracles in typing the manuscript from the sometimes almost unintelligible handwritten pages provided her by the authors. Her patience and competence was essential in allowing us to keep within the Editor's schedule. We thank also Mrs. Marie-Odile Reuter for bringing her personal touch and esthetic feeling to the otherwise rather arid figures.

We have done our best to eliminate errors from the formulae. It is not likely that all have been found. Consequently our thanks also go to the Reader for finding and communicating them to us.

Michael Danos
National Bureau of Standards, Washington, D.C.

Vincent Gillet and Monique Cauvin
Commissariat à l'Energie Atomique, CEN Saclay

Table of Contents

CHAPTER 1

Introduction

1.1. SCOPE OF THE WORK

The present work is intended to provide the methods and tools for performing actual calculations for finite many-body systems of bound relativistic constituent particles. Of course, an extensive literature exists on the subject of the many-body problem. For finite systems all these treatments are given in a non-relativistic framework. Meanwhile the need has been growing for extending these treatments to relativistic dynamics, i.e. to quantum field theory. However, that subject is always treated in the covariant framework which is well suited for exhibiting the underlying principles and for the treatment of few-body reactions. It is not easily applicable to many-body bound states.

Our aim is to cover thoroughly the methodological aspects of the relativistic many-body problem for bound states while avoiding the presentation of specific models. The many examples contained in the later part of the work are meant to give concrete illustrations of how to actually apply the methods which are given in the first part. The choice of the applications was governed by the aim to provide specific examples for the problems which arise in the context of relativistic nuclear physics with obvious extension to sub-nuclear as well as to atomic physics. Thus we shall often name, but only for convenience, the diverse fields by the mesonic or nucleonic degrees of freedom they may represent. For example, the interaction between a vector field and a spinor field here is called the vector meson-nucleon interaction. However, the methods and results are fully general and they can, of course, be applied to other situations, for example, to the gluon-quark interaction upon the introduction of the needed SU(3) recoupling coefficients.

The difficulties which have hampered the development of relativistic calculations for bound finite many-body systems, besides the problem of divergences, were of several kinds. They belong to the following families: the relativistic center of mass, the discretization of fields, and the technical difficulties of relativistic angular momentum calculus. The first two difficulties are related to the lack of a central potential as it is provided in atomic physics by the Coulomb field of a massive non-relativistic source, the nucleus. Likewise in non-relativistic nuclear physics the shell-model potential is used to provide a discrete representation. Of course in a relativistic treatment no potential can be used as a starting point of the problem. Concerning the center-of-mass motion, its elimination is linked to the restoration of translational invariance and in relativistic kinematics this leads to non-trivial many-body operators. One conventional technique is that of the Bethe-Salpeter equation, which is hard to apply to systems containing more than two fermions. Finally, the handling of angular momentum is also a formidable stumbling block in relativistic calculations. In particular, relativistic angular momentum is frequently formulated in the helicity representation. Helicity is well suited for two-body scattering problems, but it does not lend itself simply to many-body angular momentum coupling.

This work addresses itself to providing the methods and tools for treating the above problems. In it the reader will find within the framework of quantum field theory:

(a) a representation of the fields in terms of multipoles and discretized states;

(b) a consistent angular momentum technique, which may as well be used in the non-relativistic case, but whose power makes it particularly appropriate in the more difficult relativistic case;

(c) the evaluation and a tabulation of the elementary vertices of the usual interactions between bosons and fermions;

(d) the rules for systematically generating relativistic many-body matrix elements, applied for illustration purposes to systems containing up to two fermions with a meson cloud of several pions and of one heavy vector or scalar meson;

(e) the treatment of the coupling of many-body systems to an external electromagnetic field, with explicit applications to the cases above;

(f) the method for performing relativistic boosts of the non-covariant wave functions;

(g) the method for removing the center-of-mass energy and for calculating the associated matrix elements of the center-of-mass many-body operator, and again a tabulation of matrix elements derived to provide the reader with concrete examples.

1.2. ORGANIZATION AND CONTENT OF THE CHAPTERS

The basic framework of our approach is the lagrangian field theory solved in the time-independent Schrödinger picture. The opening Chapter 2 recalls the main definitions and results of quantum field theory and the forms of the basic free field solutions for the cases of 0, $\frac{1}{2}$, 1 and higher spins. The general form of the simplest interaction terms between the basic fields is also presented as well as the minimal coupling, the anomalous moment coupling, the vector-dominance coupling with an external electromagnetic field. These interactions are those which most commonly appear in nuclear physics problems. The extension to other interactions follows along similar lines. Some examples of such extensions are discussed, namely interactions involving the spin-$\frac{3}{2}$ field and interactions between Yang-Mills fields allowing application to quantum chromodynamics.

The secular problem of field theory in the Schrödinger picture is treated in Chapter 3. To actually solve this problem one requires a discretized representation of the free fields. Here, a discretization cannot be performed by the introduction of an external central potential as it is done in non-relativistic shell-model calculations, or as it exists in the relativistic hydrogen atom. The usual shell-model potential is not a relativistic concept. In a potential a system is accelerated and its different parts experience different accelerations because of retardation effects. They arise as a consequence of the transformation of the time coordinate from the laboratory to the rest system. This leads to deformations of the system, i.e. to admixture of different eigenstates, for example baryon resonances if the system is a baryon. No such effect occurs in non-relativistic mechanics for a harmonic potential. There the system can be accelerated as a whole. Analogous difficulties beset the classical relativistic model of an extended electron. Therefore, the discretization here is carried out by a unitary transformation applied to the plane wave solutions of the free field equations of motion.

All quantities can be expressed in terms of the discretized fields. In particular, the solutions of the full hamiltonian are expanded on configurations made up of these discretized fields. Therefore the secular problem is now represented by a discrete matrix equation. These equations can be solved numerically after giving a suitable truncation prescription for the Hilbert space.

Of course the solutions of the resulting finite hamiltonian matrices are unitary and non-divergent. They are, however, functions both of the input parameters, i.e., masses and coupling constants, and of the size of the retained Hilbert space. Divergences appear in this approach as the size of the retained Hilbert space is increased. This point is also discussed in Chapter 3. There a comparison is made with the time-dependent perturbation method. Finally, Chapter 3 describes a method for treating the relativistic center-of-mass problem. This method consists in introducing an auxiliary harmonic central potential for the c.m. motion, in analogy with a well-known method of the non-relativistic shell model. However, because of the reasons discussed above special care must be exercised in applying this method in the relativistic context to avoid the mixing up of eigenstates when removing the c.m. motion.

Most of the technical tools are given in Chapter 4. They include a graphical method for angular momentum recoupling, phase conventions which are uniform for all tensorial quantities, extension of the tensorial formalism to the creation and annihilation operators in the Fock space, treatment of symmetrization and anti-symmetrization of coupled tensors, etc. These techniques are central for performing the involved calculations required for relativistic bound systems. They are employed in all the examples developed in this book. Of course, they are applicable to a much wider range of problems. For example, they can be used to greatly simplify the usual non-relativistic many-body shell-model calculations. We also show how they can be extended to the treatment of SU(n) tensors, for example SU(3) of chromodynamics.

The main reason for the new phase definitions introduced in Chapter 4 is to simplify phase factors in the expressions associated with angular momentum calculus, such as in the Wigner-Eckart theorem or in the reduction and recoupling of tensorial products. It is this property of the adopted phase convention which permits the introduction of a remarkably simple graphical method for handling angular momentum problems. Thus throughout the book diagrams will replace all intermediate algebraic calculational steps and will allow the reader to obtain directly from the graphs the final expressions given in the text. The many examples presented in this book will allow the reader to acquire familiarity with the graph method.

The phase convention for creation and annihilation operators has always been a delicate matter because of the interrelationship of time reversal and hermitian conjugation. This is easily taken care of within our adopted phase convention. Consequently the graphical technique is directly applicable also to field operators and state vectors. This way the explicit symmetrization and antisymmetrization of coupled fields is no more difficult than that of the usual *c*-number wave functions.

Chapter 5 contains the expressions for the discretized basis fields in invariant tensorial form for spins 0, $\frac{1}{2}$, 1 and $\frac{3}{2}$. It also briefly presents the definition of models. The discretization of the fields involves the expansion of the fields into multipoles of good total angular momentum and it is performed by introducing harmonic oscillator wave packets in momentum space. The derivation requires substantial angular momentum computation which is given in detail using the graphical method. These basis fields are used in later chapters for constructing the many-body configurations. As examples models for light hadron systems are given together with specific tables, listing the many-body configurational states consisting of one or two nucleons with a meson cloud of several pions and one heavy meson. Of course the domains of application are much wider than these examples. For example, the present methods can be used directly for the study of the effects of mesonic degrees of freedom in all the usual aspects of the microscopic theory of nuclei, such as Hartree-Fock or BCS descriptions, Tamm-Dancoff or random-phase approximations, particle-hole or quasi-particle configuration mixing models, etc. The extension to the quark-gluon picture has been discussed in Chapters 2 and 4 and is also immediate.

Chapter 6 contains the full expressions for the matrix elements of the elementary vertices of the interactions chosen as examples in Chapter 2, except for the electromagnetic interactions which are treated in Chapter 9. Again all derivations are given in full detail. These elementary matrix elements enter the many-body hamiltonian matrix which is described in Chapter 7. In that chapter, first a systematic method for constructing properly the many-body matrix elements is presented both in Fock space and Minkowski space. Then as an example, we give the hamiltonian matrix for systems consisting of pions, nucleons and heavy mesons covering the model spaces of Chapter 5.

The details concerning the treatment of the center-of-mass problem along the lines of Chapter 3 are given in Chapter 8. The difficulties of the many-body nature of the c.m. position operator become manifest in this chapter. Here again, as an example, we generate the expressions of the c.m. pseudo-hamiltonian for the model spaces of Chapter 5.

The many-body aspects of the interaction of a relativistic system with an external electromagnetic field are developed in Chapter 9. In particular, the electron elastic scattering form factor, the magnetic dipole and electric quadrupole moments are given in detail. As already mentioned, minimal coupling, vector dominance, as well as anomalous moment interactions are included.

1.3. REFERENCES

In this book we have tried to achieve a self-contained presentation in that all needed expressions are developed from stated basic definitions. Of course, our treatment builds on the fundamental work of field theory and group theory. A minimal familiarity with these subjects is required from the reader. In order to assist in finding the background material and elucidation on the deeper points, we here give a short selection of references from the literature.

Relativistic quantum field theory has been discussed in a large number of books, practically all in the framework of the covariant renormalization theory. They differ in their relative emphasis between the presentation of concepts and the development of techniques. We mention here specifically the work of Hayward on fields of arbitrary spin, which we adopt as the basis for the treatment of spin-1 and -$\frac{3}{2}$ fields. Hayward's formulation contains a minimal number of auxiliary (ghost) fields, to be determined by gauge conditions. In this way no supplementary conditions have to be imposed and no unphysical effects arise when introducing interactions as is the case, e.g., for the Rarita-Schwinger formulation.

The mathematical basis of the angular momentum treatment is found in many books. Most of them are limited to SU(2). We use in particular the notation of Fano and Racah. Some of the references include also the treatment of SU(n), which is required in the treatment of QCD; we refer in particular to Biedenharn and Lauck.

This book is an outgrowth of our previous publication, NBS Monograph 147. The present development is much more complete and supersedes that work.

BIEDENHARN, L.C. and J.D. LAUCK, 1981, *Angular momentum in quantum physics*, vol. 8 and 9, Encycl. of Math. and its applications, ed. G.-C. Rota (Addison-Wesley, Reading, Mass.).

BJORKEN J.D. and S.D. DRELL, 1964, *Relativistic quantum mechanics* (McGraw-Hill, New York).

BJORKEN, J.D. and S.D. DRELL, 1965, *Relativistic quantum fields* (McGraw-Hill, New York).

BOGOLIUBOV, N.N. and D.V. SHIRKOV, 1959, *Introduction to the theory of quantized fields* (Interscience Publishers, New York).

DANOS, M. and V. GILLET, 1975, *Relativistic many-body bound systems* (National Bureau of Standards Monograph 147, Washington D.C.).

EDMONDS, A.R., 1957, *Angular momentum in quantum mechanics* (Princeton University Press, Princeton N.J.).

FANO, U. and G. RACAH, 1959, *Irreducible tensorial sets* (Academic Press, New York).

HAYWARD, R.W., 1976, *The dynamics of fields of higher spins* (National Bureau of Standards, Monograph 154, Washington D.C.).

ITZYKSON, C. and J.B. ZUBER, 1980, *Quantum field theory* (McGraw-Hill, New York).

LEE, T.D., 1981, *Particle physics and introduction to field theory* (Harwood Academic Publishers, Char, London, New York).

LEITE LOPES, J., 1981, *Gauge field theories* (Pergamon, Oxford).

LURIE, D., 1968, *Particles and fields* (Interscience, New York).

NORMAND, J.M., 1980, *A Lie group: rotations in quantum mechanics* (North-Holland, Amsterdam).

ROMAN, P., 1969, *Introduction to quantum field theory* (Wiley, New York).

WENTZEL, G., 1943, *Einführung in die Quantentheorie der Wellenfelder* (Franz Deuticke, Wien).

CHAPTER 2

Fields and interactions

2.1. INTRODUCTION

The general framework of quantum field theory is briefly recalled. We give the basic definitions needed in the book and the main expressions for the energy and the interactions which will be evaluated later in detail. The treatment presented here is strictly that of quantized fields and not that of classical fields. Quantum fields arise from quantization of classical fields and have a much richer structure. However, we do not attempt a complete treatment of the underlying field theory which can be found in any standard textbook on this subject.

We treat first the spin-0, spin-$\frac{1}{2}$, spin-1 fields and some of their interactions. We also discuss their interactions with the electromagnetic field for the minimal, the anomalous moment, and the vector dominance couplings. For the higher spin fields we review the comprehensive formalism of Hayward and apply it specifically to the case of the spin-$\frac{3}{2}$ field. Finally, the extension to quantum chromodynamics is discussed.

The hamiltonian of the system is a functional of the basic free fields described in sect. 2.2. It is divided into two parts

$$H = H_0 + H_{\mathrm{I}},$$

where H_0 describes the free fields and H_{I} represents the various interactions between the free fields. The free hamiltonian H_0 describes non-interacting particles and is the sum of their free field energies given in sect. 2.3. The forms of the strong interactions contained in H_{I} are given in sect. 2.4. while the electromagnetic interactions are described in sect. 2.5. The treatment of higher spins is given in sect. 2.6., and the extension to Yang-Mills theories is sketched in sect. 2.7.

Throughout we work with the metric*

$$g_{\mu\nu} = \begin{pmatrix} 1 & 0 & 0 & 0 \\ 0 & 1 & 0 & 0 \\ 0 & 0 & 1 & 0 \\ 0 & 0 & 0 & 1 \end{pmatrix}.$$

* This metric is employed for example by Lee, Hayward, Lurie, and Wentzel. It is more convenient for the angular momentum calculus than the metric $g_{ii} = -1$, $g_{00} = 1$ found for example in Bjorken and Drell, Bogoliubov and Shirkov, or Roman.

Hence the four-vector scalar product ab will be denoted by

$$ab = \boldsymbol{a} \cdot \boldsymbol{b} + a_4 b_4 = \boldsymbol{a} \cdot \boldsymbol{b} - a_0 b_0$$

with $a_4 = ia_0$ and $b_4 = ib_0$. For example

$$(x_1, x_2, x_3, x_4) \equiv (x, y, z, it).$$

Furthermore we use the conventions that repeated greek indices, μ, $\nu \ldots$, are summed over $1, 2, 3, 4$ while the latin indices, i, $j \ldots$, are summed over $1, 2, 3$.

Also we use throughout units such that $\hbar = c = 1$.

2.2. FRAMEWORK OF FIELD THEORY

2.2.1. Free fields and state vectors

We denote by φ, ϕ, ψ, Φ the basic single-particle free field solutions. We use the notation φ for the spin-0 free field, ψ for the spin-$\frac{1}{2}$ field, ϕ for the spin-1 field and Φ for the spin-$\frac{3}{2}$ field.

The free fields are solutions of the free-particle equations of motion, namely the Klein-Gordon equation for spin 0, sect. 2.3.1, the Dirac equation for spin $\frac{1}{2}$, sect. 2.3.2, and the Hayward equations for higher spin, sects. 2.3.3 and 2.6.

For quantization the free fields are expanded into suitable orthonormal bases

$$\varphi = \sum\!\!\!\!\!\!\int_i a_i \varphi_i,$$

$$\psi = \sum\!\!\!\!\!\!\int_i b_i \psi_i,$$

$$\phi = \sum\!\!\!\!\!\!\int_i A_i \phi_i,$$

$$\Phi = \sum\!\!\!\!\!\!\int_i B_i \Phi_i.$$

Quantization is achieved by ascribing the meaning of annihilation operators to the amplitudes of these expansions which together with their hermitian conjugates are postulated to fulfill commutation relations for bosons and anticommutation relations for fermions:

$$[a_i, a_j^+]_- = \delta_{ij}, \qquad [a_i^+, a_j^+]_- = 0, \qquad [a_i, a_j]_- = 0,$$

$$[b_i, b_j^+]_+ = \delta_{ij}, \qquad [b_i^+, b_j^+]_+ = 0, \qquad [b_i, b_j]_+ = 0, \qquad \text{etc.}$$

The vacuum $|0\rangle$ is defined by

$$a_i|0\rangle = 0,$$

$$b_i|0\rangle = 0, \qquad \text{etc.}$$

The complete system of interacting particles is described by state vectors $|S_n\rangle$. They are given in terms of the free field creation operators acting on the vacuum

$$|S_n\rangle = \sum_\alpha X_\alpha^n \{a^+a^+ \dots A^+A^+ \dots b^+b^+ \dots\}_\alpha |0\rangle,$$

where the sum is over basis states denoted α, each with an amplitude X_α^n. The wave functions of the system are given by the matrix elements

$$\Psi_n(\boldsymbol{r}_1t_1, \boldsymbol{r}_2t_2 \dots) = \langle 0|\varphi(1)\varphi(2)\dots\phi(k+1)\phi(k+2)\dots\psi(m+1)\psi(m+2)\dots|S_n\rangle.$$

Setting all times $t_1 = t_2 = \cdots = t$, the state vectors fulfill in the Schrödinger picture the equations of motion

$$\frac{\partial}{\partial t}|S_n^{\mathrm{S}}\rangle = -iH|S_n^{\mathrm{S}}\rangle, \tag{2.1}$$

and in the interaction picture

$$\frac{\partial}{\partial t}|S_n^{\mathrm{I}}\rangle = -iV|S_n^{\mathrm{I}}\rangle, \tag{2.2}$$

where

$$V = \mathrm{e}^{iH_0t}H_{\mathrm{I}}\mathrm{e}^{-iH_0t}.$$

These equations are not manifestly relativistically covariant, since all particles are taken at a same time t, i.e., the particle position four-vectors $(\boldsymbol{r}_it_i)$ are replaced by the three-vectors $\boldsymbol{r}_i$. Of course this is of no consequence as far as the correctness of the computed values for observables and wave functions is concerned.

The observables of the system are obtained from the state vectors as

$$\mathcal{O}_{nn'} = \langle S_n|\mathcal{O}|S_{n'}\rangle,$$

where the operators $\mathcal{O}$ are given in the chosen picture.

The interaction picture is suited for time-dependent approximations based on the expansion of the solution in powers of the interaction. On the other hand the Schrödinger picture is well adapted to the treatment of stationary states and will be employed in this book.

Before continuing the development it may be worthwhile to recall at this point the difference between classical and quantum field theory. The physically acceptable solutions of classical field theory are single-valued functions of the Minkowski variables, be it in position or momentum space. The above function $\Psi_n(\boldsymbol{r}_i, t_i)$ is evidently much more complex. Thus, for example, a system with magnetic field $\boldsymbol{B}=0$ in classical field theory requires that the vector potential be $\boldsymbol{A}=0$. In quantum field theory $\boldsymbol{B}=0$ can be achieved in infinitely many ways, for example by the "two-photon state"

$$\Psi = \phi_1\phi_2 = \left(\boldsymbol{e}_x \mathrm{e}^{ipz_1}\mathrm{e}^{-iEt_1}\right)\left(\boldsymbol{e}_{-x}\mathrm{e}^{-ipz_2}\mathrm{e}^{-iEt_2}\right),$$

where the two photons have the polarization vectors $\boldsymbol{e}_x, \boldsymbol{e}_{-x}$ and the momenta $\boldsymbol{p}, -\boldsymbol{p}$, respectively.

2.2.2. *The time-dependent treatment in the interaction picture*

Before going into the discussion of the relativistic Schrödinger secular problem we briefly recall the usual perturbative treatment in the interaction picture which is particularly useful in the context of scattering problems.

The aim there is to construct an iterative procedure to express the solutions of eq. (2.2) as a power series in V. To this end one introduces the time evolution operator $U(t, t_0)$ by the implicit equation

$$|S_n^{\mathrm{I}}(t)\rangle = U(t, t_0)|S_n^{\mathrm{I}}(t_0)\rangle .$$

Substitution in eq. (2.2) yields the Tomonaga-Schwinger equation which in its integral form is

$$U(t, t_0) = 1 - i\int_{t_0}^{t} V(t')U(t', t_0)\,\mathrm{d}t' .$$

By iterating one obtains the series in powers of V

$$U(t, t_0) = 1 - i\int_{t_0}^{t}\mathrm{d}t_1 V(t_1) + (-i)^2\int_{t_0}^{t}\mathrm{d}t_1\int_{t_0}^{t_1}\mathrm{d}t_2 V(t_1)V(t_2)$$

$$+ (-i)^3\int_{t_0}^{t}\mathrm{d}t_1\int_{t_0}^{t_1}\mathrm{d}t_2\int_{t_0}^{t_2}\mathrm{d}t_3 V(t_1)V(t_2)V(t_3) + \cdots ,$$

where $t > t_1 > t_2 > t_3 \ldots$ This expansion is the starting point for the evaluation of the S-matrix or for the construction of the diverse Green functions in terms of Feynman graphs. These iterations restore the many-time character of the treatment lost in eq. (2.2).

2.2.3. *The time-independent treatment in the Schrödinger picture*

In this volume we work in the Schrödinger picture, eq. (2.1). In that picture the stationary state vectors are characterized by the simple factorized time dependence

$$|S_n^{\mathrm{S}}(t)\rangle = \mathrm{e}^{-iE_n t}|S_n^{\mathrm{S}}\rangle . \tag{2.3}$$

Inserting this form into the equation of motion (2.1) yields

$$(H - E_n)|S_n^{\mathrm{S}}\rangle = 0\,. \tag{2.4}$$

From now on we will omit the superscript S in the Schrödinger state vector.

The eigenenergies E_n are not directly the quantities of interest. They are related to the physical mass spectrum M_n of the system, which is the observable, by the relation

$$E_n - E_{\mathrm{v}} = \sqrt{P^2 + M_n^2}\,, \tag{2.5}$$

where E_{v} is the energy of the physical vacuum and P the total momentum of the system. The physical vacuum, in contrast to the mathematical vacuum $|0\rangle$, contains the "vacuum fluctuations." The extraction of the mass spectrum M_n is discussed in sect. 3.4.1.

In order to solve the secular problem (2.4), the time-independent state vectors $|S_n\rangle$ are expanded in terms of a complete orthonormal set of basis vectors $|\alpha\rangle$

$$|S_n\rangle = \sum_\alpha X_\alpha^n |\alpha\rangle\,,$$

where the configurations $|\alpha\rangle$ are products of the creation operators for the free fields acting on the vacuum

$$|\alpha\rangle = \{a_i^+ a_j^+ \ldots A_k^+ A_l^+ \ldots b_m^+ b_n^+ \ldots\}_\alpha |0\rangle\,.$$

The bracket $\{\ \}_\alpha$ denotes a configuration constructed so as to have the conserved quantum numbers characterizing the system, among which are the total angular momentum and isospin quantum numbers. The construction of the basis $|\alpha\rangle$ later on will of course be done in terms of a discretized representation for the free fields.

Substitution in eq. (2.4) yields the secular matrix equation for the eigenmode energies E_n and state amplitudes X_α^n

$$\sum_\beta H_{\alpha\beta} X_\beta^n = E_n X_\alpha^n\,,$$

$$H_{\alpha\beta} = \langle\alpha|H|\beta\rangle\,. \tag{2.6}$$

This set of equations is of a form which is reminiscent of the Schrödinger secular problem in non-relativistic quantum mechanics. However as we shall see in the next chapter, the meaning of these equations in field theory differs somewhat from that in the non-relativistic case.

The discretization of the free fields will be treated in Chapter 3 and given in full detail in Chapter 5 after acquiring in Chapter 4 the necessary tools for handling creation and annihilation operators in the angular momentum representation. Also the discussion of the stationary secular problem of field theory (2.6) and its comparison with the time-dependent perturbative formulation will be given in Chapter 3.

2.3. FREE FIELDS

2.3.1. Spin-0 fields

For non-interacting particles of spin 0, of mass M, the lagrangian in the Heisenberg picture is

$$\mathfrak{L} = -\tfrac{1}{2}\left(\partial_\mu\varphi\,\partial_\mu\varphi + M^2\varphi^2\right),$$

where the field φ is chosen real. Through variation with respect to φ one obtains the Euler-Lagrange equations of motion, here the Klein-Gordon equation

$$\partial_\mu\partial_\mu\varphi - M^2\varphi = 0.$$

The canonical momentum field is defined by

$$\pi = \frac{\partial\mathfrak{L}}{\partial\dot{\varphi}} = \dot{\varphi}. \tag{2.7}$$

The free field hamiltonian is

$$H_0 = \int \mathrm{d}^3r\left(\pi\dot{\varphi} - \mathfrak{L}\right) = \tfrac{1}{2}\int \mathrm{d}^3r\left(\pi^2 + (\nabla\varphi)^2 + M^2\varphi^2\right).$$

The plane wave solution of the free field equation of motion, including the isospin, is

$$\begin{aligned}\varphi(\boldsymbol{r},t) &= \left(\frac{1}{2\pi}\right)^{3/2}\sum_\kappa\int \mathrm{d}^3p\,\frac{1}{\sqrt{2E}}\left(a_{p\kappa}\eta_\kappa \mathrm{e}^{i(\boldsymbol{p}\cdot\boldsymbol{r}-Et)} + a^+_{p\kappa}\eta^+_\kappa \mathrm{e}^{-i(\boldsymbol{p}\cdot\boldsymbol{r}-Et)}\right),\\ \pi(\boldsymbol{r},t) &= -i\left(\frac{1}{2\pi}\right)^{3/2}\sum_\kappa\int \mathrm{d}^3p\,\sqrt{\tfrac{1}{2}E}\left(a_{p\kappa}\eta_\kappa \mathrm{e}^{i(\boldsymbol{p}\cdot\boldsymbol{r}-Et)} - a^+_{p\kappa}\eta^+_\kappa \mathrm{e}^{-i(\boldsymbol{p}\cdot\boldsymbol{r}-Et)}\right).\end{aligned} \tag{2.8}$$

Here the index κ denotes the isospin projection in cartesian representation. For the pion field the isospin functions η_κ are the cartesian unit vectors in isospin space and hence $\eta^+_\kappa = \eta_\kappa$. They fulfill the orthonormality relations

$$\langle\eta_\kappa|\eta_{\kappa'}\rangle = \delta_{\kappa\kappa'}.$$

The boson creation and annihilation operators $a^+_{p\kappa}$, $a_{p\kappa}$ are defined for a plane wave state of linear momentum $\boldsymbol{p}$, of energy E,

$$E^2 = p^2 + M^2.$$

For the σ-meson (charge 0) the η_κ are omitted together with the index κ.

The boson operators in eq. (2.8) fulfill the commutation relations

$$\left[a_{p\kappa}, a^{+}_{p'\kappa'}\right]_{-} = \delta_{\kappa\kappa'}\delta^{3}(\boldsymbol{p}-\boldsymbol{p}'). \tag{2.9}$$

The normalization in (2.8) is chosen to achieve the equal time commutator ($t' = t$) for the free fields

$$[\varphi(\boldsymbol{r}, t), \pi(\boldsymbol{r}', t)]_{-} = i(2T+1)\delta^{3}(\boldsymbol{r}-\boldsymbol{r}'). \tag{2.10}$$

The factor $(2T+1)$ arises from the summation over the isospin index κ.

2.3.2. Spin-$\frac{1}{2}$ fields

We now consider the field ψ for spin-$\frac{1}{2}$ particles. The free field lagrangian is

$$\mathcal{L} = -\left(\bar{\psi}\gamma_{\mu}\partial_{\mu}\psi + M\bar{\psi}\psi\right),$$

with

$$\bar{\psi} = \psi^{+}\gamma_4.$$

We choose the following representation for the 4×4 γ_μ matrices

$$\gamma_i = \begin{pmatrix} 0 & -i\sigma_i \\ i\sigma_i & 0 \end{pmatrix}, \qquad \gamma_4 = \begin{pmatrix} I & 0 \\ 0 & -I \end{pmatrix},$$

with the Pauli matrices defined as

$$\sigma_1 = \begin{pmatrix} 0 & 1 \\ 1 & 0 \end{pmatrix}, \quad \sigma_2 = \begin{pmatrix} 0 & -i \\ i & 0 \end{pmatrix}, \quad \sigma_3 = \begin{pmatrix} 1 & 0 \\ 0 & -1 \end{pmatrix}, \quad I = \begin{pmatrix} 1 & 0 \\ 0 & 1 \end{pmatrix}.$$

The γ_μ matrices verify the relations

$$\gamma_\mu\gamma_\nu + \gamma_\nu\gamma_\mu = 2\delta_{\mu\nu}, \qquad \gamma_\mu^2 = 1.$$

We shall also need the matrix γ_5,

$$\gamma_5 = \gamma_1\gamma_2\gamma_3\gamma_4 = \begin{pmatrix} 0 & -I \\ -I & 0 \end{pmatrix},$$

which verifies the anticommutators

$$\gamma_5\gamma_\nu + \gamma_\nu\gamma_5 = 0.$$

The Pauli matrices σ_i fulfill

$$\sigma_i\sigma_j = i\varepsilon_{ijk}\sigma_k,$$

where ε_{ijk} is the fully antisymmetric unit tensor which vanishes unless all indices are different, and is equal to ± 1 according to the parity of the permutation $\begin{pmatrix} 1 & 2 & 3 \\ i & j & k \end{pmatrix}$. Finally, we introduce the 4×4 spin matrices $S_{\mu\nu}$ and $\sigma_{\mu\nu}$

$$S_{\mu\nu} = -\tfrac{1}{4} i [\gamma_\mu, \gamma_\nu]_- ,$$

$$\sigma_{\mu\nu} = 2 S_{\mu\nu} .$$

We have

$$S_{jk} = \tfrac{1}{4}\varepsilon_{ijk}\gamma_j\gamma_k = \tfrac{1}{2}\varepsilon_{ijk}\begin{pmatrix} \sigma_i & 0 \\ 0 & \sigma_i \end{pmatrix} = \varepsilon_{ijk} S_i .$$

They verify the commutation relations of angular momentum operators

$$[S_i, S_j]_- = i\varepsilon_{ijk} S_k .$$

We also have

$$S_{k4} = -S_{4k} = \frac{1}{2}\begin{pmatrix} 0 & \sigma_k \\ \sigma_k & 0 \end{pmatrix} .$$

A general Lorentz transformation with infinitesimal angles $\varepsilon_{\mu\nu}$ is of the form

$$L = 1 + \tfrac{1}{2} i \varepsilon_{\mu\nu} J_{\mu\nu} ,$$

with

$$J_{\mu\nu} = L_{\mu\nu} + S_{\mu\nu} .$$

Here $L_{\mu\nu} = -i(x_\mu \partial_\nu - x_\nu \partial_\mu)$ acts on the coordinates and $S_{\mu\nu}$ on the spinors.
The canonical conjugate field is

$$\pi = i(\bar{\psi}\gamma_4) = i\psi^+ .$$

The hamiltonian can now be written as

$$H_0 = \int d^3 r (\pi\dot{\psi} - \mathfrak{L}) = \int d^3 r (\bar{\psi}\gamma_i \partial_i \psi + M\bar{\psi}\psi) .$$

The field ψ is a solution of the equation of motion, i.e., the Dirac equation

$$\gamma_\mu \partial_\mu \psi + M\psi = 0 .$$

We now introduce the functions $\mathcal{U}_\sigma(\boldsymbol{p})$ and $\mathcal{V}_\sigma(\boldsymbol{p})$ where the index σ denotes the spin projection, $\sigma = \pm\frac{1}{2}$, and which are solutions of

$$(i\gamma_\mu p_\mu + M)\mathcal{U}_\sigma(\boldsymbol{p}) = 0 ,$$

$$(-i\gamma_\mu p_\mu + M)\mathcal{V}_\sigma(\boldsymbol{p}) = 0 .$$

They are four-component spinors of the form

$$\mathcal{U}_\sigma(\boldsymbol{p})=\sqrt{\frac{E+M}{2M}}\begin{pmatrix}\chi_\sigma\\ \dfrac{\boldsymbol{\sigma}\cdot\boldsymbol{p}}{E+M}\chi_\sigma\end{pmatrix}, \tag{2.11}$$

$$\mathcal{V}_\sigma(\boldsymbol{p})=\sqrt{\frac{E+M}{2M}}\begin{pmatrix}\dfrac{\boldsymbol{\sigma}\cdot\boldsymbol{p}}{E+M}\chi_{-\sigma}\\ \chi_{-\sigma}\end{pmatrix}, \tag{2.12}$$

where the spin functions are

$$\chi_{1/2}=\begin{pmatrix}1\\0\end{pmatrix},\qquad \chi_{-1/2}=\begin{pmatrix}0\\1\end{pmatrix}.$$

Note in (2.12) the usual change of the sign of the spin magnetic quantum number for the antiparticle.

We also introduce the isospin wave functions η_τ. For the nucleon they are the orthonormalized two-component functions corresponding to the two charge states

$$\eta_{1/2}=\begin{pmatrix}1\\0\end{pmatrix},\qquad \eta_{-1/2}=\begin{pmatrix}0\\1\end{pmatrix},$$

with

$$\langle\eta_\tau|\eta_{\tau'}\rangle=\delta_{\tau\tau'}.$$

Thus the creation and annihilation operators for particles of energy $E=\sqrt{p^2+M^2}$, momentum $\boldsymbol{p}$, spin and isospin projections σ and τ are denoted respectively by $b^+_{\boldsymbol{p}\sigma\tau}$, $b_{\boldsymbol{p}\sigma\tau}$. In addition we must define creation and annihilation operators for antiparticles. They are denoted by $d^+_{\boldsymbol{p}\sigma\tau}$, $d_{\boldsymbol{p}\sigma\tau}$, where the quantum numbers have the same meaning as those of the particles.

With these definitions, the plane wave expansion of the solution of the Dirac equation is

$$\psi(\boldsymbol{r},t)=\left(\frac{1}{2\pi}\right)^{3/2}\sum_{\sigma\tau}\int \mathrm{d}^3p\sqrt{\frac{M}{E}}\Big(b_{\boldsymbol{p}\sigma\tau}\mathcal{U}_\sigma(\boldsymbol{p})\eta_\tau \mathrm{e}^{i(\boldsymbol{p}\cdot\boldsymbol{r}-Et)}$$

$$+d^+_{\boldsymbol{p}\sigma\tau}\mathcal{V}_\sigma(\boldsymbol{p})\eta_\tau \mathrm{e}^{-i(\boldsymbol{p}\cdot\boldsymbol{r}-Et)}\Big). \tag{2.13}$$

Note that the spin and isospin wave functions in the antiparticle part of the field (2.12) are not transposed. The transposed isospin wave function for spin 0, eq. (2.8), and later on for spin 1, arises from writing the creation part as the hermitian conjugate of the annihilation, thus making the fields real. Here the transposed spin and isospin functions χ^+ and η^+ will appear in the hermitian conjugate ψ^+ and also in $\bar{\psi}=\psi^+\gamma_4$.

With the chosen normalization the spinors obey the following relations (α is the spinor index): orthogonality,

$$\sum_{\alpha} \mathfrak{U}^{+}_{\sigma\alpha} \mathfrak{U}_{\sigma'\alpha} = \frac{E}{M} \delta_{\sigma\sigma'},$$

$$\sum_{\alpha} \mathfrak{V}^{+}_{\sigma\alpha} \mathfrak{V}_{\sigma'\alpha} = \frac{E}{M} \delta_{\sigma\sigma'};$$

completeness,

$$\sum_{\sigma} \left(\mathfrak{U}^{+}_{\sigma\alpha} \mathfrak{U}_{\sigma\alpha'} - \mathfrak{V}^{+}_{\sigma\alpha} \mathfrak{V}_{\sigma\alpha'} \right) = (\gamma_4)_{\alpha\alpha'} = \begin{cases} \delta_{\alpha\alpha'} & \text{for } \alpha = 1,2, \\ -\delta_{\alpha\alpha'} & \text{for } \alpha = 3,4. \end{cases}$$

It follows from these relations that the normalized field ψ, eq. (2.13), and its canonical conjugate field π fulfill the equal time anticommutation relations

$$\left[\psi_{\alpha}(\boldsymbol{r}, t), \pi_{\beta}(\boldsymbol{r}', t) \right]_{+} = i(2T+1)(2S+1) \delta_{\alpha\beta} \delta^3(\boldsymbol{r} - \boldsymbol{r}'), \tag{2.14}$$

where the factor $(2T+1)(2S+1)$ comes from the summation over the charge and the spin states.

2.3.3. *Spin-1 fields*

We now consider the field ϕ for particles of spin 1. As one can see from the fact that component by component there are 8 Maxwell equations (see eqs. (2.20) for $M = 0$), the general treatment requires an 8-component framework. This will be discussed in more generality in sect. 2.6 for higher spin fields. Here we use the known case of the spin-1 field to introduce the higher spin field description of Hayward. In this description the lagrangian is more general than the usual one written in terms of the antisymmetric field strength tensor $F_{\mu\nu}$, which contains a gauge freedom only for the case $M = 0$. Here the lagrangian has a gauge freedom also for $M \neq 0$. We shall choose the Lorentz gauge, eq. (2.17), in order to simplify the equations. As a consequence the Hayward lagrangian reduces to the usual form in terms of $F_{\mu\nu}$ and the equations of motion yield for $M = 0$ the well-known form of the Maxwell equations.

The spin-1 field is described by spinors with 8 components. The upper and lower components have opposite parity, i.e. they have the transformation properties of a polar four-vector, V, and of an axial four-vector, A,

$$\phi = \begin{pmatrix} \phi_1 \\ \cdot \\ \cdot \\ \phi_4 \\ \phi_5 \\ \cdot \\ \cdot \\ \phi_8 \end{pmatrix} = \begin{pmatrix} V_x \\ \cdot \\ \cdot \\ iV_0 \\ iA_x \\ \cdot \\ \cdot \\ A_0 \end{pmatrix}. \tag{2.15}$$

In other words, under space reflection and rotation, the four parts of the field $\boldsymbol{V}, V_0, \boldsymbol{A}, A_0$ which are chosen real transform respectively as a vector, a scalar, a pseudo-vector and a pseudo-scalar. This separation will later on allow to carry out conveniently angular momentum coupling calculations.

The lagrangian for the free field ϕ in the Heisenberg picture is

$$\mathfrak{L} = -\tfrac{1}{2}\left(\bar{\phi}\gamma_\mu \overleftarrow{\partial}_\mu \gamma_\nu \overrightarrow{\partial}_\nu \phi + M^2 \bar{\phi}\phi\right),$$

where

$$\bar{\phi} = \phi^+ \gamma_4,$$

and where the arrows indicate on which field the operator acts. In contrast with Hayward, we use real instead of complex potential fields, i.e. $\boldsymbol{V}^* = \boldsymbol{V}$, $V_0^* = V_0$, $\boldsymbol{A}^* = \boldsymbol{A}$, $A_0^* = A$. This requires the factor $\frac{1}{2}$ in the definition of the lagrangian.

Here the 8×8 matrices γ_μ are given by

$$\gamma_1 = \left(\begin{array}{cccc|cccc} \cdot & \cdot & \cdot & -1 & \cdot & \cdot & \cdot & \cdot \\ \cdot & \cdot & \cdot & \cdot & \cdot & \cdot & -1 & \cdot \\ \cdot & \cdot & \cdot & \cdot & \cdot & 1 & \cdot & \cdot \\ -1 & \cdot & \cdot & \cdot & \cdot & \cdot & \cdot & \cdot \\ \hline \cdot & \cdot & \cdot & \cdot & \cdot & \cdot & \cdot & 1 \\ \cdot & \cdot & 1 & \cdot & \cdot & \cdot & \cdot & \cdot \\ \cdot & -1 & \cdot & \cdot & \cdot & \cdot & \cdot & \cdot \\ \cdot & \cdot & \cdot & \cdot & 1 & \cdot & \cdot & \cdot \end{array}\right),$$

$$\gamma_2 = \left(\begin{array}{cccc|cccc} \cdot & \cdot & \cdot & \cdot & \cdot & \cdot & 1 & \cdot \\ \cdot & \cdot & \cdot & -1 & \cdot & \cdot & \cdot & \cdot \\ \cdot & \cdot & \cdot & \cdot & -1 & \cdot & \cdot & \cdot \\ \cdot & -1 & \cdot & \cdot & \cdot & \cdot & \cdot & \cdot \\ \hline \cdot & \cdot & -1 & \cdot & \cdot & \cdot & \cdot & \cdot \\ \cdot & \cdot & \cdot & \cdot & \cdot & \cdot & \cdot & 1 \\ 1 & \cdot & \cdot & \cdot & \cdot & \cdot & \cdot & \cdot \\ \cdot & \cdot & \cdot & \cdot & \cdot & 1 & \cdot & \cdot \end{array}\right),$$

$$\gamma_3 = \left(\begin{array}{cccc|cccc} \cdot & \cdot & \cdot & \cdot & \cdot & -1 & \cdot & \cdot \\ \cdot & \cdot & \cdot & \cdot & 1 & \cdot & \cdot & \cdot \\ \cdot & \cdot & \cdot & -1 & \cdot & \cdot & \cdot & \cdot \\ \cdot & \cdot & -1 & \cdot & \cdot & \cdot & \cdot & \cdot \\ \hline \cdot & 1 & \cdot & \cdot & \cdot & \cdot & \cdot & \cdot \\ -1 & \cdot & \cdot & \cdot & \cdot & \cdot & \cdot & \cdot \\ \cdot & \cdot & \cdot & \cdot & \cdot & \cdot & \cdot & 1 \\ \cdot & \cdot & \cdot & \cdot & \cdot & \cdot & 1 & \cdot \end{array}\right),$$

$$\gamma_4 = \left(\begin{array}{cccc|cccc} 1 & \cdot & \cdot & \cdot & \cdot & \cdot & \cdot & \cdot \\ \cdot & 1 & \cdot & \cdot & \cdot & \cdot & \cdot & \cdot \\ \cdot & \cdot & 1 & \cdot & \cdot & \cdot & \cdot & \cdot \\ \cdot & \cdot & \cdot & -1 & \cdot & \cdot & \cdot & \cdot \\ \hline \cdot & \cdot & \cdot & \cdot & -1 & \cdot & \cdot & \cdot \\ \cdot & \cdot & \cdot & \cdot & \cdot & -1 & \cdot & \cdot \\ \cdot & \cdot & \cdot & \cdot & \cdot & \cdot & -1 & \cdot \\ \cdot & \cdot & \cdot & \cdot & \cdot & \cdot & \cdot & 1 \end{array}\right).$$

We shall also need the γ_5 matrix,

$$\gamma_5 = \gamma_1\gamma_2\gamma_3\gamma_4 = \begin{pmatrix} 0 & -I \\ -I & 0 \end{pmatrix}.$$

They fulfill the relations

$$\gamma_\mu\gamma_\nu + \gamma_\nu\gamma_\mu = 2\delta_{\mu\nu}, \qquad \gamma_5\gamma_\mu + \gamma_\mu\gamma_5 = 0, \qquad \gamma_\mu^2 = \gamma_5^2 = 1.$$

Using the above definitions we get for the matrix $\gamma_\mu \partial_\mu$ the following detailed expression

$$\gamma_\mu \partial_\mu = \left(\begin{array}{ccc|c|ccc|c} \partial_4 & \cdot & \cdot & -\partial_1 & \cdot & -\partial_3 & \partial_2 & \cdot \\ \cdot & \partial_4 & \cdot & -\partial_2 & \partial_3 & \cdot & -\partial_1 & \cdot \\ \cdot & \cdot & \partial_4 & -\partial_3 & -\partial_2 & \partial_1 & \cdot & \cdot \\ \hline -\partial_1 & -\partial_2 & -\partial_3 & -\partial_4 & \cdot & \cdot & \cdot & \cdot \\ \hline \cdot & \partial_3 & -\partial_2 & \cdot & -\partial_4 & \cdot & \cdot & \partial_1 \\ -\partial_3 & \cdot & \partial_1 & \cdot & \cdot & -\partial_4 & \cdot & \partial_2 \\ \partial_2 & -\partial_1 & \cdot & \cdot & \cdot & \cdot & -\partial_4 & \partial_3 \\ \hline \cdot & \cdot & \cdot & \cdot & \partial_1 & \partial_2 & \partial_3 & \partial_4 \end{array}\right).$$

When $\gamma_\mu \partial_\mu$ acts on an 8-component vector we get

$$\gamma_\mu \partial_\mu \begin{pmatrix} \boldsymbol{F} \\ F_4 \\ \boldsymbol{G} \\ G_4 \end{pmatrix} = \begin{pmatrix} \partial_4 \boldsymbol{F} - \operatorname{grad} F_4 + \operatorname{rot} \boldsymbol{G} \\ -\operatorname{div} \boldsymbol{F} - \partial_4 F_4 \\ -\operatorname{rot} \boldsymbol{F} + \operatorname{grad} G_4 - \partial_4 \boldsymbol{G} \\ \operatorname{div} \boldsymbol{G} + \partial_4 G_4 \end{pmatrix}. \tag{2.16}$$

Since we are concerned with fields of definite parity and with parity conserving interactions, we now drop the axial four-vector part of the field ϕ, eq. (2.15), i.e., we put the four last components $\phi_5 \ldots \phi_8$ equal to zero,

$$\phi = \begin{pmatrix} \boldsymbol{\phi} \\ \phi_4 \\ 0 \\ 0 \end{pmatrix}.$$

This simplication is not permissible if the system contains magnetic monopoles since then $\operatorname{div} \boldsymbol{B} = 0$, eq. (2.20), is not fulfilled.

The Hayward lagrangian is invariant under the gauge transformation

$$\phi \to \phi + \gamma_\mu \partial_\mu \Lambda,$$

where the scalar field Λ obeys the Klein-Gordon equation with mass M. We use this gauge freedom to work in the Lorentz gauge. This gauge fixes the time-like component of the field in terms of the space-like components

$$\operatorname{div} \boldsymbol{\phi} = -\partial_4 \phi_4. \tag{2.17}$$

With these assumptions we note that the operator $\gamma_\mu \partial_\mu$ applied to the four-vector field generates a six-vector quantity where we recognize the definition of the Maxwell electric and magnetic fields,

$$\gamma_\mu \partial_\mu \begin{pmatrix} \boldsymbol{\phi} \\ \phi_4 \\ 0 \\ 0 \end{pmatrix} = \begin{pmatrix} \partial_4 \boldsymbol{\phi} - \operatorname{grad} \phi_4 \\ 0 \\ -\operatorname{rot} \boldsymbol{\phi} \\ 0 \end{pmatrix} = \begin{pmatrix} i\boldsymbol{E} \\ 0 \\ -\boldsymbol{B} \\ 0 \end{pmatrix}. \tag{2.18}$$

We see that in the case $M \neq 0$, as in the Maxwell theory, ϕ represents the "potential" fields and $\gamma_\mu \partial_\mu \phi$ the "strength" fields.

The canonical momentum field is

$$\bar{\pi} = \frac{\partial \mathcal{L}}{\partial \dot{\phi}} = i \begin{pmatrix} \partial_4 \boldsymbol{\phi} - \operatorname{grad} \phi_4 \\ \operatorname{div} \boldsymbol{\phi} + \partial_4 \phi_4 \\ 0 \\ 0 \end{pmatrix} = \begin{pmatrix} \bar{\boldsymbol{\pi}} \\ \bar{\pi}_4 \\ 0 \\ 0 \end{pmatrix}. \tag{2.19}$$

and

$$\pi = \frac{\partial \mathcal{L}}{\partial \dot{\bar{\phi}}} = \bar{\pi}^+ \gamma_4.$$

Here we recognize the relation between the canonical field and the electric field $\boldsymbol{E}$ as given in (2.18)

$$\bar{\pi} = -\boldsymbol{E}.$$

In the Lorentz gauge (2.17)

$$\bar{\pi}_4 = 0.$$

Hence, together with the reality character of ϕ we have

$$\bar{\pi} = \pi^{\mathrm{T}}, \qquad \bar{\pi}^{+} = \bar{\pi}^{\mathrm{T}}.$$

We also can rewrite the lagrangian in terms of the $\boldsymbol{E}$ and $\boldsymbol{B}$ fields given in eq. (2.18). This way we obtain, using (2.17)

$$\mathcal{L} = -\tfrac{1}{2}\left((\partial_4\boldsymbol{\phi})^2 - 2\partial_4\boldsymbol{\phi}\,\mathrm{grad}\,\phi_4 + (\mathrm{grad}\,\phi_4)^2 + (\mathrm{rot}\,\boldsymbol{\phi})^2 + M^2\bar{\phi}\phi\right)$$
$$= \tfrac{1}{2}\left(\boldsymbol{E}^2 - \boldsymbol{B}^2 - M^2\bar{\phi}\phi\right).$$

Indeed, for the case of vanishing mass this lagrangian is that of classical electrodynamics.

The equations of motion derived from the above lagrangian by variation with respect to $\bar{\phi}$ are

$$\gamma_\mu\partial_\mu\gamma_\nu\partial_\nu\phi - M^2\phi = 0,$$

which writes as

$$\gamma_\mu\partial_\mu\begin{pmatrix} i\boldsymbol{E} \\ 0 \\ -\boldsymbol{B} \\ 0 \end{pmatrix} = \begin{pmatrix} \partial_t\boldsymbol{E} - \mathrm{rot}\,\boldsymbol{B} \\ -i\,\mathrm{div}\,\boldsymbol{E} \\ -i\,\mathrm{rot}\,\boldsymbol{E} - i\partial_t\boldsymbol{B} \\ -\mathrm{div}\,\boldsymbol{B} \end{pmatrix} = M^2\begin{pmatrix} \boldsymbol{\phi} \\ \phi_4 \\ 0 \\ 0 \end{pmatrix}. \tag{2.20}$$

We recognize for the case $M = 0$ the Maxwell equations in the absence of external electromagnetic sources.

The free field hamiltonian is, with the help of eq. (2.20) after integrating by parts for treating the term $M^2\bar{\phi}\phi$,

$$H_0 = \int \mathrm{d}^3r\,(\pi\dot{\phi} - \mathcal{L}) = \int \mathrm{d}^3r\,\tfrac{1}{2}\left(-(\partial_4\boldsymbol{\phi})^2 + (\mathrm{grad}\,\phi_4)^2 + (\mathrm{rot}\,\boldsymbol{\phi})^2 + M^2\bar{\phi}\phi\right)$$
$$= \int \mathrm{d}^3r\,\tfrac{1}{2}\left(\boldsymbol{E}^2 + \boldsymbol{B}^2 + M^2\boldsymbol{\phi}^+\boldsymbol{\phi} - M^2\phi_4^+\phi_4\right),$$

which again is the expression of classical electrodynamics for $M = 0$. For $M \neq 0$, H_0 is still a positive definite quantity.

The free field plane wave solutions are thus, for the three (space-like) vector components ϕ_1, ϕ_2, ϕ_3 of the field, introducing the isospin projection index k,

$$\boldsymbol{\phi}(\boldsymbol{r}, t) = \left(\frac{1}{2\pi}\right)^{3/2} \sum_k \int \mathrm{d}^3 p \, \frac{N}{\sqrt{2E}} \left(A_{\boldsymbol{p}\boldsymbol{d}k} \boldsymbol{d} \eta_k \mathrm{e}^{i(\boldsymbol{p}\cdot\boldsymbol{r} - Et)} + A^+_{\boldsymbol{p}\boldsymbol{d}k} \boldsymbol{d} \eta^+_k \mathrm{e}^{-i(\boldsymbol{p}\cdot\boldsymbol{r} - Et)} \right), \tag{2.21}$$

where $A_{\boldsymbol{p}\boldsymbol{d}k}$ and $A^+_{\boldsymbol{p}\boldsymbol{d}k}$ are respectively the annihilation or creation boson operators for a spin-1 particle of linear momentum $\boldsymbol{p}$ and polarization $\boldsymbol{d}$. Here η_k are the cartesian isospin unit vectors for the isovector fields, and $\eta = 1$ for the isoscalar fields.

The (time-like) scalar component $\phi_4(\boldsymbol{r}, t)$ is related to the vector components $\boldsymbol{\phi}$ by the Lorentz condition (2.17),

$$\phi_4(\boldsymbol{r}, t) = -i \int \mathrm{d}t \, \mathrm{div} \, \boldsymbol{\phi}(\boldsymbol{r}, t),$$

which entails, in agreement with eq. (2.15)

$$\phi_4^+(\boldsymbol{r}, t) = -\phi_4(\boldsymbol{r}, t).$$

The normalization constant N in (2.21) is determined by the equal time commutation property which must be fulfilled by the field $\boldsymbol{\phi}$ and its conjugate $\boldsymbol{\pi}$

$$[\boldsymbol{\phi}(\boldsymbol{r}, t), \boldsymbol{\pi}(\boldsymbol{r}, t)]_- = i(2S+1)(2T+1)\delta^3(\boldsymbol{r} - \boldsymbol{r}'),$$

where the factors come from the summation over the three spin states ($S = 1$), two transverse and one longitudinal, and over the isospin states ($T = 0$ for the ω meson, $T = 1$ for the ρ meson). The condition on the field commutators to be unity for each mode yields for the normalization in eq. (2.21) as will be shown in Chapter 5

$$N = 1, \qquad \text{for the transverse modes},$$

$$N = E/M, \qquad \text{for the longitudinal mode}.$$

For the massless case, $M = 0$, the longitudinal mode is absent and hence the factor $2S + 1 = 3$ in the commutator relation is replaced by 2 and $N = 1$. The massless case is not an analytic limit of the massive case, i.e. it cannot be achieved as the limit $M \to 0$. This can be seen from the normalization of the longitudinal mode above. The special treatment for $M = 0$ is sketched in Appendix A.1. for QED. In short, in the massless case, the free field solutions, or in the terminology of mathematics, the solutions of the homogeneous equation, do not form a complete set owing to the absence of the longitudinal modes. These modes arise in the massless case only as the solution to the inhomogeneous equation, i.e. in the presence of sources. Therefore when dealing with massless vector fields, such as photons, the Coulomb part of the field solution of the inhomogeneous equation must be added to the treatment, for example by adding explicitly to the hamiltonian a Coulomb interaction between the charged particles, as discussed in Appendix A1.

2.4. HADRONIC INTERACTIONS

The interaction hamiltonian H_I contains hadronic and electromagnetic interactions. We collect in this section a list of the simplest hadronic interactions between nucleons, pions, scalar or vector mesons. Their forms are obtained by imposing all the known invariances of strong interactions to the simplest hermitian Lorentz-invariant products of the fields and if necessary of their derivatives. The electromagnetic interactions are treated later. The expressions are defined so that the coupling constants are dimensionless.

All interactions are written as normal products of the quantized fields, denoted as usual by $:\varphi\varphi\ldots:$. The normal products are defined by the prescription that in each term the operators must be re-ordered such that all annihilators are on the right, and that no contractions are to be performed between these operators.

The pion-nucleon interaction consists of a pseudo-scalar term

$$H_{\pi N}^{PS} = ig_{\pi N}^{PS} \int d^3r : \bar{\psi}\gamma_5 \boldsymbol{\tau}\psi\boldsymbol{\varphi}:, \tag{2.22}$$

and a pseudo-vector term, where M is the nucleon mass,

$$H_{\pi N}^{PV} = -ig_{\pi N}^{PV} \frac{1}{2M} \int d^3r : \bar{\psi}\gamma_\mu\gamma_5 \boldsymbol{\tau}\psi\, \partial_\mu \boldsymbol{\varphi}:, \tag{2.23}$$

where for Lorentz invariance all the fields and their derivatives under the integration symbol are to be taken at the same space-time point (point-vertex interaction). The relative sign between the pseudo-scalar and pseudo-vector terms is arbitrarily chosen so that for small momenta the two expressions after integration by parts coincide. The intrinsic parity of the pion field $\boldsymbol{\varphi}$ is odd and parity conservation is insured by γ_5. Charge symmetry is obtained by coupling the integrand into a scalar quantity in isospin space. Furthermore, the combination of the isospin operator $\boldsymbol{\tau}$ with the isospin wave functions of the fields into an isospin rotationally invariant product insures charge conservation, as will be shown later in sect. 6.3.1. When working in an angular momentum representation these interactions are likewise coupled into a quantity invariant under space rotation to insure angular momentum conservation. Finally, the phase factor i is introduced for hermiticity. For example,

$$(i\bar{\psi}\gamma_5\boldsymbol{\tau}\psi\boldsymbol{\varphi})^+ = -i\boldsymbol{\varphi}^+\psi^+\boldsymbol{\tau}\gamma_5\gamma_4\psi = i\psi^+\gamma_4\boldsymbol{\tau}\gamma_5\psi\boldsymbol{\varphi} = i\bar{\psi}\gamma_5\boldsymbol{\tau}\psi\boldsymbol{\varphi},$$

owing to the commutation properties of the quantities involved and the reality of the boson field.

The interaction between the neutral scalar meson σ and the nucleon is, from similar considerations,

$$H_{\sigma N} = g_{\sigma N} \int d^3r : \bar{\psi}\psi\varphi^{(\sigma)} :, \tag{2.24}$$

where $\varphi^{(\sigma)}$ is real.

The vector interactions between the vector meson and the nucleon are, for the neutral ω-meson

$$H_{\omega N}^{V} = -ig_{\omega N}^{V} \int d^3r : \bar{\psi}\gamma_\mu\psi\phi_\mu^{(\omega)} :, \tag{2.25}$$

and for the charged ρ-meson

$$H_{\rho N}^{V} = -ig_{\rho N}^{V} \int d^3r : \bar{\psi}\gamma_\mu\tau\psi\phi_\mu^{(\rho)} : .$$

Here again the phase factor i insures hermiticity of the expression. Recall that these mesons are vector fields and not pseudo-vector fields. Hence, they have intrinsic negative parity.

The tensor interaction between the vector meson and the nucleon is, for the ω-meson, where M is the nucleon mass

$$H_{\omega N}^{T} = -g_{\omega N}^{T} \frac{1}{2M} \int d^3r : \bar{\psi}\sigma_{\nu\mu}\left(\partial_\nu\phi_\mu^{(\omega)} - \partial_\mu\phi_\nu^{(\omega)}\right)\psi :, \tag{2.26}$$

and a similar expression for the ρ-meson.

The interaction between pions requires a four-field vertex for parity conservation, coupled into an isospin scalar for charge symmetry and charge conservation

$$H_\pi = g_\pi \int d^3r : \varphi\varphi\varphi\varphi : . \tag{2.27}$$

The interaction of the σ-meson with the pions likewise is

$$H_{\sigma\pi} = g_{\sigma\pi}(2M_\sigma) \int d^3r : \varphi^{(\sigma)}\varphi\varphi : .$$

The interaction of the ρ-meson with the pions requires two pion fields since the $\mathcal{G}$-parity of the ρ-meson is even and that of the pion is odd. Thus,

$$H_{\rho\pi} = -ig_{\rho\pi} \int d^3r : \phi_\mu^{(\rho)}\left(\varphi\overleftrightarrow{\partial}_\mu\tau\varphi\right) : . \tag{2.28}$$

As in the case of the pion-nucleon interaction this form insures charge conservation since the operator is a scalar in isospin space. The symmetrized derivative $\overleftrightarrow{\partial}_\mu = (\overrightarrow{\partial}_\mu - \overleftarrow{\partial}_\mu)$ generates a conserved four-vector current for the spin-0 field. The quantity $\phi_\mu\overleftrightarrow{\partial}_\mu$ is a proper scalar since ϕ_μ is a vector.

The interaction of the ω-meson with the pions is more complicated. The ω-meson has odd $\mathcal{G}$-parity and thus it requires a vertex with three pion fields. The simplest form incorporating all the necessary invariances is

$$H_{\omega\pi} = ig_{\omega\pi}\left(\frac{1}{2M_\omega}\right)^2 \int \mathrm{d}^3r : \left(\varepsilon_{ijk}\phi_i^{(\omega)}\partial_j\partial_k\right)(\varphi\varphi\varphi) : . \tag{2.29}$$

Here a projection is performed on total pion isospin 0 in the three-dimensional isospin space. The symmetry properties of the symbol $\varepsilon_{\lambda\mu\nu}$ are defined as follows. The product of the three pion fields has negative intrinsic parity. Thus the tensorial product $\varepsilon_{\lambda\mu\nu}\phi_\lambda\partial_\mu\partial_\nu$ must be a pseudo-scalar quantity. This can be achieved only by a quantity of the form $\boldsymbol{A}\cdot(\boldsymbol{B}\times\boldsymbol{C})$. Hence λ, μ, $\nu \neq 4$ is required. Likewise, the charge antisymmetrization arising from projection on $T=0$ in the second parentheses generates a state odd under charge conjugation (pseudo-scalar in isospin space), which fits the charge conjugation properties of the ω-meson. The differential operators $\partial_i\partial_j$ act on the pion fields according to

$$\begin{aligned}\partial_i\partial_j\varphi(1)\varphi(2)\varphi(3) = {} & (\partial_i\varphi(1))(\partial_j\varphi(2))\varphi(3) \\ & - (\partial_i\varphi(1))\varphi(2)(\partial_j\varphi(3)) + \varphi(1)(\partial_i\varphi(2))(\partial_j\varphi(3)),\end{aligned} \tag{2.30}$$

so as to yield a non-vanishing result. This would arise if the expression were symmetrized in the indices i, j, since the product of the pion fields has mixed symmetry in space and isospace.

2.5. ELECTROMAGNETIC INTERACTIONS

2.5.1. The vector current interaction

We restrict ourselves here to those terms which one obtains by the minimal substitution in the free field lagrangian. Of course, a large number of terms arises from the minimal substitution in the interactions containing derivatives. These can be obtained in a straightforward manner and treated along the lines of the examples we have retained. The anomalous moment interaction with the nucleon is described in sect. 2.5.4.

The minimal coupling prescription consists in the replacement $\boldsymbol{p} \to \boldsymbol{p} - e\boldsymbol{A}$ in the lagrangian and in ascribing no anomalous moment to the particle fields.

For photon emission or absorption processes the electromagnetic interaction is chosen linear in the photon field A_μ,

$$H_{\mathrm{EM}} = -\int \mathrm{d}^3r : J_\mu A_\mu : .$$

The currents J_μ for the interacting charged particles are of the following forms. In the case of real spin-0 fields

$$J_\mu = \tfrac{1}{2} ie\varphi\tau_z \overleftrightarrow{\partial}_\mu \varphi \tag{2.31}$$

and of spin-$\frac{1}{2}$ fields

$$J_\mu = ie\bar{\psi}\gamma_\mu \tfrac{1}{2}(1+\tau_z)\psi. \tag{2.32}$$

For spin-1 fields we have

$$J_\mu = -\tfrac{1}{2}ie\left(\bar{\phi}\gamma_\mu\gamma_4\tau_z\bar{\pi} - \pi\gamma_4\gamma_\mu\tau_z\phi\right). \tag{2.33}$$

In all three forms the isospin operator τ_z measures the charge, and insures that neutral particles have no interaction.

Apart from the interaction hamiltonian H_{EM}, another quantity of interest is the form factor $G_\mu(\boldsymbol{q})$ related to the electron scattering cross section at momentum transfer $\boldsymbol{q}$. Its expression is

$$G_\mu(\boldsymbol{q}) = \int \mathrm{d}^3 r : J_\mu(\boldsymbol{r})\mathrm{e}^{i\boldsymbol{q}\cdot\boldsymbol{r}} : .$$

2.5.2. *The long wavelength limit*

The long wavelength limit is of special importance for static properties. It also does apply to transition processes as long as the momentum transfer is small enough so that $qR \ll 1$, where R is the radius of the interaction region.

The interaction with real photons in the transverse gauge

$$A_4 = 0$$

reduces to

$$H_{\text{EM}} = -\int \mathrm{d}^3 r \boldsymbol{J}\cdot\boldsymbol{A}.$$

Taking the absorption part, the electric field in the transverse gauge is,

$$\boldsymbol{E} = -\dot{\boldsymbol{A}}^{(-)} = i\omega \boldsymbol{A}^{(-)}$$

and the interaction can be expressed in terms of this field $\boldsymbol{E}$,

$$H_{\text{EM}} = \frac{i}{\omega}\int \mathrm{d}^3 r \boldsymbol{J}\cdot\boldsymbol{E}.$$

In the long wavelength limit, for a given multipolarity L in the multipole expansion of the electric field $\boldsymbol{E}$, the magnetic multipole $\boldsymbol{E}_{\mathfrak{M}}(L, qr)$ behaves as $(qr)^L$ while the electric multipole $\boldsymbol{E}_{\mathcal{E}}(L, qr)$ behaves as $(qr)^{L-1}$.

We first consider electric multipolarities for which the long wavelength limit yields the well-known Siegert theorem. There the general form for the electric field of multipole L is

$$\boldsymbol{E} = -\nabla V_L + (\text{higher order terms in powers of } qr),$$

with

$$V_L(\boldsymbol{r}) = a_L Y_{L0}(\hat{\boldsymbol{r}})(qr)^L. \tag{2.34}$$

Herewith, integrating by parts

$$H_{\mathrm{EM}} = -\frac{i}{\omega}\int \mathrm{d}^3 r \boldsymbol{J}\cdot\nabla V_L = -\frac{i}{\omega}\int \mathrm{d}^3 r (\operatorname{div}\boldsymbol{J}) V_L .$$

Using the current conservation relation between the current and the charge density ρ,

$$\operatorname{div}\boldsymbol{J} = -\dot{\rho},$$

this expression yields a real term. Hence together with the identical emission term we obtain

$$H_{\mathrm{EM}} = \int \mathrm{d}^3 r \rho V_L . \tag{2.35}$$

This is the Siegert theorem.

The charge density $\rho = -iJ_4$ is given in terms of the charged particle fields by the expressions (2.31), (2.32) and (2.33). The electric multipoles of the electric field are given by the general form (2.34). For example, for the quadrupole moment, this form is

$$Q = V_2 = 2z^2 - x^2 - y^2 .$$

Consider now magnetic multipolarities. We limit ourselves to the important case of the magnetic dipole interaction. Again we look for a relation analogous to the Siegert theorem. Namely, from

$$\boldsymbol{B} = \operatorname{rot}\boldsymbol{A},$$

by expanding $\boldsymbol{B}$ around $r = 0$, in the limit $q \to 0$, we get the identity

$$\boldsymbol{A} = \tfrac{1}{2}\boldsymbol{r}\times\boldsymbol{B}_0 + \boldsymbol{q}\times\boldsymbol{r}f(qr),$$

where $\boldsymbol{B}_0$ is the magnetic field at $r = 0$. Herewith we have for the interaction with a spinless particle, dropping the subscript 0 from $\boldsymbol{B}_0$, and dropping the higher order terms in qr,

$$H_{\mathrm{EM}} = -\int \mathrm{d}^3 r \boldsymbol{J}\cdot\boldsymbol{A} = -\frac{e}{M}\langle \boldsymbol{p}\cdot\boldsymbol{A}\rangle = \frac{-e}{2M}\langle \boldsymbol{p}\cdot(\boldsymbol{r}\times\boldsymbol{B})\rangle$$

$$= \frac{e}{2M}\langle(\boldsymbol{r}\times\boldsymbol{p})\cdot\boldsymbol{B}\rangle = \frac{e\boldsymbol{B}}{2M}\cdot\langle \boldsymbol{l}\rangle,$$

where $\langle \boldsymbol{l}\rangle$ means the expectation value of the orbital angular momentum of the system. Thus we have

$$H_{\text{EM}} = \frac{e}{2M} \boldsymbol{B} \cdot \langle \boldsymbol{l} \rangle = \boldsymbol{B} \cdot \boldsymbol{\mu},$$

which defines the magnetic moment $\boldsymbol{\mu}$. The expectation value $\langle \boldsymbol{l} \rangle$ is to be evaluated by integration over a normalized quantity which has the sign of the charge, i.e., the charge density. Hence the magnetic moment is

$$\boldsymbol{\mu} = -\frac{i}{2M} \int \mathrm{d}^3 r J_4 \boldsymbol{l}. \tag{2.36}$$

This expression is valid for spin-0 fields. More generally, the full magnetic operator is of the form $\boldsymbol{\mu} = \mu_0(\boldsymbol{l} + \boldsymbol{\sigma})$ where $\boldsymbol{\sigma}$ is the spin operator of the field and $\mu_0 = e/2M$. This yields in the case of spin-1 fields of total angular momentum $\boldsymbol{j} = \boldsymbol{l} + \boldsymbol{s}$

$$\boldsymbol{\mu} = -\frac{i}{2M} \int \mathrm{d}^3 r J_4 \boldsymbol{j}. \tag{2.37}$$

In the case of spin-$\frac{1}{2}$ fields, still including no anomalous part ($\boldsymbol{\sigma} = 2\boldsymbol{s}$)

$$\boldsymbol{\mu} = -\frac{i}{2M} \int \mathrm{d}^3 r J_4 (\boldsymbol{j} + \boldsymbol{s}). \tag{2.38}$$

In fact, the fields used in the actual calculations will not be strictly free fields but are partly dressed. In the case of non-zero spin one should therefore allow for an anomalous (effective) magnetic intrinsic moment. This effect modifies the magnetic dipole coupling constant $\mu_0 = e/2M$. This correction is particularly important in the case of the neutron and proton for which we use modified forms introducing Landé factors g. For the proton

$$\boldsymbol{\mu} = -\frac{i}{2M} \int \mathrm{d}^3 r J_4 (\boldsymbol{j} + (g_{\mathrm{P}} - 1)\boldsymbol{s}), \tag{2.39}$$

and for the neutron

$$\boldsymbol{\mu} = -\frac{i}{2M} g_{\mathrm{n}} \int \mathrm{d}^3 r J_4^{(\mathrm{n})} \boldsymbol{s}, \tag{2.40}$$

where the neutron current $J_4^{(\mathrm{n})}$ is

$$J_4^{(\mathrm{n})} = ie \bar{\psi} \gamma_4 \tfrac{1}{2}(1 - \tau_z) \psi.$$

The anomalous moment interaction can also be derived from the Pauli term $S_{\mu\nu} F_{\mu\nu}$, which will be treated below in sect. 2.5.4 for the general case, $q \neq 0$.

2.5.3. *The vector dominance interaction*

In addition to the current interaction, a neutral vector meson (the ω-meson or the ρ_0 meson) may convert into a photon.

The simplest gauge invariant Lorentz scalar which is linear in both the photon and vector meson fields is

$$L_{\mathrm{VD}} = g_{\mathrm{VD}} \int \mathrm{d}^4 x\, F_{\mu\nu}(x) \phi_{\mu\nu}(x),$$

where g_{VD} is a coupling constant and where in terms of the photon vector potential $A_\mu(x)$ and of the field $\phi_\mu(x)$ of the neutral vector meson we have

$$F_{\mu\nu}(x) = \partial_\mu A_\nu(x) - \partial_\nu A_\mu(x),$$

$$\phi_{\mu\nu}(x) = \partial_\mu \phi_\nu(x) - \partial_\nu \phi_\mu(x).$$

Since we use the Lorentz gauge for the vector meson field,

$$\partial_\mu \phi_\mu = 0,$$

the interaction can be simplified by integrating by parts

$$L_{\mathrm{VD}} = -g_{\mathrm{VD}} \int \mathrm{d}^4 x \left(A_\nu \partial_\mu \phi_{\mu\nu} - A_\mu \partial_\nu \phi_{\mu\nu} \right)$$

$$= -2 g_{\mathrm{VD}} \int \mathrm{d}^4 x\, A_\nu \left(\partial_\mu \partial_\mu \phi_\nu - \partial_\mu \partial_\nu \phi_\mu \right)$$

$$= -2 g_{\mathrm{VD}} M_{\mathrm{V}}^2 \int \mathrm{d}^4 x\, A_\nu(x) \phi_\nu(x).$$

In the first term of the second line we have used the equation of motion for spin 1, eq. (2.20), to replace the d'alembertian by the meson mass, while the second term vanishes owing to the Lorentz condition. The replacement of the d'alembertian by the vector meson mass M_{V} here is correct since the basic field ϕ used in our expansions obeys the free field equation of motion.

Thus the vector dominance interaction hamiltonian for the neutral vector field is of the form

$$H_{\mathrm{VD}} = g_{\mathrm{VD}} \left(2 M_{\mathrm{V}}^2 \right) \int \mathrm{d}^3 r\, A_\mu(\boldsymbol{r}) \phi_\mu(\boldsymbol{r}), \tag{2.41}$$

where the coupling constant g_{VD} may be different for the ρ_0 or ω interactions.

2.5.4. *The anomalous moment interaction*

In addition to the vector current interaction obtained by minimal substitution in sect. 2.5.1., there exists also the Pauli tensor moment interaction. It is needed in particular for the description of the anomalous magnetic moment. We already have introduced the anomalous magnetic moment interaction for the nucleon in the long wavelength limit in sect. 2.5.2.

This interaction arises when the basic fields are assumed to be partially dressed. It exists for all particles, both charged and neutral, which have spin greater than zero. The general form of this interaction is for all fields, where M is the particle mass,

$$H_a = g_a \frac{1}{2M} \int d^3r \, T_{\mu\nu} F_{\mu\nu} ,$$

where g_a is a coupling constant, $F_{\mu\nu}$ is the field strength tensor and where the spin tensor is, for example for the spin-$\frac{1}{2}$ field,

$$T_{\mu\nu} = \bar{\psi} S_{\mu\nu} C\psi = -\tfrac{1}{4} i \bar{\psi} [\gamma_\mu, \gamma_\nu]_- C\psi .$$

In this last expression C is an appropriate charge conserving isospin operator. In Chapter 9 we shall treat the cases of the anomalous moment interaction of the nucleon and of the nucleon-delta photoproduction.

2.6. HIGHER SPIN FIELDS

2.6.1. General framework

We have surveyed the familiar treatments of the spin-0 and spin-$\frac{1}{2}$ fields and of their interactions. The spin-1 field formalism in sect. 2.3.3 has already introduced the reader to the Hayward formulation. We now extend this formulation to the description of higher spin fields and sketch its general features. In fact, the developments follow the lines of sect. 2.3.3 and very few new aspects arise. As a concrete example, we shall emphasize the important case of the spin-$\frac{3}{2}$ field and of its interactions.

Let us consider a field of spin $S > \frac{1}{2}$. In a non-relativistic framework its description requires $2S+1$ components. In the relativistic case additional components are required to account for (i) antiparticles, (ii) the existence of two kind of fields, the potential fields and the strength fields, (iii) admixtures by a general Lorentz transformation of spin $(S-1)$ fields. This way the number of field components is

$$n = 2 \times 2 \times ((2S+1) + (2S-1)) = 16\, S .$$

For example for the case $S = 1$ treated in sect. 2.3.3 we had a sixteen-component field which we had broken into two 8-component fields, viz. the potential fields (V, A) and the strength fields (E, B), eqs. (2.15) and (2.18). For integer spin fields the particle-antiparticle symmetry allowed us to set some of the components equal to zero as it appeared in (2.18). Furthermore a gauge freedom associated with the presence of a $S = 0$ admixture, namely the scalar time-like component ϕ_4, enabled us to introduce a gauge condition between the field components, here the Lorentz gauge of eq. (2.17).

In the arbitrary spin-S case, in complete analogy with the $S = 1$ case above, the spin $(S-1)$ solutions have indefinite norm, i.e. they are ghost fields and they are associated with a gauge freedom. Also, one has $2S-1$ gauge conditions at one's disposal which may be used, for example, to eliminate the spin $S-1$ components. This elimination is not Lorentz invariant, as examplified by the transverse gauge in electrodynamics. Other gauges, like the covariant Lorentz gauge (2.17), may be used to fix, but of course not to eliminate, the ghost fields.

The free spin-S field Ψ obeys component by component the Klein-Gordon equation

$$\left(\partial_\mu \partial_\mu - M^2\right)\Psi_i = 0 .$$

As in the Dirac formalism the number of components introduced above allows to linearize this equation by means of anticommuting matrices Γ_μ

$$(\Gamma_\mu \partial_\mu + M)\Psi = 0 . \tag{2.42}$$

Writing the field Ψ in terms of two $8S$-component fields Φ and η

$$\Psi = \begin{pmatrix} \Phi \\ \eta \end{pmatrix},$$

the matrices Γ_μ can be chosen of the purely off-diagonal form

$$\Gamma_\mu = \begin{pmatrix} 0 & \gamma_\mu \\ \gamma_\mu & 0 \end{pmatrix},$$

where the γ_μ are $8S \times 8S$ anticommuting matrices obeying the usual Dirac algebra

$$[\gamma_\mu, \gamma_\nu]_+ = 2\delta_{\mu\nu} . \tag{2.43}$$

The Dirac-like equation (2.42) can then be written as the coupled set of equations

$$\gamma_\mu \partial_\mu \Phi = -M\eta , \qquad \gamma_\mu \partial_\mu \eta = -M\Phi . \tag{2.44}$$

The components of Φ play the role of the potential fields (2.15) of the spin-1 case. Introducing the field Θ, the components of which were the strength fields in (2.16) and (2.18),

$$\Theta = -M\eta = \gamma_\mu \partial_\mu \Phi , \tag{2.45}$$

we have

$$\gamma_\mu \partial_\mu \Theta = M^2 \Phi . \tag{2.46}$$

We obtain by substitution the free Klein-Gordon equation for the field Φ, as in (2.20),

$$\left(\gamma_\mu \partial_\mu \gamma_\nu \partial_\nu - M^2\right)\Phi = 0 .$$

The field Θ fulfills the same Klein-Gordon equation.

When introducing the interaction with an electromagnetic field A_μ by the minimal substitution as in sect. 2.5, $\partial_\mu \to \partial_\mu - ieA_\mu$, (2.44) and (2.46) read

$$\gamma_\mu(\partial_\mu - ieA_\mu)\Phi = \Theta,$$

$$\gamma_\mu(\partial_\mu - ieA_\mu)\Theta = M^2\Phi.$$

Eliminating Θ we obtain

$$\left(-(\partial_\mu - ieA_\mu)^2 + ie\sigma_{\mu\nu}F_{\mu\nu} + M^2\right)\Phi = 0, \tag{2.47}$$

with

$$2i\sigma_{\mu\nu} = [\gamma_\mu, \gamma_\nu]_-, \tag{2.48}$$

A similar equation is obeyed by the field Θ. From eq. (2.47) we see that Φ (and Θ) are not Klein-Gordon fields, i.e. fields which obey upon minimal substitution the full Klein-Gordon equation, like the pion field φ

$$\left(-(\partial_\mu - ieA_\mu)^2 + M^2\right)\varphi = 0.$$

The above equations of motion can be derived from the Hayward lagrangian. Employing complex fields this lagrangian is

$$\mathfrak{L} = \bar{\Theta}\Theta - M^2\bar{\Phi}\Phi$$

with

$$\bar{\Phi} = \Phi^+\gamma_4, \tag{2.49}$$

$$\bar{\Theta} = \Theta^+\gamma_4. \tag{2.50}$$

Note that the factor $\frac{1}{2}$ in the lagrangians of the spin-0 and spin-1 fields and of their currents in the previous sections arose from the use of real fields in contrast to the complex fields employed here. This lagrangian can be expressed in terms of only the Φ field using the relation

$$\bar{\Theta} = -\bar{\Phi}\overleftarrow{\partial}_\mu\gamma_\mu.$$

Then

$$\mathfrak{L} = -\bar{\Phi}\overleftarrow{\partial}_\mu\gamma_\mu\gamma_\nu\overrightarrow{\partial}_\nu\Phi - M^2\bar{\Phi}\Phi.$$

The conjugate momentum is

$$\pi = \frac{\partial\mathfrak{L}}{\partial\dot{\Phi}} = i\bar{\Phi}\overleftarrow{\partial}_\mu\gamma_\mu\gamma_4.$$

We also define

$$\bar{\pi} = \frac{\partial\mathfrak{L}}{\partial\dot{\bar{\Phi}}} = i\gamma_4\gamma_\mu\overrightarrow{\partial}_\mu\Phi.$$

The hamiltonian is then in terms of the Φ field

$$H_0 = \int \mathrm{d}^3 r \left(\pi \dot{\Phi} + \dot{\bar{\Phi}} \bar{\pi} - \mathcal{L} \right)$$

$$= \int \mathrm{d}^3 r \left(-\bar{\Phi} \overleftarrow{\partial}_\mu \gamma_\mu \gamma_4 \overrightarrow{\partial}_4 \Phi - \bar{\Phi} \overleftarrow{\partial}_4 \gamma_4 \gamma_\mu \overrightarrow{\partial}_\mu \Phi + \bar{\Phi} \overleftarrow{\partial}_\mu \gamma_\mu \gamma_\nu \overrightarrow{\partial}_\nu \Phi + M^2 \bar{\Phi} \Phi \right).$$

The $8S$-component field Φ is now broken up into component fields U, V, u and v, as in eq. (2.15)

$$\Phi = \begin{pmatrix} U \\ u \\ V \\ v \end{pmatrix}, \tag{2.51}$$

and similarly for Θ. Here U and V have $2S+1$ components each, while u and v have $2S-1$ components. They transform under space rotation as spin-S and spin-$(S-1)$ spinors respectively. As mentioned above a gauge freedom is associated with the presence of these spin-$(S-1)$ fields u and v, i.e. the physics of the theory is unchanged under the gauge transformation

$$\Phi \to \Phi + \gamma_\mu \partial_\mu \Lambda,$$

provided the scalar field Λ obeys the free Klein-Gordon equation with mass M.

The $8S \times 8S$ matrices γ_μ acting on the fields are written as

$$\gamma_\mu \begin{pmatrix} U \\ u \\ V \\ v \end{pmatrix} = \begin{pmatrix} A_\mu & B_\mu \\ C_\mu & D_\mu \end{pmatrix} \begin{pmatrix} U \\ u \\ V \\ v \end{pmatrix},$$

in analogy to the representation of the Dirac γ-matrices in terms of the Pauli σ-matrices. Here A, B, C and D are $4S \times 4S$ matrices. We have*

$$\gamma_4 = \begin{pmatrix} K & 0 \\ 0 & -K \end{pmatrix}, \tag{2.52}$$

$$\gamma_k = i \begin{pmatrix} -\hat{\tau}_k & -\sigma_k \\ \sigma_k & \hat{\tau}_k \end{pmatrix}, \qquad k = 1,2,3, \tag{2.53}$$

$$\gamma_5 = \gamma_1 \gamma_2 \gamma_3 \gamma_4 = \begin{pmatrix} 0 & -I \\ -I & 0 \end{pmatrix},$$

with

$$K \begin{pmatrix} U \\ u \end{pmatrix} = \begin{pmatrix} U \\ -u \end{pmatrix}, \qquad K^2 = 1,$$

$$\hat{\tau}_k = K \tau_k = -\tau_k K.$$

* Here, the notation σ_k and τ_k refers to the space-like and the time-like character of the sub-matrices. In this context they must not be confused with the spin and isospin matrices.

The matrices σ_k and τ_k are chosen such that the γ_μ fulfill the anticommutation relations (2.43). Their form for the case $S=\frac{3}{2}$ is given below. For the general case we refer the reader to Hayward.

In analogy with the spin matrices of sect. 2.3.2 we also introduce the matrices S_k and T_k

$$S_k = S\sigma_k, \tag{2.54}$$

$$T_k = S\tau_k. \tag{2.55}$$

They are written in blocks

$$S_k\begin{pmatrix} U \\ u \end{pmatrix} = \begin{pmatrix} a_k & b_k \\ c_k & d_k \end{pmatrix}\begin{pmatrix} U \\ u \end{pmatrix}, \tag{2.56}$$

where a_k is a $(2S+1)\times(2S+1)$ matrix, d_k is a $(2S-1)\times(2S-1)$ matrix and b_k, c_k are non-square matrices of dimension $(2S+1)\times(2S-1)$. Furthermore, S_k is diagonal in spin ($b_k = c_k = 0$), while T_k is purely off-diagonal in spin ($a_k = d_k = 0$). In words, the S_k have the properties of spin matrices and conserve the spin while the T_k change the spin by one unit.

The Θ field is defined such that a general Lorentz transformation with infinitesimal angles $\varepsilon_{\mu\nu}$ acting on it is of the form

$$L = 1 + \tfrac{1}{2}i\varepsilon_{\mu\nu}J_{\mu\nu}, \tag{2.57}$$

with

$$J_{\mu\nu} = L_{\mu\nu} + S_{\mu\nu}.$$

Here $L_{\mu\nu}$ acts on the coordinates and $S_{\mu\nu}$ on the spinors. The latter are given by

$$S_{ij} = \varepsilon_{ijk}\begin{pmatrix} S_k & 0 \\ 0 & S_k \end{pmatrix}, \tag{2.58}$$

$$S_{k4} = \begin{pmatrix} 0 & S_k \\ S_k & 0 \end{pmatrix}. \tag{2.59}$$

For a Φ field, related to the Θ field by eq. (2.45), the Lorentz transformation then is of the form

$$\bar{L} = 1 + \tfrac{1}{2}i\varepsilon_{\mu\nu}\bar{J}_{\mu\nu}, \tag{2.60}$$

with

$$\bar{J}_{\mu\nu} = L_{\mu\nu} + \bar{S}_{\mu\nu}.$$

We have

$$\bar{S}_{ij} = S_{ij},$$

but

$$\bar{S}_{k4} = \sigma_{k4} - S_{k4} = \begin{pmatrix} \tau_k & \hat{\sigma}_k - S_k \\ \hat{\sigma}_k - S_k & \tau_k \end{pmatrix}, \tag{2.61}$$

where

$$\hat{\sigma}_k = K\sigma_k = \sigma_k K.$$

The usual five covariants can be constructed with the γ-matrices. Recalling the definitions (2.49) and (2.50), they are respectively the scalar, vector, tensor, pseudo-vector and pseudo-scalar forms

$$\begin{aligned} S &= \bar{\Phi}\Phi, \\ V_\mu &= -i\left(\bar{\Phi}\gamma_\mu\Theta + \bar{\Theta}\gamma_\mu\Phi\right), \\ T_{\mu\nu} &= \bar{\Phi}\sigma_{\mu\nu}\Phi, \\ A_\mu &= i\left(\bar{\Phi}\gamma_\mu\gamma_5\Theta + \bar{\Theta}\gamma_\mu\gamma_5\Phi\right), \\ P &= i\bar{\Phi}\gamma_5\Phi. \end{aligned}$$

These covariants are then used for constructing the interactions as in sect. 2.4. These general results are now written for the case of the spin-$\frac{3}{2}$ field.

2.6.2. *Spin-$\frac{3}{2}$ fields*

As an example, we consider the case of the spin-$\frac{3}{2}$ field. Its potential field Φ has $8S = 12$ components, eq. (2.51). Namely, U and V have 4 components each and they have the rotation transformation properties of a spin $\frac{3}{2}$, while u and v have 2 components each and the properties of a spin $\frac{1}{2}$. The K, S_k and T_k matrices are then 6×6. Their values are given in Hayward. We shall only need later on the components

$$S_z = \left(\begin{array}{cccc|cc} \frac{3}{2} & 0 & 0 & 0 & & \\ 0 & \frac{1}{2} & 0 & 0 & & 0 \\ 0 & 0 & -\frac{1}{2} & 0 & & \\ 0 & 0 & 0 & -\frac{3}{2} & & \\ \hline & & 0 & & \frac{1}{2} & 0 \\ & & & & 0 & -\frac{1}{2} \end{array}\right), \tag{2.62}$$

$$T_z = \left(\begin{array}{cccc|cc} & & & & 0 & 0 \\ & & 0 & & \sqrt{2} & 0 \\ & & & & 0 & \sqrt{2} \\ & & & & 0 & 0 \\ \hline 0 & \sqrt{2} & 0 & 0 & & 0 \\ 0 & 0 & \sqrt{2} & 0 & & \end{array}\right). \tag{2.63}$$

As in eq. (2.56) we have indicated how they are broken up into sub-matrices associated with the $S=\frac{3}{2}$ and $(S-1)=\frac{1}{2}$ parts of the fields.

We now turn to the form of the spinors $\mathcal{U}_m(\boldsymbol{p})$ and $\mathcal{V}_m(\boldsymbol{p})$ entering the field Φ, eq. (2.51), for positive and negative energy solutions respectively. Recall that they have $8S=12$ components each. According to the gauge freedom discussed above, we shall choose a gauge such that in the Φ field the $S=\frac{1}{2}$ components vanish. This gauge is the analogue of the transverse gauge of electrodynamics. It is different from the Lorentz gauge we used in the spin-1 case. In order to achieve this gauge we proceed as follows: we begin with a positive energy state, completely aligned along Oz in its rest system. There, in the notation of eq. (2.51) its 12-component spinor is $(\boldsymbol{p}=0)$

$$\mathcal{U}_{m=3/2}(0)=\begin{pmatrix}\chi_{m=3/2}\\ u=0\\ V=0\\ v=0\end{pmatrix},$$

where χ_m are the four-component spin-$\frac{3}{2}$ wave functions normalized according to

$$\langle\chi_m|\chi_{m'}\rangle=\delta_{mm'}.$$

We then apply a boost along the z-axis to a momentum p_z, i.e. the integrated form of eq. (2.60) with (2.61). It conserves helicity, hence

$$\mathcal{U}_{m=3/2}(p_z)=\sqrt{\frac{E+M}{2M}}\begin{pmatrix}\chi_{m=3/2}\\ 0\\ \dfrac{-p_z}{E+M}\chi_{m=3/2}\\ 0\end{pmatrix}.$$

The other m states are obtained by means of the m-lowering operator S_-

$$S_-=\begin{pmatrix}S_x-iS_y & 0\\ 0 & S_x-iS_y\end{pmatrix},$$

to yield

$$\mathcal{U}_m(p_z)=\sqrt{\frac{E+M}{2M}}\begin{pmatrix}\chi_m\\ 0\\ \dfrac{-p_z}{E+M}\chi_m\\ 0\end{pmatrix}.$$

These spinors obey the normalization, for given p_z, with a summation over the spinor index $\alpha=1,2,\ldots,12$

$$\langle\mathcal{U}_m|\mathcal{U}_{m'}\rangle=\sum_\alpha\mathcal{U}^+_{m\alpha}\mathcal{U}_{m'\alpha}=\frac{E}{M}\delta_{mm'}.$$

Note that the relative sign between the large and small components differs from that of the spin-$\frac{1}{2}$ case. This can be understood from the difference between the forms of the boost for the $\frac{1}{2}$-field, sect. 2.3.2 and for the Φ field, eq. (2.61). It is the Θ-field spinors which have the relative sign of the spin-$\frac{1}{2}$ case, see eq. (2.59). All quantities will be expressed in terms of the Φ-field, using the relation (2.45).

For a boost of arbitrary direction it is seen from eq. (2.61) that the spinors are obtained by the replacement of p_z by $\boldsymbol{\sigma}\cdot\boldsymbol{p}$, where $\boldsymbol{\sigma}$ are the usual non-relativistic spin-$\frac{3}{2}$ matrices, in agreement with (2.62) and (2.54)

$$\sigma_z = \begin{pmatrix} 1 & 0 & 0 & 0 \\ 0 & \frac{1}{3} & 0 & 0 \\ 0 & 0 & -\frac{1}{3} & 0 \\ 0 & 0 & 0 & -1 \end{pmatrix}. \tag{2.64}$$

Hence the general form of the spinors is, for the positive energy solutions

$$\mathcal{U}_m(\boldsymbol{p}) = \sqrt{\frac{E+M}{2M}} \begin{pmatrix} \chi_m \\ 0 \\ \dfrac{-\boldsymbol{\sigma}\cdot\boldsymbol{p}}{E+M}\chi_m \\ 0 \end{pmatrix}, \tag{2.65}$$

and similarly for the antiparticle solutions

$$\mathcal{V}_m(\boldsymbol{p}) = \sqrt{\frac{E+M}{2M}} \begin{pmatrix} \dfrac{-\boldsymbol{\sigma}\cdot\boldsymbol{p}}{E+M}\chi_{-m} \\ 0 \\ \chi_{-m} \\ 0 \end{pmatrix}, \tag{2.66}$$

with in the latter the usual sign change of the spin magnetic quantum numbers in the spin wave functions.

It is important to note that these spinors were generated by applying first a boost in the maximal helicity state and then a rotation in the boosted system. This way no spin $S = \frac{1}{2}$ components [$u = 0$, $v = 0$ in eq. (2.51)] arose. Conversely, when boosting our $m = \pm\frac{1}{2}$ states back into the rest system the resulting states will be mixtures of $S = \frac{3}{2}$ and $S = \frac{1}{2}$ components, as seen from the form of (2.61). This procedure is not that of the Wigner rotation where the m-lowering operation from the maximum helicity $m = S$ states is done by the sequence of operations: (i) boost into the rest system, (ii) perform the m-lowering operation by S_-, (iii) boost back to the initial system. In the Wigner rotation the spinors of the Φ field would be mixtures of S and $(S-1)$ components. Having only spin-S components the present formulation will lend itself directly to the conventional tensorial calculus, which is not the case for the helicity formulation based on Wigner rotations.

The plane wave expansion of the higher spin Φ fields then is, omitting isopin,

$$\Phi(\boldsymbol{r}, t) = \left(\frac{1}{2\pi}\right)^{3/2} \sum_m \int \mathrm{d}^3p \frac{1}{\sqrt{2E}} \Big(B_{pm} \mathcal{U}_m(\boldsymbol{p}) \mathrm{e}^{i(\boldsymbol{p}\cdot\boldsymbol{r} - Et)}$$

$$+ D^+_{pm} \mathcal{V}_m(\boldsymbol{p}) \mathrm{e}^{-i(\boldsymbol{p}\cdot\boldsymbol{r} - Et)} \Big). \qquad (2.67)$$

For half-integer spin, B is the particle annihilation operator and D^+ the antiparticle creation operator. We note the difference of normalization in (2.13) and (2.67) which arises from the fact that the spin-$\frac{1}{2}$ field is a solution of a first-order differential equation while in the general case the field Φ is a solution of a second-order equation. The fields are quantized by postulating the anticommutation relations

$$\left[B_{pm}, B^+_{p'm'} \right]_+ = \delta_{mm'} \delta^3(\boldsymbol{p} - \boldsymbol{p}'),$$

$$\left[D_{pm}, D^+_{p'm'} \right]_+ = \delta_{mm'} \delta^3(\boldsymbol{p} - \boldsymbol{p}'). \qquad (2.68)$$

For integer spin, D^+ is replaced by B^+ in (2.67) and commutation relations are postulated.

In the plane wave expansion (2.67), the components of different $\boldsymbol{p}$ are related to each other not by a Lorentz transformation alone as in the cases of spin $S \leqslant 1$. Here also a gauge transformation restoring the transverse gauge is required. Finally, we do not write the expressions for the strength field Θ since all quantities can be written in terms of the field Φ by means of the relation (2.45).

The interactions of a spin-S field with a pion field φ are, the pseudo-scalar interaction (M_S is the mass of the spin-S particle)

$$H^{\mathrm{PS}}_{\pi S} = i g^{\mathrm{PS}}_{\pi S} (2M_S) \int \mathrm{d}^3r : \overline{\Phi} \gamma_5 \boldsymbol{\tau} \Phi \varphi :,$$

and the pseudo-vector interaction

$$H^{\mathrm{PV}}_{\pi S} = -i g^{\mathrm{PV}}_{\pi S} \frac{1}{2M_S} \int \mathrm{d}^3r : \left(\overline{\Phi} \gamma_\mu \gamma_5 \boldsymbol{\tau} \Theta + \overline{\Theta} \gamma_\mu \gamma_5 \boldsymbol{\tau} \Phi \right) \partial_\mu \varphi :,$$

where again the relative sign is arbitrarily chosen so that for small momenta the two expressions coincide. These equations become (2.22) and (2.23) for the $S = \frac{1}{2}$ case. Indeed the comparison of the Dirac equation, sect. 2.3.2, with the relations (2.44)–(2.46) shows that in the case of spin $\frac{1}{2}$, Θ and Φ are identical within a norm M.

Likewise, the interaction of a spin-S field with the ω vector meson is constructed with the current V_μ above

$$H_{\omega S} = g_{\omega S} \int \mathrm{d}^3r : V_\mu \phi^{(\omega)}_\mu :,$$

and similarly for the ρ-meson, introducing as usual the isospin operator $\boldsymbol{\tau}$

$$H_{\rho S} = -ig_{\rho S}\int \mathrm{d}^3r : \left(\bar{\Phi}\gamma_\mu \boldsymbol{\tau}\Theta + \bar{\Theta}\gamma_\mu \boldsymbol{\tau}\Phi\right)\phi_\mu^{(\rho)} :,$$

in full analogy with the $S=\frac{1}{2}$ case, eq. (2.25) and (2.26).

We consider in particular the important case of the $\pi N\Delta$ vertex, where Δ is the $S=\frac{3}{2}$, $t=\frac{3}{2}$ resonance when taken as a basic field. For angular momentum and parity conservation this interaction must be of the form

$$H_{\pi N\Delta} = g_{\pi N\Delta}\sqrt{\frac{1}{2M_\Delta}}\int \mathrm{d}^3r : \left(\bar{\Phi} D_\mu \gamma_5 \boldsymbol{C}\psi\, \partial_\mu \boldsymbol{\varphi} - \left(\partial_\mu \boldsymbol{\varphi}\right)\bar{\psi}\boldsymbol{C}^{+}\gamma_5 D_\mu^{+}\Phi\right): \quad (2.69)$$

where ψ is the spin-$\frac{1}{2}$ field, $\boldsymbol{C}$ is an operator which changes the isospin from $\frac{1}{2}$ to $\frac{3}{2}$ and conserves the charge and D_μ is a four-vector matrix which must obey the following relation for the interaction to be Lorentz invariant:

$$S_{\mu\nu}^{3/2} D_\lambda - D_\lambda S_{\mu\nu}^{1/2} = iD_\nu \delta_{\mu\lambda} - iD_\mu \delta_{\nu\lambda},$$

where $S_{\mu\nu}^{3/2}$ and $S_{\mu\nu}^{1/2}$ are the spin matrices for the $\frac{3}{2}$ and $\frac{1}{2}$ fields respectively. The matrices D_λ in this expression are found to be of the form (from Hayward)

$$D_k = \begin{pmatrix} 0 & F_k \\ G_k & 0 \\ -F_k & 0 \\ 0 & -G_k \end{pmatrix}, \qquad D_4 = i\sqrt{\tfrac{9}{8}}\begin{pmatrix} 0 & 0 \\ 0 & -I \\ 0 & 0 \\ I & 0 \end{pmatrix}, \quad (2.70)$$

where in D_4 the unit matrices I are 2×2 and where for example in D_z

$$F_z = \begin{pmatrix} 0 & 0 \\ 1 & 0 \\ 0 & 1 \\ 0 & 0 \end{pmatrix}, \qquad G_z = \sqrt{8}\begin{pmatrix} 1 & 0 \\ 0 & -1 \end{pmatrix}. \quad (2.71)$$

Finally, the part $J_\mu A_\mu$ of the interaction with the electromagnetic field linear in the photon field A_μ, as discussed in sect. 2.5.1, is obtained with the current

$$J_\mu = eV_\mu.$$

This result can also be checked from (2.47), keeping only the terms which are linear in A_μ.

2.7. EXTENSION TO YANG-MILLS THEORIES

In this work the examples will be constructed with interacting fields for particles having an intrinsic symmetry of at most SU(2). However, the present methods and results can be extended to particles with intrinsic symmetry groups SU(n), for example SU(3) in the case of quantum chromodynamics (QCD).

In particular they are applicable to Yang-Mills type gauge theories. These theories describe fermions (quarks,...) interacting via vector gauge fields (gluons,...), and they contain the following assumptions: (i) the gauge fields obey a SU(n) Lie algebra (SU(3) in QCD); (ii) the fields interact via minimal coupling, i.e. $\partial_\mu \to \partial_\mu - i\varepsilon A_{\mu\varepsilon}$ where A_μ is the vector gauge field and ε is a charge tensor (colour tensor in QCD); (iii) consequently the gauge fields have self-interactions with point vertices containing up to four fields; (iv) if the theory has exact conservation laws (conservation of colour in QCD), the vector gauge fields have vanishing rest mass.

For these theories, the expansions for the fermion fields are not changed except for the replacement of the SU(2) isospin wave functions by the appropriate SU(n) functions. The expansions of the spin-1 gauge fields are also unchanged except for the introduction of the SU(n) algebra.

However, for the case of vanishing mass (e.g. for the gluons) the longitudinal solutions require a special treatment. Namely, only the transverse modes of the solutions for the free spin-1 field, as given above, are valid for massless particles. For the longitudinal modes the massless case is not an analytic limit $M \to 0$ of the massive case. This is evident from the normalization of the longitudinal mode of the massive spin-1 field given in sect. 2.3.3. The physical reason for this is the fact that for massless particles only the extremal helicity states can propagate as free particles. The homogeneous equations of motion yield only transverse solutions. A general solution of the inhomogeneous equations (i.e. the equations with sources) is obtained by adding to the transverse solutions a particular solution of the inhomogeneous equations. That means that to the transverse field components of sect. 2.3.3 one must add the contributions from the longitudinal modes by means of a special treatment. One possible way, along the lines of the derivation of the Coulomb potential in electrodynamics, is as follows. The fermion and the transverse boson sector of the Hilbert space can be treated in the general framework of this book. The longitudinal boson sector is treated by Feynman techniques to yield a potential acting between the particles of the first sector. A complete discussion of this problem is, however, outside the scope of this work.

Once the contribution of the longitudinal vector bosons has been taken into account the discretization procedure of Chapter 5 is applicable to Yang-Mills theories and a secular problem can be formulated. Likewise, all the recoupling techniques, graphs and formulae of the following chapters can be simply generalized to the case of SU(n). This will be discussed in sect. 4.3.5.

The interactions of the Yang-Mills hamiltonian are of two kinds: the fermion-vector boson interactions and the 3- and 4-boson interactions. The former are exactly of the form (2.25) generalized from SU(2) to SU(n) and are given in sect. 6.5. The 3- and 4-boson interactions can likewise be easily evaluated along the lines of Chapter 6. The only difference is the replacement of the SU(2) recoupling coefficients in the isospin parts by the appropriate SU(n) recoupling coefficients.

CHAPTER 3

The relativistic secular problem

3.1. INTRODUCTION

In relativistic quantum field theory the stationary states of finite many-body bound states are best described by a time-independent treatment in the Schrödinger picture, as is the case in a non-relativistic theory. In this chapter we set up the Schrödinger relativistic secular problem. We discuss its physical meaning and the related difficulty of the spurious center-of-mass motion arising from the truncation of the Hilbert space.

The actual solution of the Schrödinger problem, eq. (2.4), requires a discrete representation of the Hilbert space. To that end one needs a discretized form for the relativistic free fields used in the construction of the many-body configurations. In sect. 3.2. the method of discretization is presented for the case of a spin-0 field. This procedure will be applied in full detail to all fields in Chapter 5 after acquiring the necessary techniques in Chapter 4.

The secular problem is set up in sect. 3.3. where also a discussion is given of the truncation procedure in the relativistic context. In that section the meaning of stationary solutions having components with different numbers of particles is explained. The differences between the time-dependent perturbation treatment in the interaction picture and the time-independent Schrödinger framework are also discussed in this section.

The problem of the center-of-mass motion which arises from the truncation of the Hilbert space is more involved in the relativistic case than in the non-relativistic case. A method for its treatment is described in sect. 3.4. It is based on the separation in the solution of the c.m. modes by the introduction of an harmonic oscillator pseudo-hamiltonian for the c.m. motion. This requires the construction of the c.m. position operator for a system of relativistic particles, which is done in sect. 3.4.2. This operator has a full N-body character which requires a specific procedure for the evaluation of its many-body matrix elements.

3.2. FREE FIELD DISCRETIZATION

The discretization of relativistic fields is somewhat different from the non-relativistic case. Namely, in non-relativistic theories one can use an arbitrary potential to generate a complete set of eigenstates with discrete energy eigenvalues. The solutions of the true hamiltonian then can be expanded in terms of these discrete states. In a relativistic theory one cannot use the non-relativistic concept of an arbitrary potential. Consequently one has to start from the free-field solutions themselves for constructing the basis states. The way to achieve a discrete secular problem then is to form wave packets over the continuous variable $\boldsymbol{p}$, both over the direction $\hat{\boldsymbol{p}}$ of the momentum and the energy $E=\sqrt{p^2+M^2}$. In particular the latter is achieved by applying a unitary transformation on the field operators, i.e.,

$$a_\nu^+ = \int \mathrm{d}E f_\nu(E) a_E^+ ,$$

where the energy weight functions $f_\nu(E)$ with discrete indices ν form an orthonormal set. A convenient choice is, for example, the set of harmonic oscillator functions.

In other words, the discretization of the basis free fields and of their creation and annihilation operators is carried out by going from the plane wave expansions of sect. 2.2, having continuous quantum numbers to expansions into wave packets specified by discrete quantum numbers. With the choice of a harmonic oscillator basis they will have the meaning of the number of mean energy quanta and of angular momentum.

We limit ourselves at this stage to the simple case of spin 0 which will be used to show the principle of the discretization method. The procedure is essentially the same for the other fields except for couplings associated with spin and will be developed in Chapter 5.

Thus the problem at hand is to go from the plane wave representation of the free field solution in the Schrödinger picture eq. (2.8), setting $t=0$ and omitting the isospin at this stage,

$$\varphi(\boldsymbol{r}) = \int \mathrm{d}^3p \left(a_p \varphi(\boldsymbol{p},\boldsymbol{r}) + a_p^+ \varphi^+(\boldsymbol{p},\boldsymbol{r})\right), \tag{3.1}$$

to the discretized representation

$$\varphi(\boldsymbol{r}) = \sum_{\nu l m} \left(a_{\nu l m}\varphi_{\nu l m}(\boldsymbol{r}) + a_{\nu l m}^+ \varphi_{\nu l m}^+(\boldsymbol{r})\right). \tag{3.2}$$

Here in (3.1) the boson operators a_p, a_p^+ fulfill the commutation relations (2.9), and in eq. (3.2) ν is the radial quantum number, l the angular momentum, and m its projection along the quantization axis.

In order to carry out this discretization, first the multipole expansion of the plane wave is introduced in (2.8) as

$$\mathrm{e}^{i\boldsymbol{p}\cdot\boldsymbol{r}} = 4\pi \sum_{lm} i^l Y_{lm}^*(\hat{\boldsymbol{p}}) Y_{lm}(\hat{\boldsymbol{r}}) j_l(pr),$$

and new multipole creation and annihilation operators a_{plm}^+, a_{plm} are defined by carrying out the integration over the direction $\hat{\boldsymbol{p}}$ of the linear momentum,

$$a_{plm}=\int \mathrm{d}^2 p\, Y_{lm}(\hat{\boldsymbol{p}})\,a_{\boldsymbol{p}}.$$

This unitary transformation leads to the commutation property

$$\left[a_{plm},a^{+}_{p'l'm'}\right]_{-}=\delta_{ll'}\delta_{mm'}\frac{\delta(p-p')}{p^2}.$$

Next, the basis states of continuous index $p=|\boldsymbol{p}|$ are replaced by basis states of discrete indices ν. To that end a complete set of orthogonal functions $f_{\nu l}(p)$ is introduced, defined to obey the relations

$$\int p^2 \mathrm{d}p\, f_{\nu l}(p) f_{\nu' l}(p)=\delta_{\nu\nu'},$$

$$\sum_{\nu} f_{\nu l}(p) f_{\nu l}(p')=\frac{\delta(p-p')}{p^2}.$$

Then, discretized creation and annihilation operators $a^{+}_{\nu lm}$ and $a_{\nu lm}$ are defined by integrating over the value p of the linear momentum,

$$a_{\nu lm}=\int p^2 dp\, f_{\nu l}(p)\,a_{plm}. \tag{3.3}$$

The inverse transform is

$$a_{plm}=\sum_{\nu} f_{\nu l}(p)\,a_{\nu lm}.$$

The new discretized operators obey the commutation relations

$$\left[a_{\nu lm},a^{+}_{\nu' l'm'}\right]_{-}=\delta_{\nu\nu'}\delta_{ll'}\delta_{mm'}.$$

The substitution of the discretized operators (3.3) into the plane wave expansion of the field (3.1) yields the discretized expansion

$$\varphi(\boldsymbol{r})=\sqrt{\tfrac{1}{2}}\sum_{\nu lm} i^l g_{\nu l}(r)\left(a_{\nu lm}Y_{lm}(\hat{\boldsymbol{r}})+(-)^l a^{+}_{\nu lm}Y^{*}_{lm}(\hat{\boldsymbol{r}})\right),$$

where the functions $g_{\nu l}(r)$ represent radial wave packets

$$g_{\nu l}(r)=\left(\frac{2}{\pi}\right)^{1/2}\int p^2 \mathrm{d}p\,\frac{1}{\sqrt{E}} f_{\nu l}(p)\, j_l(pr), \tag{3.4}$$

with $E=\sqrt{p^2+M^2}$. Likewise, the conjugate field (2.8) in its discretized form is

$$\pi(\boldsymbol{r})=-i\sqrt{\tfrac{1}{2}}\sum_{\nu lm} i^l h_{\nu l}(r)\left(a_{\nu lm}Y_{lm}(\hat{\boldsymbol{r}})-(-)^l a^+_{\nu lm}Y^*_{lm}(\hat{\boldsymbol{r}})\right), \tag{3.5}$$

with the radial functions

$$h_{\nu l}(r)=\left(\frac{2}{\pi}\right)^{1/2}\int p^2\,\mathrm{d}p\sqrt{E}\,f_{\nu l}(p)\,j_l(pr). \tag{3.6}$$

Using the relation

$$\sum_\nu g_{\nu l}(r)h_{\nu l}(r')=\frac{2}{\pi}\sum_\nu\int p^2\,\mathrm{d}p\int p'^2\,\mathrm{d}p'\sqrt{\frac{E'}{E}}\,j_l(pr)\,j_l(p'r')f_{\nu l}(p)f_{\nu l}(p')$$

$$=\frac{\delta(r-r')}{r^2}, \tag{3.7}$$

one may verify that the commutation relations (2.10) between the fields $\varphi(\boldsymbol{r})$ and $\pi(\boldsymbol{r})$ are indeed fulfilled.

As already mentioned, one convenient set of orthonormal functions $f_{\nu l}(p)$ is of course that of the radial eigenstates of the harmonic oscillator. Thus we take

$$f_{\nu l}(p)=\alpha^{3/2}c_{\nu l}\exp\left(-\tfrac{1}{2}\alpha^2p^2\right)(\alpha p)^l L_{\nu-1}^{l+1/2}(\alpha^2p^2), \tag{3.8}$$

where α specifies the scale, $\nu=1,2,3\ldots$ etc is the the radial quantum number and

$$c_{\nu l}=\frac{1}{\pi^{1/4}}\left[\frac{2^{\nu+l+1}(\nu-1)!}{(2\nu+2l-1)!!}\right]^{1/2}.$$

The harmonic oscillator excitation quantum number is $N=2\nu+l-2$.

3.3 THE SECULAR PROBLEM

The solution state vectors $|S_n\rangle$ given by eq. (2.1) are expanded as sums over discretized many-particle configurations $|\alpha\rangle$,

$$|S_n\rangle=\sum_\alpha X^n_\alpha|\alpha\rangle=\sum_\alpha X^n_\alpha\sum_A C_{\alpha A}|A\rangle \tag{3.9}$$

with

$$\langle\alpha|\beta\rangle=\delta_{\alpha\beta}.$$

Here $|A\rangle$ is a product of one-particle creation operators acting on the vacuum in the discretized basis defined above and the transformation $C_{\alpha A}$ generates an orthonormal set of configurations $|\alpha\rangle$ having the desired quantum numbers of the system. Owing to the emission and absorption processes contained in H_{I} the configurations $|\alpha\rangle$ in a given state $|S_n\rangle$ may contain different numbers of particles. More precisely they have different numbers of bosons and of fermion-antifermion pairs but of course the same conserved baryon number.

The unknown amplitudes X_α^n and energies E_n are the solutions of eq. (2.4)

$$\sum_\beta \langle\alpha|H_0 + H_{\mathrm{I}}|\beta\rangle X_\beta^n = E_n X_\alpha^n . \tag{3.10}$$

Note that the discretized configurations $|\alpha\rangle$ made of wave packets over the particle momenta $\boldsymbol{p}_i$ are not eigenstates of the free hamiltonian H_0, and the off-diagonal matrix elements $\langle\alpha|H_0|\beta\rangle$ must be computed. Furthermore, the observables are not directly the eigenvalues E_n but the related mass spectrum M_n as discussed in eq. (2.5).

A "model" is specified by the choice of the form of the hamiltonian H, of the values of its parameters, and of the retained Hilbert space, i.e. the choice of the configurations $|\alpha\rangle$. The lowest eigenstate $|\mathrm{v}\rangle$ with the quantum numbers of the vacuum is the " vacuum state" of the model. All energies have to be measured from the energy E_{V} of that vacuum. The analysis of the validity of a model requires a discussion which will depend on the situation under study. More generally one can use as "models" all the well-known approximations from non-relativistic physics of the problem (3.10): perturbations, generalizations of Hartree-Fock or BCS theories, Tamm-Dancoff or random-phase treatments, etc.

The divergence of interacting field theories appears here as a divergent dependence of the solutions upon the dimension of the truncated Hilbert space. This situation is in common with the non-relativistic shell configurational treatments of nuclei, or with the relativistic field theories in the usual time-dependent formalism. In the latter case, renormalization is used to introduce "physical" masses and coupling constants. In the application of the present treatment, the discussion must in particular bear on whether the use of effective masses and coupling constants for the partially dressed free fields can yield a proper description of the system within the retained configuration space.

We now discuss a possible truncation criterion. We may classify the configurations $|\alpha\rangle$ according to the energy number $\mathcal{E}_\alpha$ which contains the total rest mass of the constituent particles together with the sum of the "energies" of the oscillator modes making up the basis states $|A\rangle$ in the expansion (3.9):

$$\mathcal{E}_\alpha = M_1 + M_2 + \cdots + \tfrac{1}{2}(N_1\omega_1 + N_2\omega_2 + \cdots), \tag{3.11}$$

where M_i are the mass parameters of the constituent particles i and the numbers N_i are the oscillator quantum numbers of their states defined at the end of sect. 3.2. Note that the number of "phonons" $N_1 + N_2 + \cdots = N$ is the same for all the basis states $|A\rangle$ associated with a given configuration $|\alpha\rangle$ since the transformation $C_{\alpha A}$ entails only summations over the single-particle magnetic quantum numbers m_i. The

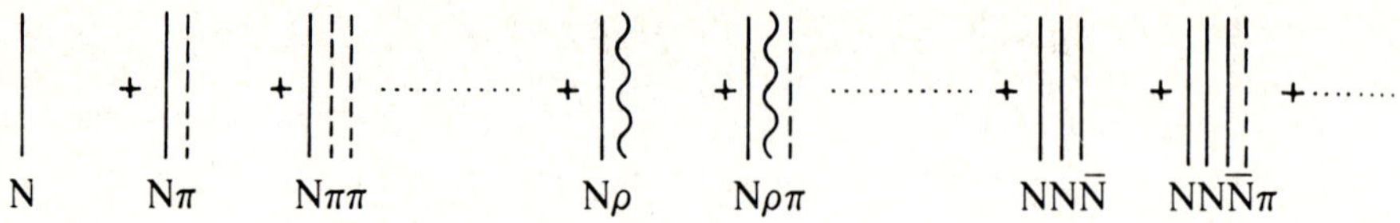

Fig. 3.1.

"scale energies" ω_i are related to the size parameter of the chosen representation functions (3.8) and are an input parameter of the model. For example in order to have pion wave packets extending over the size of the nucleon, the energy scale ω_i of the representation is roughly of the order of 140 MeV. For a nucleon in a nucleus this "scale" energy will approximately be of the order of magnitude of the shell-model energies, from 15 MeV for light nuclei to 8 MeV for heavy nuclei. The factor $\frac{1}{2}$ in (3.11) reflects the fact that H_0 contains only the kinetic and no potential energy.

One can use this quantity $\mathcal{E}_\alpha$ to specify the truncation prescription as discussed in more detail in Chapter 5. The simplest prescription is: retain all configurations with $\mathcal{E}_\alpha \leqslant E_{\text{truncation}}$. Of course, one may use other truncation prescriptions, such as: retain only certain subclasses of configurations in that space.

In order to discuss the general characteristics of the solutions of the secular problem (3.10) let us consider the example of a model of the "physical nucleon", in which it is pictured as containing components with 0, 1, 2... bosons (π, ρ, ω...) and likewise arbitrary numbers of baryon-antibaryon pairs in addition to one baryon, as illustrated in fig. 3.1.

The absorption and emission of quanta of the basic fields by the interaction hamiltonian H_{I} are depicted by vertices of the kind of fig. 3.2. The secular matrix looks schematically like fig. 3.3, if we limit ourselves to a hamiltonian having only pion-nucleon and pion-pion interactions and to configurations having at most 3 pions in the meson cloud. In contrast to the non-relativistic secular problem the energy matrix is made up of matrix elements where the number of interacting particles may change as a result of the interaction.

Thus, the solution of the secular problem of fig. 3.3 contains an undefined number of particles. In the time-dependent description the number of particles would fluctuate.

Here, the meaning of a "stationary" solution is one in which the relative amplitudes of the various components, with different numbers of particles, do not

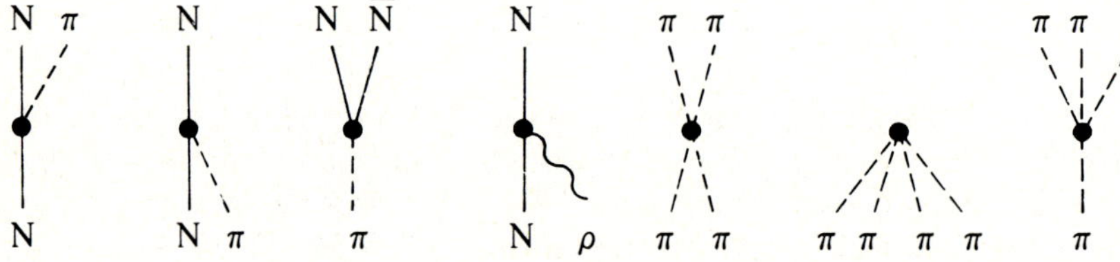

Fig. 3.2.

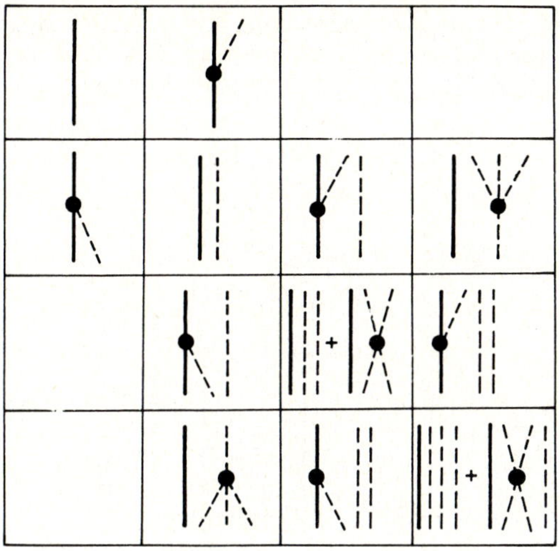

Fig. 3.3.

change in time. They all evolve together: they have the common time factor exp$(-iEt)$, eq. (2.3). We now show in fig. 3.4 how this is achieved microscopically.

At time t the state vector $|S(t)\rangle$ is a configuration mixture of the components $|\alpha\rangle$, namely $|N\rangle$, $|N\pi\rangle$, $|N\pi\pi\rangle$, $|N\pi\pi\pi\rangle$. They have amplitudes X_α $(\alpha = 0, 1, 2, 3)$. After the time Δt this mixture is unchanged, although a number of processes have taken place, i.e., bosons have been created or absorbed. We have from eq. (2.1)

$$|S(t+\Delta t)\rangle = |S(t)\rangle - i\Delta t H|S(t)\rangle .$$

Introducing the expansion (3.9) together with (2.3) we get

$$X_\alpha e^{-iE\Delta t} \simeq X_\alpha(1 - iE\Delta t) = X_\alpha - i\Delta t \sum_\beta \langle\alpha|H|\beta\rangle X_\beta ,$$

which are the relations between the amplitudes given by the secular problem.

Assuming for simplicity H_0 to be diagonal with eigenvalues $\varepsilon_0, \varepsilon_1, \varepsilon_2 \ldots$ and also assuming that H_I contains only terms which change the pion number by one, these relations read in detail

$$EX_0 = \varepsilon_0 X_0 + \langle 0|H_\mathrm{I}|1\rangle X_1 ,$$

$$EX_1 = \langle 1|H_\mathrm{I}|0\rangle X_0 + \varepsilon_1 X_1 + \langle 1|H_\mathrm{I}|2\rangle X_2 , \text{ etc} \ldots$$

Graphically, see fig. 3.4, when going from t to $t + \Delta t$ the component with no pions, $\alpha = 0$, either goes into itself via the free field hamiltonian H_0 with the amplitude $\varepsilon_0 X_0$, or into the one-pion component, $\beta = 1$, via an interaction H_I creating one

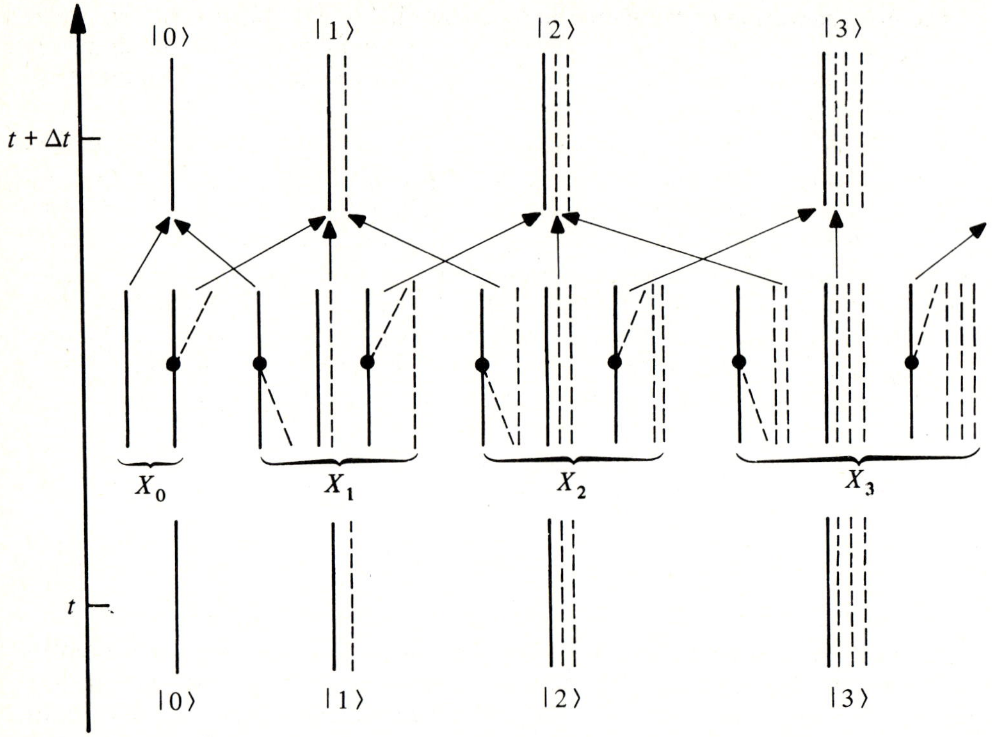

Fig. 3.4.

boson with the amplitude $\langle 1|H_I|0\rangle X_0$. Likewise, the component with one pion, $\alpha = 1$, may either go to $\beta = 0$ via the absorption of a boson with amplitude $\langle 0|H_I|1\rangle X_1$, or go into itself with amplitude $\varepsilon_1 X_1$, or into the component $\beta = 2$ with amplitude $\langle 2|H_I|1\rangle X_1$. In order for the state vector $|S\rangle$ to be stationary the amplitude flows, indicated by arrows in fig. 3.4, must be such that they leave all the ratios X_α/X_β unchanged. This is precisely the meaning of the linear relations contained in the secular equations (3.10).

We conclude this section by reviewing some of the differences between these stationary solutions and the usual perturbation solutions of the time-dependent treatment of sect. 2.2.2. In the time-dependent treatment one selects particular graphs which are either evaluated as such or summed to all orders. The integrations over the graph loop energies are either carried out to infinity (including renormalization, if needed) or up to some cut-off energy. The difficulties associated with the evaluation of high-order diagrams and their interconnected loops limit generally the calculation to processes with intermediate states containing very few particles. On the other hand in the time-independent Schrödinger formulation one selects a configuration space introducing a truncation prescription on physical grounds, and the computation of the many-body matrix elements requires only the evaluation of elementary vertices between the configurations $|\alpha\rangle$.

The diagonalization of the hamiltonian of the time-independent treatment corresponds, in the time-dependent treatment, to the summation of all the graphs of arbitrarily high order and arbitrary topology which can be constructed with the basis states contained in the retained configuration space.

A useful feature of the Schrödinger picture is that unitarity of the solutions obtained by diagonalization of the secular problem (3.10) is guaranteed, owing to the hermiticity of the hamiltonian matrix. It would not be so in the time-dependent formulation where unitarity has to be restored with special techniques.

The difficulties encountered in the construction of the hamiltonian matrix are essentially those of the many-body problem, viz., they arise from the complexity of symmetrizing or antisymmetrizing states containing many particles, having good parity, isospin and angular momentum. In this aspect of the problem, relativity introduces no complications above the non-relativistic case.

3.4. THE CENTER OF MASS

3.4.1. The mass spectrum

When working in a truncated Hilbert space, the solutions of the secular problem are each a mixture of states with different center-of-mass (c.m.) motion. This fact and its consequences may be neglected when treating heavy systems. However, for the calculation of light systems the c.m. motion must be accounted for. This entails two distinct aspects: the definition of the c.m. operator in relativity, and the extraction of the effect of the c.m. motion on the solutions.

In order to obtain solutions with a well-defined c.m. motion, a standard procedure in non-relativistic shell-model treatments consists in adding an artificial energy operator to the hamiltonian,

$$\mathcal{H} = H + H_{\mathrm{c.m.}}$$

with

$$H_{\mathrm{c.m.}} = \tfrac{1}{2}\zeta(P^2 + \Omega^2 R^2)\,. \tag{3.12}$$

Here $\boldsymbol{P}$ and $\boldsymbol{R}$ are the c.m. momentum and position operators. This hamiltonian is used only as a device to split apart the solutions of $\mathcal{H}$ into groups each with a given c.m. motion, without ascribing a physical meaning to it. This splitting increases with increasing values of the parameter ζ. The choice of the harmonic oscillator parameter Ω in the non-relativistic case is arbitrary. It determines the velocity of the c.m. motion. Non-relativistically this procedure is exact except for the inaccuracies introduced by the truncation of the Hilbert space.

The solutions of the pseudo-hamiltonian $\mathcal{H}$ for the non-relativistic case

$$\mathcal{H}\Phi_n(\zeta, \Omega) = \underline{E}_n \Phi_n(\zeta, \Omega)$$

are of the form, owing to the fact that H and $H_{\text{c.m.}}$ commute,

$$\Phi_n(\zeta, \Omega) = \phi_c \chi_a, \qquad \underline{E}_n = E_c + E_a, \tag{3.13}$$

where ϕ_c is an eigenstate of the c.m. hamiltonian (3.12) and χ_a an eigenstate for the intrinsic motion of the system.

A similar method can be used in the relativistic case. There the c.m. momentum is still

$$\boldsymbol{P} = \boldsymbol{p}_1 + \boldsymbol{p}_2 + \boldsymbol{p}_3 \dots,$$

and the evaluation of the matrix elements of P^2 on the relativistic basis state vectors presents no particular difficulty. However, as will be demonstrated in the next section, in relativistic kinematics the c.m. coordinate $\boldsymbol{R}$ takes on the more complicated form

$$\boldsymbol{R} = \frac{(\Sigma_i E_i \boldsymbol{r}_i)}{\Sigma_j E_j}, \tag{3.14}$$

where the E_i are the free energies of the particles making up the configuration,

$$E_i = \left(p_i^2 + M_i^2\right)^{1/2}.$$

This form of the c.m. position operator entails that in the relativistic case $H_{\text{c.m.}}$ and H do not commute. The solutions therefore do not simply factorize, as in eq. (3.13). However, owing to completeness they can in principle be written in the general form

$$\Phi_n(\zeta, \Omega) = \sum K_{ca}^{n} \phi_c \chi_a. \tag{3.15}$$

Here the ϕ_c are the set of wave functions of the c.m. motion while the χ_a are the set of wave functions for the internal motion. Since the magnitude of the mixing in (3.15) is associated with the size of the non-commuting term $\zeta \Omega^2 R^2$ in the pseudo-hamiltonian, the expression (3.15) goes into (3.13) when $\zeta \Omega^2 \to 0$. Thus we assume that with a suitable choice of the parameters ζ and Ω, the low-energy eigenstates of $\mathcal{H}$ are of the form

$$\Phi_n(\zeta, \Omega) \simeq \phi_0 \chi_n, \tag{3.16}$$

where ϕ_0 describes a c.m. motion in the harmonic oscillator state 1s. The accuracy of the assumption (3.16) as well as the independence of the results as a function of the c.m. parameters ζ and Ω must be discussed in the context of the numerical solutions.

We finally turn to the extraction of the mass spectrum M_n of the solutions. The masses M_n are related to the eigenvalues of the pseudo-hamiltonian $\mathcal{H}$, denoted $\underline{E}_n$, by an expression similar to (2.5). Here we have

$$E_n = \underline{E}_n - \langle \Phi_n | H_{\text{c.m.}} | \Phi_n \rangle ,$$

$$E_\text{v} = \underline{E}_\text{v} - \langle \Phi_\text{v} | H_{\text{c.m.}} | \Phi_\text{v} \rangle ,$$

where Φ_v is the lowest eigenstate of $\mathcal{H}$ for a system with the quantum numbers of the vacuum. Thus the mass spectrum M_n is defined by

$$\hat{E}_n = E_n - E_\text{v} = \langle \Phi_n | \sqrt{P^2 + M_n^2}\, | \Phi_n \rangle .$$

Recall that the explicit form of ϕ_0 and χ_n separately are not necessarily known. With our assumption of a suitable choice of the parameters ζ and Ω, so as to achieve a slow c.m. motion, we have $P^2 \ll M_n^2$. The square root can be expanded

$$\hat{E}_n = \langle \Phi_n | M_n + \frac{P^2}{2M_n} + \cdots | \Phi_n \rangle .$$

Upon inversion the mass spectrum of the system is

$$M_n \simeq \hat{E}_n \left(1 - \frac{\langle \Phi_n | P^2 | \Phi_n \rangle}{2\hat{E}_n^2} + \cdots \right) .$$

These masses M_n are the physical quantities of interest in the relativistic context and they replace the energy eigenvalues of the non-relativistic problem.

3.4.2. Relativistic kinematics of the center of mass

We now demonstrate the validity of eq. (3.14), namely that the c.m. coordinate $\boldsymbol{R}$ for a system of relativistic particles i of energies E_i, all considered at an equal time, is given by

$$\boldsymbol{R} = \frac{\Sigma_i E_i \boldsymbol{r}_i}{\Sigma_j E_j} .$$

The discussion shall be carried out by considering first the case of a single particle which decays into two others, then the case of two distinct particles.

In the case of a decaying particle the c.m. trajectory is given by the trajectory of the initial particle and we simply must compute the geometrical relations between that trajectory and the trajectories of the daughter particles.

In a given reference system S, see fig. 3.5, considering only the projections on the x-t plane and assuming that the trajectory of the initial particle goes through the origin, the c.m. trajectory is

$$x = V_x t \tag{3.17}$$

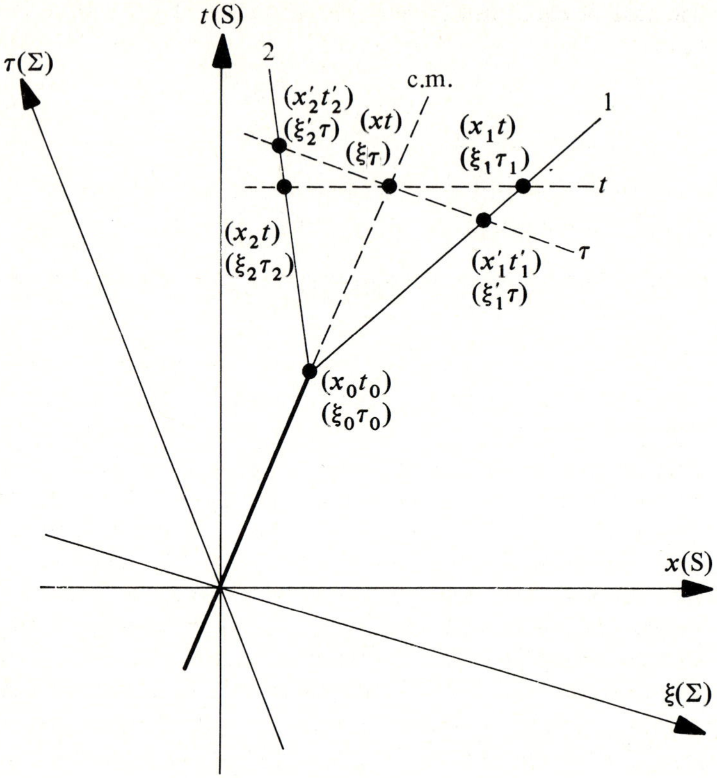

Fig. 3.5.

with

$$V_x = P_x/E,$$

$$E = \sqrt{P_x^2 + P_y^2 + P_z^2 + M^2},$$

where M is the mass of the parent particle and $\boldsymbol{P}$ its momentum.

At the point (x_0, t_0) the particle M decays into the two daughter particles of mass M_1 and M_2 the positions of which at time t are

$$x_1 = x_0 + v_{1_x}(t_1 - t_0),$$

$$x_2 = x_0 + v_{2_x}(t_2 - t_0),$$

$$v_{1_x} = \frac{p_{1_x}}{E_1}, \qquad E_1 = \sqrt{p_{1_x}^2 + p_{1_y}^2 + p_{1_z}^2 + M_1^2},$$

$$v_{2_x} = \frac{p_{2_x}}{E_2}, \qquad E_2 = \sqrt{p_{2_x}^2 + p_{2_y}^2 + p_{2_z}^2 + M_2^2},$$

$$E_1 + E_2 = E, \qquad \boldsymbol{p}_1 + \boldsymbol{p}_2 = \boldsymbol{P}. \tag{3.18}$$

Consider the case of equal time $t_1 = t_2 = t$. The ratio of the distances to the c.m. coordinate x given in eq. (3.17) is, see fig. 3.5,

$$\frac{\Delta x_1}{\Delta x_2} = \frac{x_1 - x}{x_2 - x} = \frac{\dfrac{p_{1_x}}{E_1} - \dfrac{p_{1_x} + p_{2_x}}{E_1 + E_2}}{\dfrac{p_{2_x}}{E_2} - \dfrac{p_{1_x} + p_{2_x}}{E_1 + E_2}} = -\frac{E_2}{E_1}, \tag{3.19}$$

which demonstrates the result (3.14), and which for the non-relativistic limit goes into the usual relation

$$\frac{\Delta x_1}{\Delta x_2} = -\frac{M_2}{M_1}, \qquad \text{or } x = \frac{M_1 x_1 + M_2 x_2}{M_1 + M_2}.$$

Now consider the general case of unequal times $t_1' \neq t_2'$. To that end we simply go into another system Σ, fig. 3.5, where the c.m. and the two daughter particles have coordinates ξ, ξ_1 and ξ_2 corresponding to x, x_1 and x_2 with now however non-equal times τ, τ_1 and τ_2. Nevertheless these three world points still retain the character that $(\xi\tau)$ is the position of the c.m. for the particles M_1 and M_2 at positions $(\xi_1\tau_1)$ and $(\xi_2\tau_2)$ moving now with velocities v_1' and v_2'. On the other hand the point $(\xi\tau)$ is also the c.m. of the particles M_1 and M_2 at the equal time τ in Σ, i.e., at the positions $(\xi_1'\tau)$ and $(\xi_2'\tau)$ as shown on fig. 3.5. Conversely going back to the initial system S, the world point (xt), corresponding to $(\xi\tau)$ in Σ, is also the c.m. of the particles at $(x_1't_1')$ and $(x_2't_2')$, corresponding to $(\xi_1'\tau)$ and $(\xi_2'\tau)$ in Σ respectively. Thus we see that the c.m. in relativity is associated with a whole family of particle positions with space-like separations, i.e., all the positions obtained at the crossing points of any straight lines containing (xt) with the particle trajectories. To each of these straight lines corresponds a Lorentz transform into a reference system in which the three points have equal times. Thus again we find in the equal time system the same relations (3.19) which demonstrates in the case of a decaying particle the general validity of (3.14)*.

We now consider the case of two distinct particles. This yields the following problem: Given two particles with arbitrary 4-momenta $(\boldsymbol{p}_1 E_1)$ and $(\boldsymbol{p}_2 E_2)$ in S, find the center of mass if the particles have the positions $(\boldsymbol{r}_1 t_1)$ and $(\boldsymbol{r}_2 t_2)$.

The c.m. energy and momentum are again given by eqs. (3.18). Restricting ourselves again to the x-t plane, see fig. 3.6, one can find the crossing of the particle world lines by drawing lines through the points $(x_1 t_1)$ and $(x_2 t_2)$ with slopes

$$v_{1_x} = \frac{p_{1_x}}{E_1}, \qquad v_{2_x} = \frac{p_{2_x}}{E_2}.$$

* For this demonstration the separation between the world points was taken to be space-like, which is sufficient for our single-time treatment. In covariant treatments, as for example in the Bethe-Salpeter equations, also time-like separations arise. The present demonstration can easily be extended to include that case.

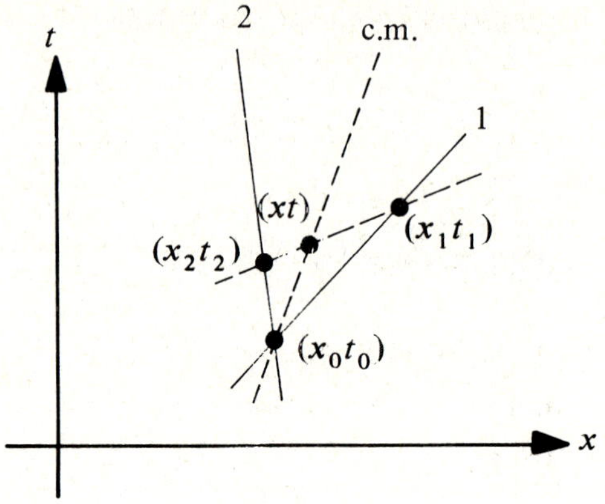

Fig. 3.6.

Let us call the crossing point (x_0t_0). In contrast to the above example of the decay of a particle, here the point (x_0t_0) has no physical meaning since in general the world lines of the particles do not cross, i.e., the projections of the world lines on the x-t, y-t and z-t planes cross at different times. We now can construct the trajectory of the c.m. in the x-t plane as it goes through the point (x_0t_0) with the slope

$$V_x = \frac{P_x}{E}.$$

Then, the c.m. corresponding to the given points (x_1t_1) and (x_2t_2) is given by the crossing of the c.m. trajectory with the line connecting the two points. After some algebra we find, considering the x-t plane, for the c.m. coordinates

$$x = \frac{(x_1 - x_2)(E_1x_1 + E_2x_2) - (t_1 - t_2)(p_{1_x}x_1 + p_{2_x}x_2)}{(x_1 - x_2)E - (t_1 - t_2)P_x}, \tag{3.20}$$

$$t = \frac{(x_1 - x_2)(E_1t_1 + E_2t_2) - (t_1 - t_2)(p_{1_x}t_1 + p_{2_x}t_2)}{(x_1 - x_2)E - (t_1 - t_2)P_x}. \tag{3.21}$$

For the y-t plane, we have likewise

$$y = \frac{(y_1 - y_2)(E_1y_1 + E_2y_2) - (t_1 - t_2)(p_{1_y}y_1 + p_{2_y}y_2)}{(y_1 - y_2)E - (t_1 - t_2)P_y},$$

$$t = \frac{(y_1 - y_2)(E_1t_1 + E_2t_2) - (t_1 - t_2)(p_{1_y}t_1 - p_{2_y}t_2)}{(y_1 - y_2)E - (t_1 - t_2)P_y},$$

and, for the z-t plane

$$z=\frac{(z_1-z_2)(E_1z_1+E_2z_2)-(t_1-t_2)(p_{1_z}z_1+p_{2_z}z_2)}{(z_1-z_2)E-(t_1-t_2)P_z},$$

$$t=\frac{(z_1-z_2)(E_1t_1+E_2t_2)-(t_1-t_2)(p_{1_z}t_1+p_{2_z}t_2)}{(z_1-z_2)E-(t_1-t_2)P_z}.$$

As discussed above we need only the case $t_1=t_2$, for which we immediately verify that $t=t_1=t_2$ and that the c.m. position vector (xyz) has indeed the form (3.14).*

Finally, there holds for the c.m. position x

$$\left(\frac{\partial}{\partial x_1}+\frac{\partial}{\partial x_2}\right)x=1,$$

which can be verified by direct calculation from (3.20). Thus, formally, the commutation relations $[P_x, x]=-i$ are fulfilled.

This procedure for finding the c.m. coordinates can be generalized to more particles, for example in analogy to the Jacobi coordinates of non-relativistic kinematics. Thus one first defines the c.m. for particles 1 and 2, x_{12}, t_{12}, by (3.20), (3.21). Now one adds the third particle by replacing in these two expressions $x_1\to x_{12}, x_2\to x_3, t_1\to t_{12}, t_2\to t_3$. This way one obtains $x_{12,3}, t_{12,3}$ and so on. As to be expected, the general expressions are cumbersome. However in the present work we are interested only in the equal time case $(t_1=t_2=t_3\dots)$ for which the expression for the c.m. coordinate is simply (using the notation $E_{12}=E_1+E_2$)

$$x_{12,3}(t_1=t_2=t_3)=\frac{E_{12}x_{12}+E_3x_3}{E_{12}+E_3}=\frac{(E_1+E_2)\dfrac{E_1x_1+E_2x_2}{E_1+E_2}+E_3x_3}{E_1+E_2+E_3}$$

$$=\frac{E_1x_1+E_2x_2+E_3x_3}{E_1+E_2+E_3},$$

as asserted in (3.14).

To re-write in relativistic kinematics the familiar expression for the c.m. and relative coordinate separation

$$e^{i(\boldsymbol{p}_1\cdot\boldsymbol{r}_1+\boldsymbol{p}_2\cdot\boldsymbol{r}_2)}=e^{i(\boldsymbol{P}\cdot\boldsymbol{R}+\boldsymbol{p}\cdot\boldsymbol{r})},$$

* Note that the c.m. coordinates for two equal-mass particles are not given by the often used ansatz $t=\frac{1}{2}(t_1+t_2)$, $x=\frac{1}{2}(x_1+x_2)$, see for example Lurie, p. 421.

we need in the equal time case the following relations:

$$\boldsymbol{P} = \boldsymbol{p}_1 + \boldsymbol{p}_2, \qquad \boldsymbol{p} = \left(\frac{\boldsymbol{p}_1}{E_1} - \frac{\boldsymbol{p}_2}{E_2}\right)\frac{E_1E_2}{E_1 + E_2},$$

$$\boldsymbol{R} = \frac{E_1\boldsymbol{r}_1 + E_2\boldsymbol{r}_2}{E_1 + E_2}, \qquad \boldsymbol{r} = \boldsymbol{r}_1 - \boldsymbol{r}_2,$$

and the inverse formula

$$\boldsymbol{p}_1 = \boldsymbol{P}\frac{E_1}{E_1 + E_2} + \boldsymbol{p}, \qquad \boldsymbol{p}_2 = \boldsymbol{P}\frac{E_2}{E_1 + E_2} - \boldsymbol{p},$$

$$\boldsymbol{r}_1 = \boldsymbol{R} + \boldsymbol{r}\frac{E_2}{E_1 + E_2}, \qquad \boldsymbol{r}_2 = \boldsymbol{R} - \boldsymbol{r}\frac{E_1}{E_1 + E_2}.$$

They all yield the non-relativistic formulae by the replacement $E_i \to M_i$.

CHAPTER 4

Techniques and conventions

4.1. INTRODUCTION

In this chapter we develop, with attention to full detail, the mathematical concepts and calculational techniques which will be used throughout this book. The techniques are constructed around the concept of the invariants. These are the quantities which do not depend on the choice of the coordinate system, or, more generally, which are invariant under the transformations of the symmetry groups of the physical system. All quantities of the theory, i.e. fields, state vectors, operators, etc., will be written as invariants. Once the different invariants have been defined no vector coupling coefficients, i.e. 3-*j* coefficients, and no magnetic quantum numbers need to be written. All calculations will be performed exclusively in terms of invariants.

We begin the development by introducing in sect. 4.2, a consistent phase convention. It is defined so as to avoid the appearance of complicated phases in the successive recoupling steps and to simplify the evaluation of matrix elements. In order to conform to that phase convention, the Wigner reduced matrix elements must be redefined. In this new phase convention they will be called "invariant matrix elements". Most of the required invariant matrix elements needed are collected in this section.

In sect. 4.3, we describe a graphical calculus for handling angular momentum problems. This method replaces lengthy recoupling and reduction calculations by the drawing of a single graph, needing very few rules. The graph is then read off to yield directly the final expression of the complete calculation without any additional phase or factor. In this respect the present graphical method is economical, eliminates sources of errors and gives access to the calculation of quantities of great complexity. A few examples, chosen for their general interest, are treated in detail. The generalisation of the graphical calculus to SU(n) tensors is discussed in sect. 4.3.4.

We then proceed to the definition of the invariants which represent wave functions, state vectors, and operators. This is done in sect. 4.4. With these invariants some applications are given in detail in sect. 4.5. They concern vector algebra and vector calculus needed later on.

The expressions of Fock-space operators and their matrix elements in the present framework is given in sect. 4.6. The usual phase difficulty associated with time reversal of tensors in the occupation number representation is eliminated. Namely, the particles and the antiparticles, i.e. the tensorial forms of the creation and the annihilation operators, are handled identically. Finally, the relations between Fock-space many-body state vectors and the corresponding wave functions in Minkowski space are given in order to introduce correctly in both spaces the fractional parentage coefficients. The particle-hole conjugation is treated in the appendix A.2.

4.2. TENSORIAL SETS AND INVARIANT MATRIX ELEMENTS

4.2.1. Standard and contrastandard tensors

The Hilbert space vectors, in this work, are always defined as irreducible contra-standard tensors. More precisely the $(2J+1)$ components $\psi_M^{[J]}(\theta\varphi)$ of a vector are chosen so as to transform under a space rotation according to (J can be integer or half-integer)

$$\psi_M^{[J]}(\theta\varphi) = \sum_{M'} \psi_{M'}^{[J]}(\theta'\varphi')\mathcal{D}_{M'M}^{J}(\alpha\beta\gamma).$$

Contrastandard sets are denoted with a superscript $[J]$ with square parentheses. Standard tensor components on the other hand transform according to

$$\psi_M^{(J)}(\theta\varphi) = \sum_{M'} \psi_{M'}^{(J)}(\theta'\varphi')\mathcal{D}_{M'M}^{J*}(\alpha\beta\gamma).$$

Note that standard sets are distinguished from contrastandard sets by a superscript (J) with round parentheses.

Here the rotation matrix is defined as in Fano-Racah and Edmonds

$$\mathcal{D}_{M'M}^{J}(\alpha\beta\gamma) = \mathrm{e}^{iM'\gamma} d_{M'M}^{J}(\beta)\mathrm{e}^{iM\alpha},$$

where the coordinate system Ox′y′z′ is obtained from Oxyz by a rotation through the Euler angles α, β and γ in that order.

The fundamental property of standard and contrastandard sets is that the quantity

$$I = \sum_{M} \psi_M^{(J)}\phi_M^{[J]}$$

is an invariant, i.e. is independent of the orientation of the coordinate system.

The definitions are now completed by the convention that the phase between standard and contrastandard sets is fixed as

$$\psi_M^{(J)}(\theta\varphi) = D_y(\pi)\psi_M^{[J]}(\theta\varphi) = (-)^{J+M}\psi_{-M}^{[J]}(\theta\varphi), \tag{4.1}$$

where $D_y(\pi)$ is a rotation operator for a rotation by an angle π around the y-axis. This conjugation operation is identical to the Wigner time-reversal operation $\mathcal{T}$

$$\psi_M^{(J)}(\theta\varphi) = \mathcal{T}\psi_M^{[J]}(\theta\varphi)$$

with

$$\mathcal{T}^2 = (-)^{2J}.$$

Hermitian conjugation is written as

$$\psi_M^{[J]+} = (-)^{J+M} \tilde{\psi}_{-M}^{[J]}, \tag{4.2}$$

which defines the tilde operation discussed in detail below. We also have

$$\psi_M^{[J]++} = \psi_M^{[J]}.$$

Hermitian conjugation and time-reversal are related by

$$\psi_M^{[J]+} = \mathcal{T}\tilde{\psi}_M^{[J]}.$$

These two operations differ only by the tilde operation which in fact represents the charge conjugation operation $\mathcal{C}$, since hermitian conjugation is the product $\mathcal{C}\mathcal{T}$.

The tilde operation depends on the character of the particular tensor ψ. This dependence can be seen already on two examples which we now discuss: the spherical harmonics and the spin-$\frac{1}{2}$ wave functions. The tilde operation for other cases, in particular those involving operators, will be discussed when they arise.

The usual spherical harmonics transform as contrastandard sets. They have the hermitian conjugation property

$$Y_{lm}^+ = (-)^m Y_{l-m}.$$

Comparison with (4.2) shows that the tilde operation here means

$$\tilde{Y}_{lm} = (-)^l Y_{lm}.$$

Thus it is convenient to introduce modified spherical harmonics by the phase convention

$$Y_m^{[l]} = (-i)^l Y_{lm}. \tag{4.3}$$

Our choice of the phase $(-i)^l$ instead of $(+i)^l$ is dictated by the form of the vectorial product discussed below, eq. (4.10). These redefined functions, which we shall call the contrastandard spherical harmonics, fulfill

$$Y_m^{[l]+} = (-)^{l+m} Y_{-m}^{[l]}.$$

Hence, from (4.2),

$$\tilde{Y}_m^{[l]} = Y_m^{[l]},$$

i.e., these tensors are said to be self-conjugate. Likewise the standard spherical harmonics given by (4.1) from these $Y_m^{[l]}$ also are self-conjugate. Thus, when dealing with contrastandard and standard spherical harmonics, or any other self-conjugate tensors, a tilde is not required in the hermitian conjugation (4.2) and will be omitted.

Not all tensors can be chosen self-conjugate. In those cases the tilde must be retained. A case of this kind is provided by our second example. Let us consider the spin $s = \frac{1}{2}$ wave functions, $\chi_m^{[s]}$. They are the contrastandard components of a set of rank $2s + 1$, defined to verify (4.2) and orthonormalized according to

$$\langle \chi_m^{[s]} | \chi_{m'}^{[s]} \rangle = \delta_{mm'} .$$

In matrix representation

$$\chi_m^{[s]+} = \chi_m^{[s]*\mathrm{T}},$$

where T means transpose. Comparison with (4.2) gives

$$\tilde{\chi}_m^{[s]} = (-)^{s-m} \chi_m^{[s]*\mathrm{T}},$$

which defines the meaning of the tilde operation in that case. This relation in fact is valid for all half-integer spin wave functions.

We collect for convenience all the relations deduced from eqs. (4.1) and (4.2) between standard and contrastandard quantities in table 4.1. We will use these relations in order to rewrite all expressions in terms of contrastandard tensors only.

TABLE 4.1
Relations between standard and contrastandard tensors

$\psi_M^{(J)} = (-)^{J+M} \psi_{-M}^{[J]}$	$\psi_M^{[J]} = (-)^{J-M} \psi_{-M}^{(J)}$
$\psi_M^{(J)} = (-)^{2J} \tilde{\psi}_M^{[J]+}$	$\psi_M^{[J]} = \tilde{\psi}_M^{(J)+}$
$\psi_M^{(J)+} = (-)^{J+M} \psi_{-M}^{[J]+}$	$\psi_M^{[J]+} = (-)^{J-M} \psi_{-M}^{(J)+}$
$\psi_M^{(J)+} = (-)^{2J} \tilde{\psi}_M^{[J]}$	$\psi_M^{[J]+} = \tilde{\psi}_M^{(J)}$
$\tilde{\psi}_M^{(J)} = \psi_M^{[J]+}$	$\tilde{\psi}_M^{[J]} = (-)^{2J} \psi_M^{(J)+}$
$\tilde{\psi}_M^{(J)} = (-)^{J+M} \tilde{\psi}_{-M}^{[J]}$	$\tilde{\psi}_M^{[J]} = (-)^{J-M} \tilde{\psi}_{-M}^{(J)}$
$\tilde{\psi}_M^{(J)+} = \psi_M^{[J]}$	$\tilde{\psi}_M^{[J]+} = (-)^{2J} \psi_M^{(J)}$
$\tilde{\psi}_M^{(J)+} = (-)^{J+M} \tilde{\psi}_{-M}^{[J]+}$	$\tilde{\psi}_M^{[J]+} = (-)^{J-M} \tilde{\psi}_{-M}^{(J)+}$
$\psi_M^{(J)} = (-)^{J+M} \tilde{\psi}_{-M}^{(J)+}$	$\psi_M^{[J]} = (-)^{J+M} \tilde{\psi}_{-M}^{[J]+}$
$\tilde{\psi}_M^{(J)} = (-)^{J-M} \psi_{-M}^{(J)+}$	$\tilde{\psi}_M^{[J]} = (-)^{J-M} \psi_{-M}^{[J]+}$
$\psi_M^{(J)+} = (-)^{J+M} \tilde{\psi}_{-M}^{(J)}$	$\psi_M^{[J]+} = (-)^{J+M} \tilde{\psi}_{-M}^{[J]}$
$\tilde{\psi}_M^{(J)+} = (-)^{J-M} \psi_{-M}^{(J)}$	$\tilde{\psi}_M^{[J]+} = (-)^{J-M} \psi_{-M}^{[J]}$

We conclude this section by introducing as an abbreviation the compact notation

$$\hat{r}_m^{[l]} = Y_m^{[l]}(\hat{\boldsymbol{r}}) = Y_m^{[l]}(\theta\varphi)$$

and similarly for the momentum space

$$\hat{p}_m^{[l]} = Y_m^{[l]}(\hat{\boldsymbol{p}}).$$

This notation will often be used. We will also use the notation of rank-3 Hilbert tensors and write $r^{[1]}, e^{[1]}, p^{[1]}, \nabla^{[1]}, \sigma^{[1]}, \tau^{[1]}$, etc.

4.2.2. Invariant products

Both contrastandard and standard tensors can be coupled with the usual vector coupling coefficients to yield tensors of good total angular momentum. The Fano-Racah notation is used throughout:

$$\psi_M^{[I]}(1,2) = [\phi^{[J]}(1)\phi^{[K]}(2)]_M^{[I]} = \sum_{M_J} (JKM_JM_K|IM)\phi_{M_J}^{[J]}(1)\phi_{M_K}^{[K]}(2).$$

Owing to the phase convention, this product retains all the properties of a contrastandard tensorial set under rotation, time reversal, and hermitian conjugation. For example,

$$[\psi^{[J]}\psi^{[K]}]_M^{[I]+} = (-)^{I+M}[\tilde{\psi}^{[J]}\tilde{\psi}^{[K]}]_{-M}^{[I]}. \tag{4.4}$$

An example of coupled tensors are the tensorial multipoles involving coupling of a spin wave function with a spherical harmonic written as

$$Y_{SlM}^{[J]}(\hat{\boldsymbol{r}}) = [\chi^{[S]}Y^{[l]}(\hat{\boldsymbol{r}})]_M^{[J]}.$$

Of particular importance will be the case of spin $S=\frac{1}{2}$ and $S=1$ for which the following abbreviated forms are used (note that the coupling order adopted here for these tensors is spin-orbit and not orbit-spin)

$$Y_{lm}^{[j]}(\hat{\boldsymbol{r}}) = Y_{1/2lm}^{[j]}(\hat{\boldsymbol{r}}) = [\chi^{[1/2]}Y^{[l]}(\hat{\boldsymbol{r}})]_m^{[j]}, \tag{4.5}$$

$$\boldsymbol{Y}_{lM}^{[J]}(\hat{\boldsymbol{r}}) = Y_{1lM}^{[J]}(\hat{\boldsymbol{r}}) = [e^{[1]}Y^{[l]}(\hat{\boldsymbol{r}})]_M^{[J]}. \tag{4.6}$$

The spin-1 wave functions $e_m^{[1]}$ are the contrastandard unit vectors defined in terms of the usual cartesian unit vectors $\boldsymbol{e}_x$, $\boldsymbol{e}_y$ and $\boldsymbol{e}_z$ as

$$e_1^{[1]} = -\sqrt{\tfrac{1}{2}}\,(-i\boldsymbol{e}_x + \boldsymbol{e}_y),$$

$$e_0^{[1]} = -i\boldsymbol{e}_z,$$

$$e_{-1}^{[1]} = -\sqrt{\tfrac{1}{2}}\,(i\boldsymbol{e}_x + \boldsymbol{e}_y).$$

It can be checked along the lines of the previous section that these unit vectors $e_M^{[1]}$ are self-conjugate. Hence the vector spherical harmonics are self-conjugate

$$\tilde{\boldsymbol{Y}}_{lM}^{[J]} = \boldsymbol{Y}_{lM}^{[J]}.$$

This is not the case for the spin-$\frac{1}{2}$ spherical harmonics for which holds

$$\tilde{Y}_{lm}^{[j]} = \left[\tilde{\chi}^{[1/2]} Y^{[l]}\right]_m^{[j]}.$$

The invariant product of two tensors is defined as the scalar

$$\left[\phi^{[j]}\varphi^{[j]}\right]^{[0]} = \sum_m (jj - mm|00)\,\phi_{-m}^{[j]}\varphi_m^{[j]} = \sum_m \frac{(-)^{j+m}}{\hat{j}}\phi_{-m}^{[j]}\varphi_m^{[j]},$$

with the values of the vector coupling coefficients

$$(jj - mm|00) = \frac{(-)^{j+m}}{\hat{j}},$$

and with the notation

$$\hat{j} = \sqrt{2j+1}\,.$$

This last notation will be used throughout.

Comparing the above form with eq. (4.1) one sees that it can be written as

$$\hat{j}\left[\phi^{[j]}\varphi^{[j]}\right]^{[0]} = \sum_m \phi_m^{(j)}\varphi_m^{[j]},$$

which shows that indeed it is an invariant.

This way, when writing for a general state of a physical system, which of course must be independent of the orientation of the coordinate system

$$\psi = \sum_m a_m \Phi_m,$$

the amplitudes a_m must form a standard set when the wave functions Φ_m form a contrastandard set to yield an invariant product

$$\Psi = \sum_m a_m^{(j)}\Phi_m^{[j]} = \sum_m (-)^{j+m} a_{-m}^{[j]}\Phi_m^{[j]} = \hat{j}\left[a^{[j]}\Phi^{[j]}\right]^{[0]}. \tag{4.7}$$

Using the coupled form of the invariants only contrastandard sets appear in the expressions as shown by the right-hand side of this equation. This will be done throughout this book.

The product of a contrastandard tensorial component $\phi_M^{[J]}$ with its hermitian conjugate is given by the expansion

$$\phi_M^{[J]+}\phi_M^{[J]}=\sum_K(-)^{J+M}(JJ-MM|K0)\big[\tilde{\phi}^{[J]}\phi^{[J]}\big]_0^{[K]}.$$

An important application of this relation is the normalization integral of a wave function $\boldsymbol{\Phi}_M^{[J]}$ which reads

$$\sum\!\!\!\!\!\!\int \Phi_M^{[J]+}\Phi_M^{[J]}=\frac{1}{\hat{J}}\sum\!\!\!\!\!\!\int\big[\tilde{\Phi}^{[J]}\Phi^{[J]}\big]^{[0]}=\frac{1}{\hat{J}}\big[\Phi^{[J]}|\Phi^{[J]}\big]=1,$$

since the integrals and summations project on $K=0$. This equation defines our notation for the invariant norm matrix element

$$\big[\Phi^{[J]}|\Phi^{[J]}\big]=\hat{J}. \tag{4.8}$$

Note that in the invariant norm matrix element we omit the tilde as it is redundant with the bra notation.

The invariant triple product of three tensors plays also an important role, as it will appear in the Wigner-Eckart theorem when calculating invariant matrix elements in the next sections. It is defined by

$$\big[\phi^{[j]}\phi^{[k]}\phi^{[l]}\big]^{[0]}=\Big[\big[\phi^{[j]}\phi^{[k]}\big]^{[l]}\phi^{[l]}\Big]^{[0]}=\Big[\phi^{[j]}\big[\phi^{[k]}\phi^{[l]}\big]^{[j]}\Big]^{[0]},$$

i.e., it is associative and it verifies relatively to a circular permutation of its factors

$$(-)^{2k}\big[\phi^{[j]}\phi^{[k]}\phi^{[l]}\big]^{[0]}=(-)^{2l}\big[\phi^{[k]}\phi^{[l]}\phi^{[j]}\big]^{[0]}$$
$$=(-)^{2j}\big[\phi^{[l]}\phi^{[j]}\phi^{[k]}\big]^{[0]},$$

and more generally

$$\big[\phi^{[j]}\phi^{[k]}\phi^{[l]}\big]^{[0]}=(-)^{j+k-l}\big[\phi^{[k]}\phi^{[j]}\phi^{[l]}\big]^{[0]},\qquad \text{etc.}$$

In fact the invariant triple product defines the Wigner 3-j symbol by

$$\big[\phi^{[j]}\phi^{[k]}\phi^{[l]}\big]^{[0]}=(-)^{j-k+l}\sum_{m_jm_km_l}\begin{pmatrix} j & k & l\\ m_j & m_k & m_l\end{pmatrix}\phi_{m_j}^{[j]}\phi_{m_k}^{[k]}\phi_{m_l}^{[l]}.$$

Hermitian conjugation reads

$$\big[\phi^{[j]}\phi^{[k]}\phi^{[l]}\big]^{[0]+}=\big[\tilde{\phi}^{[j]}\tilde{\phi}^{[k]}\tilde{\phi}^{[l]}\big]^{[0]}.$$

4.2.3. Cartesian vectors

The contrastandard spherical components of a vector $\boldsymbol{V}$ with cartesian components V_x, V_y, V_z are defined like the spherical unit vectors

$$V_1^{[1]} = -\sqrt{\tfrac{1}{2}}\,(-iV_x + V_y),$$

$$V_0^{[1]} = -iV_z,$$

$$V_{-1}^{[1]} = -\sqrt{\tfrac{1}{2}}\,(iV_x + V_y).$$

These spherical components are self-conjugate.

The scalar product of two vectors

$$\boldsymbol{A}\cdot\boldsymbol{B} = A_xB_x + A_yB_y + A_zB_z = A_1^{[1]}B_{-1}^{[1]} - A_0^{[1]}B_0^{[1]} + A_{-1}^{[1]}B_1^{[1]}$$

can be written in the invariant form

$$\boldsymbol{A}\cdot\boldsymbol{B} = \hat{1}[A^{[1]}B^{[1]}]^{[0]}. \tag{4.9}$$

Using the numerical values of the vector coupling coefficients $(11mm'|1\,m+m')$, one finds that for the vector product

$$\boldsymbol{C} = \boldsymbol{A}\times\boldsymbol{B},$$

there holds

$$C_M^{[1]} = \sqrt{2}\,[A^{[1]}B^{[1]}]_M^{[1]}. \tag{4.10}$$

This result explains the choice of the phase $(-i)^l$ in the definition (4.3), since the phase $(+i)^l$ would have led to the inverted order of the factors in the right-hand side of (4.10).

For the particular case of the angular momentum operator $\boldsymbol{J}$, the commutation relations in contrastandard form read

$$[J_1^{[1]}, J_{-1}^{[1]}] = iJ_0^{[1]},$$

$$[J_1^{[1]}, J_0^{[1]}] = iJ_1^{[1]},$$

$$[J_{-1}^{[1]}, J_0^{[1]}] = -iJ_{-1}^{[1]},$$

or equivalently in the coupled form (4.10)

$$[J^{[1]}J^{[1]}]_M^{[1]} = i\sqrt{\tfrac{1}{2}}\,J_M^{[1]}.$$

Also, from the scalar product (4.9), we have

$$\boldsymbol{J}^2 = \hat{1}[J^{[1]}J^{[1]}]^{[0]}.$$

4.2.4. Invariant matrix elements

Consider the evaluation of the matrix element of an operator $\Omega_\mu^{[\lambda]}$ between the initial wave function $\Psi_M^{[I]}$ and the final wave function $\Phi_m^{[J]+}$. To that end we first perform the reduction, using the completeness relations of the vector coupling coefficients

$$\Phi_m^{[J]+}\Omega_\mu^{[\lambda]}\Psi_M^{[I]} = (-)^{J+m}\tilde{\Phi}_{-m}^{[J]}\Omega_\mu^{[\lambda]}\Psi_M^{[I]} = (-)^{J+m}\sum_{RS}\left[\tilde{\Phi}^{[J]}\left[\Omega^{[\lambda]}\Psi^{[I]}\right]^{[R]}\right]^{[S]}$$

$$\times (JR - m\, M_S + m|SM_S)(\lambda I\mu M|R\, M_S + m)\,.$$

Next, upon integrating over angles and summing over spin indices the sum over R, S reduces to a single term, $R = J$ and $S = 0$, which enforces $\mu + M = m$. Recalling the definition of the Wigner 3-j symbol

$$\begin{pmatrix} j & k & l \\ m & n & p \end{pmatrix} = \frac{(-)^{j-k-p}}{\hat{l}}(jkmn|l - p)\,,$$

the matrix element becomes

$$\oint \Phi_m^{[J]+}\Omega_\mu^{[\lambda]}\Psi_M^{[I]} = (-)^{J+m}(-)^{J-\lambda+I}\begin{pmatrix} J & \lambda & I \\ -m & \mu & M \end{pmatrix}\oint\left[\tilde{\Phi}^{[J]}\Omega^{[\lambda]}\Psi^{[I]}\right]^{[0]}. \tag{4.11}$$

This is the Wigner-Eckart theorem given in terms of an integral over an invariant triple product. The meaning of the phases is obvious: the first phase is associated with the re-writing of the hermitian conjugate according to table 4.1 while the second is associated with the definition of the invariant triple product in terms of the Wigner 3-j coefficients, sect. 4.2.2.

We now introduce the notation

$$\left[\Phi^{[J]}|\Omega^{[\lambda]}|\Psi^{[I]}\right] = \oint\left[\tilde{\Phi}^{[J]}\Omega^{[\lambda]}\Psi^{[I]}\right]^{[0]}, \tag{4.12}$$

and we call this quantity the invariant matrix element. It is an invariant triple product with all its properties. In the notation (4.12) for the invariant matrix element we drop the tilde as it was done for the invariant norm matrix element (4.8), since it is redundant with the bra notation.

The invariant matrix element is related to the usual Wigner reduced matrix element by

$$\left[\Phi^{[J]}|\Omega^{[\lambda]}|\Psi^{[I]}\right] = (-)^{J+\lambda-I}\langle\Phi^{[J]}\|\Omega^{[\lambda]}\|\Psi^{[I]}\rangle\,.$$

The calculation of matrix elements requires the knowledge of the invariant matrix elements of very few operators which we collect now. These invariants are obtained from their defining equations (4.11) and (4.12):

$$[\psi^{[I]}|1^{[0]}|\psi^{[I]}] = [\psi^{[I]}|\psi^{[I]}] = \sqrt{2I+1} = \hat{I}, \tag{4.13}$$

$$[\chi^{[1/2]}|\sigma^{[1]}|\chi^{[1/2]}] = [\tfrac{1}{2}|\sigma|\tfrac{1}{2}] = i\sqrt{6}, \tag{4.14}$$

$$[\psi^{[I]}|J^{[1]}|\psi^{[I]}] = i\hat{I}\sqrt{I(I+1)}, \tag{4.15}$$

$$[Y^{[l_1]}|Y^{[l_2]}|Y^{[l_3]}] = [l_1|l_2|l_3] = (-)^{(l_1+l_2+l_3)/2}\frac{\hat{l}_1\hat{l}_2\hat{l}_3}{\sqrt{4\pi}}\begin{pmatrix} l_1 & l_2 & l_3 \\ 0 & 0 & 0 \end{pmatrix} \geqslant 0. \tag{4.16}$$

In these relations $\sigma^{[1]}$ and $J^{[1]}$ are the spherical contrastandard self-conjugate forms of the Pauli spin matrices and of the angular momentum operators defined from their cartesian components according to sect. 4.2.3.

The invariant matrix elements of the position operator $\boldsymbol{r}$ and of the momentum operator $\boldsymbol{p}$ are also often needed. Denoting an orbital wave function of a complete orthonormal set

$$\varphi_{\alpha m}^{[l]}(\boldsymbol{r}) = F_{\alpha l}(r)Y_m^{[l]}(\hat{\boldsymbol{r}}),$$

the invariant matrix elements of the radius vectors in position and momentum spaces

$$r_m^{[1]} = \sqrt{\frac{4\pi}{3}}\, rY_m^{[1]}(\hat{\boldsymbol{r}}),$$

$$p_m^{[1]} = -i\nabla_m^{[1]},$$

are

$$[\varphi_\alpha^{[l]}|r^{[1]}|\varphi_\beta^{[l-1]}] = \sqrt{l}\,\langle F_{\alpha l}|r|F_{\beta l-1}\rangle$$

$$= \sqrt{l}\int r^2\,\mathrm{d}r\, F_{\alpha l}(r)rF_{\beta l-1}(r), \tag{4.17}$$

$$[\varphi_\alpha^{[l]}|r^{[1]}|\varphi_\beta^{[l+1]}] = \sqrt{l+1}\,\langle F_{\alpha l}|r|F_{\beta l+1}\rangle, \tag{4.18}$$

$$[\varphi_\alpha^{[l]}|p^{[1]}|\varphi_\beta^{[l-1]}] = i\sqrt{l}\left\langle F_{\alpha l}\left|\frac{l-1}{r} - \frac{\partial}{\partial r}\right|F_{\beta l-1}\right\rangle, \tag{4.19}$$

$$[\varphi_\alpha^{[l]}|p^{[1]}|\varphi_\beta^{[l+1]}] = -i\sqrt{l+1}\left\langle F_{\alpha l}\left|\frac{l+2}{r} + \frac{\partial}{\partial r}\right|F_{\beta l+1}\right\rangle. \tag{4.20}$$

The radial functions $F_{\alpha l}(r)$ will frequently be normalized spherical Bessel functions.

In this particular case, setting

$$\varphi_p^{[l]}(\boldsymbol{r}) = \sqrt{\frac{2}{\pi}}\, j_l(pr) Y^{[l]}(\hat{\boldsymbol{r}}),$$

the relations

$$j_{l+1}(x) = \left(\frac{l}{x} - \frac{\partial}{\partial x}\right) j_l(x),$$

$$j_{l-1}(x) = \left(\frac{l+1}{x} + \frac{\partial}{\partial x}\right) j_l(x), \tag{4.21}$$

$$\frac{2}{\pi}\int r^2 \mathrm{d}r\, j_l(pr)\, j_l(p'r) = \frac{\delta(p-p')}{pp'} \tag{4.22}$$

yield for the gradient operator the result,

$$\left[\varphi_p^{[\lambda]}\middle|\, \nabla^{[1]} \middle| \varphi_q^{[l]}\right] = \alpha_{\lambda l} \frac{\delta(p-q)}{p^2} p, \tag{4.23}$$

with the factor $\alpha_{\lambda l}$ defined by

$$\alpha_{\lambda l} = \begin{cases} +\sqrt{\lambda+1} & \text{if } l = \lambda + 1 \\ -\sqrt{\lambda} & \text{if } l = \lambda - 1 \\ 0 & \text{otherwise,} \end{cases} \tag{4.24}$$

or

$$\alpha_{\lambda l} = (l - \lambda)\sqrt{\tfrac{1}{2}(\lambda + l + 1)}\,, \qquad \text{with } |\lambda - l| \leqslant 1\,.$$

For example the evaluation of the action of the gradient operator on the pion multipole function $Y^{[l]}(\hat{\boldsymbol{r}}) g_{\nu l}(r)$, eq. (3.4), yields, collecting all the above expressions,

$$\left[Y^{[\lambda]}\middle|\, \nabla^{[1]} \middle| Y^{[l]}\right] g_{\nu l}(r) = \alpha_{\lambda l} k_{\nu l \lambda}(r)$$

with

$$k_{\nu l \lambda}(r) = \left(\frac{2}{\pi}\right)^{1/2} \int p^2 \mathrm{d}p \frac{p}{\sqrt{E}} f_{\nu l}(p)\, j_\lambda(pr)\,.$$

The evaluation of the radial integrals of the position operators r and r^2 on wave functions expanded into spherical Bessel functions can be performed using the relations

$$\frac{2}{\pi}\int r^2 \mathrm{d}r\, j_l(pr) r j_{l+1}(p'r) = \frac{l}{p'}\frac{\delta(p-p')}{pp'} - \frac{\mathrm{d}}{\mathrm{d}p'}\frac{\delta(p-p')}{pp'},$$

$$\frac{2}{\pi}\int r^2 \mathrm{d}r\, j_l(pr) r^2 j_l(p'r) = \frac{(l+2)^2}{pp'}\frac{\delta(p-p')}{pp'}$$

$$+(l+2)\left(\frac{1}{p'}\frac{\mathrm{d}}{\mathrm{d}p} + \frac{1}{p}\frac{\mathrm{d}}{\mathrm{d}p'}\right)\frac{\delta(p-p')}{pp'} + \frac{\mathrm{d}}{\mathrm{d}p}\frac{\mathrm{d}}{\mathrm{d}p'}\frac{\delta(p-p')}{pp'},$$

and integrating the resulting expressions by parts to transfer the derivatives from the delta functions.

4.3. THE GRAPHICAL RECOUPLING METHOD

4.3.1. Recoupling graphs

The calculation of a general invariant matrix element consists in expressing it in terms of the simple invariant matrix elements of the previous section, after proper recoupling of the angular momenta. This task is greatly systematized and simplified by the use of a graphical method. To each recoupling corresponds a graphical representation and each graphical representation is associated with an analytical expression. The rules of the diagrammatic method used throughout the text are now collected.

A system of two coupled tensors

$$\psi^{[c]} = [\phi^{[a]}\phi^{[b]}]^{[c]} \tag{4.25}$$

is represented by the graph of fig. 4.1a. Each of the two tensors is represented by a horizontal line labelled by its angular momentum. The coupling to total angular momentum c is represented by the bracket labelled c. Note that the order of the lines from top to bottom corresponds to the order of the tensors in (4.25) from left to right. Sometimes, in order to simplify the graphical display the two coupled lines of fig. 4.1a may be merged into a single line as shown in fig. 4.1b. It is then necessary to remember that this line, denoted with the total angular momentum c, in fact represents two coupled tensors.

Considering four tensors, the basic recoupling transformation is

$$[[\phi^{[a]}\phi^{[b]}]^{[c]}[\phi^{[d]}\phi^{[e]}]^{[f]}]^{[g]} = \sum_{hi}\begin{bmatrix} a & b & c \\ d & e & f \\ h & i & g \end{bmatrix} [[\phi^{[a]}\phi^{[d]}]^{[h]}[\phi^{[b]}\phi^{[e]}]^{[i]}]^{[g]}, \tag{4.26}$$

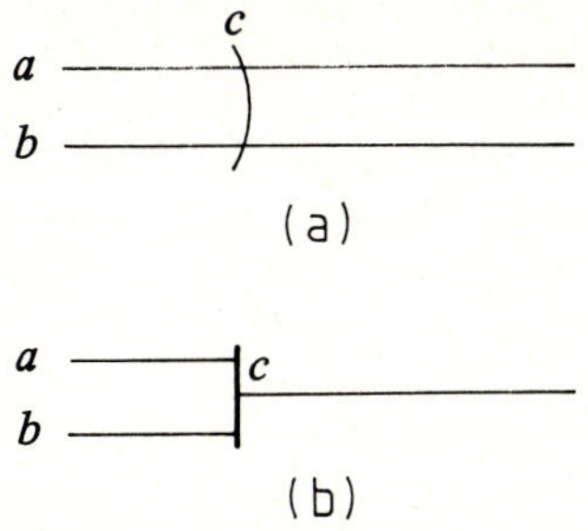

Fig. 4.1.

where the square symbol is related to the ordinary 9-*j* coefficient by

$$\begin{bmatrix} a & b & c \\ d & e & f \\ h & i & g \end{bmatrix} = \hat{c}\hat{f}\hat{h}\hat{i} \begin{Bmatrix} a & b & c \\ d & e & f \\ h & i & g \end{Bmatrix}.$$

This transformation is represented by the basic diagram of fig. 4.2, where summation over all new quantum numbers appearing in the diagram is implied. In the present case the summation must go over h and i.

Recoupling transformations for three tensors, viz., of the 6-*j* type, are represented by the same diagram with a dashed line representing a mock tensor of rank one, $I = 0$, as shown in fig. 4.3 for the case

$$\left[\phi^{[a]}\left[\phi^{[b]}\phi^{[c]}\right]^{[d]}\right]^{[g]} = \sum_e \begin{bmatrix} a & 0 & a \\ b & c & d \\ e & c & g \end{bmatrix} \left[\left[\phi^{[a]}\phi^{[b]}\right]^{[e]}\phi^{[c]}\right]^{[g]}.$$

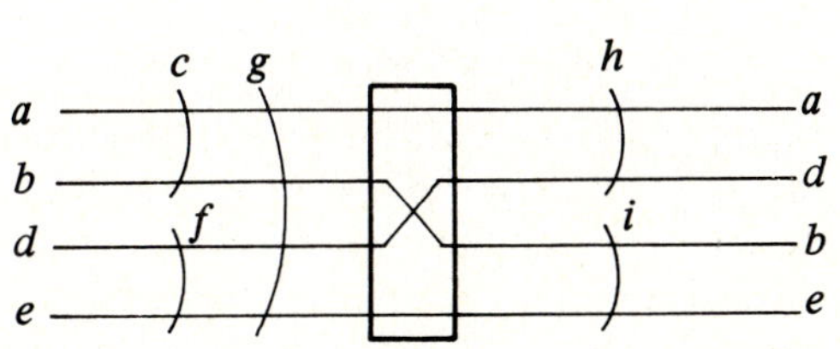

Fig. 4.2.

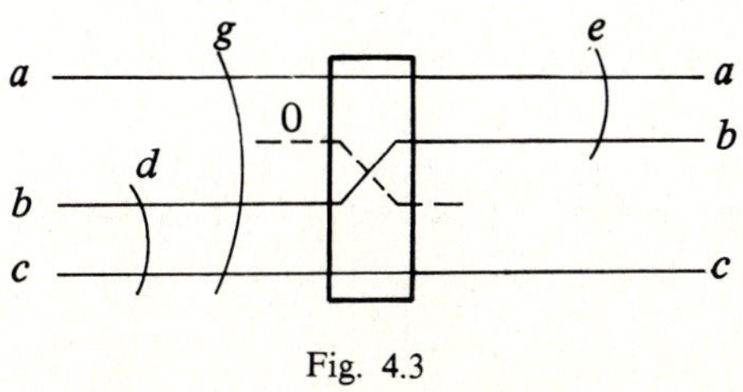

Fig. 4.3

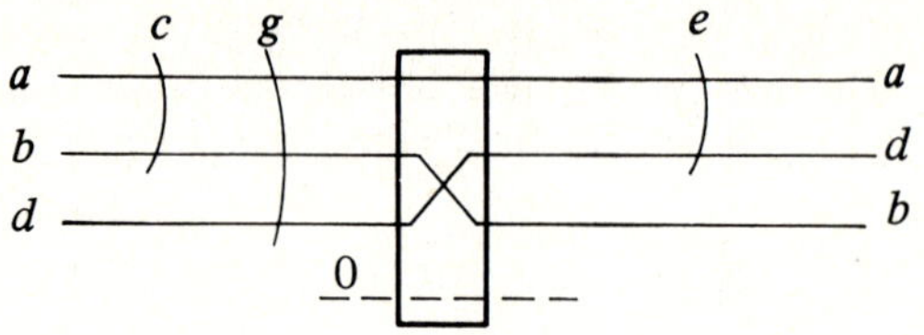

Fig. 4.4.

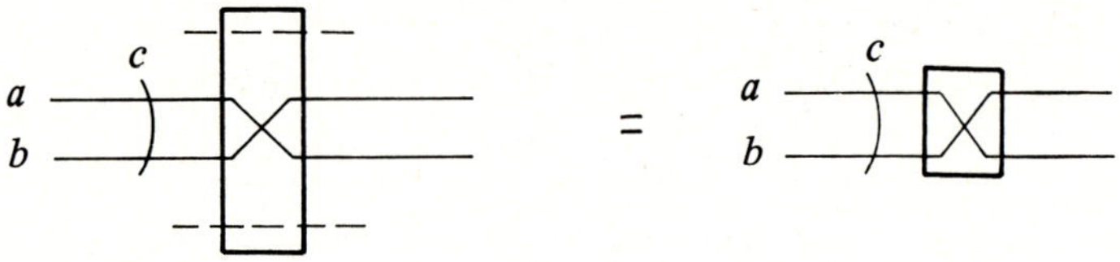

Fig. 4.5.

The three other possible recouplings of 3 tensors are obtained by permutation of the dashed line associated with the mock tensor of rank one, as shown for example in fig. 4.4.

The interchange in the coupling order of two tensors is represented by the simplified diagram on the right of fig. 4.5 which of course corresponds to the special value of the recoupling appearing on the left of fig. 4.5

$$\begin{bmatrix} 0 & a & a \\ b & 0 & b \\ b & a & c \end{bmatrix} = (-)^{a+b-c}.$$

The expressions for the recoupling coefficients having one or more zeroes are listed in table 4.2, where the usual notation is employed for the 6-j coefficients.

TABLE 4.2
Recoupling coefficients with one zero

$$\begin{bmatrix} e & f & b \\ c & d & b \\ a & a & 0 \end{bmatrix} = (-)^{f+c+a+b}\hat{a}\hat{b}\begin{Bmatrix} e & f & b \\ d & c & a \end{Bmatrix}$$

$$\begin{bmatrix} 0 & a & a \\ b & d & c \\ b & f & e \end{bmatrix} = \begin{bmatrix} d & b & c \\ a & 0 & a \\ f & b & e \end{bmatrix} = (-)^{f+c+a+b}\hat{f}\hat{c}\begin{Bmatrix} e & f & b \\ d & c & a \end{Bmatrix}$$

$$\begin{bmatrix} b & d & c \\ 0 & a & a \\ b & f & e \end{bmatrix} = \begin{bmatrix} a & 0 & a \\ d & b & c \\ f & b & e \end{bmatrix} = (-)^{e+d+a+b}\hat{f}\hat{c}\begin{Bmatrix} e & f & b \\ d & c & a \end{Bmatrix}$$

$$\begin{bmatrix} a & a & 0 \\ e & f & b \\ c & d & b \end{bmatrix} = \begin{bmatrix} a & e & c \\ a & f & d \\ 0 & b & b \end{bmatrix} = (-)^{f+c+a+b}\frac{\hat{c}\hat{d}}{\hat{a}}\begin{Bmatrix} e & f & b \\ d & c & a \end{Bmatrix}$$

$$\begin{bmatrix} e & f & b \\ a & a & 0 \\ c & d & b \end{bmatrix} = \begin{bmatrix} e & a & c \\ f & a & d \\ b & 0 & b \end{bmatrix} = (-)^{e+d+a+b}\frac{\hat{c}\hat{d}}{\hat{a}}\begin{Bmatrix} e & f & b \\ d & c & a \end{Bmatrix}$$

TABLE 4.2 (continued)

Recoupling coefficients with two zeroes

$$\begin{bmatrix} e & 0 & e \\ 0 & f & f \\ e & f & b \end{bmatrix} = \begin{bmatrix} e & 0 & e \\ b & f & e \\ f & f & 0 \end{bmatrix} = \begin{bmatrix} e & b & f \\ 0 & f & f \\ e & e & 0 \end{bmatrix} = 1$$

$$\begin{bmatrix} 0 & e & e \\ f & 0 & f \\ f & e & b \end{bmatrix} = \begin{bmatrix} 0 & e & e \\ f & b & e \\ f & f & 0 \end{bmatrix} = \begin{bmatrix} b & e & f \\ f & 0 & f \\ e & e & 0 \end{bmatrix} = (-)^{e+f-b}$$

$$\begin{bmatrix} e & e & 0 \\ e & b & f \\ 0 & f & f \end{bmatrix} = \begin{bmatrix} b & e & f \\ e & e & 0 \\ f & 0 & f \end{bmatrix} = \frac{1}{\hat{e}\hat{e}}$$

$$\begin{bmatrix} e & e & 0 \\ b & e & f \\ f & 0 & f \end{bmatrix} = \begin{bmatrix} e & b & f \\ e & e & 0 \\ 0 & f & f \end{bmatrix} = \frac{(-)^{e+f-b}}{\hat{e}\hat{e}}$$

$$\begin{bmatrix} 0 & e & e \\ f & f & 0 \\ f & b & e \end{bmatrix} = \begin{bmatrix} 0 & f & f \\ e & f & b \\ e & 0 & e \end{bmatrix} = \begin{bmatrix} e & e & 0 \\ f & 0 & f \\ b & e & f \end{bmatrix} = \begin{bmatrix} e & f & b \\ e & 0 & e \\ 0 & f & f \end{bmatrix} = \frac{\hat{b}}{\hat{e}\hat{f}}$$

$$\begin{bmatrix} e & 0 & e \\ f & f & 0 \\ b & f & e \end{bmatrix} = \begin{bmatrix} e & f & b \\ 0 & f & f \\ e & 0 & e \end{bmatrix} = \begin{bmatrix} e & 0 & e \\ e & f & b \\ 0 & f & f \end{bmatrix} = \begin{bmatrix} e & e & 0 \\ 0 & f & f \\ e & b & f \end{bmatrix} = (-)^{e+f-b}\frac{\hat{b}}{\hat{e}\hat{f}}$$

Recoupling coefficients with three zeroes

$$\begin{bmatrix} e & f & b \\ 0 & 0 & 0 \\ e & f & b \end{bmatrix} = \begin{bmatrix} e & 0 & e \\ f & 0 & f \\ b & 0 & b \end{bmatrix} = \begin{bmatrix} 0 & 0 & 0 \\ e & f & b \\ e & f & b \end{bmatrix} = \begin{bmatrix} 0 & e & e \\ 0 & f & f \\ 0 & b & b \end{bmatrix} = 1$$

$$\begin{bmatrix} e & f & b \\ e & f & b \\ 0 & 0 & 0 \end{bmatrix} = \begin{bmatrix} e & e & 0 \\ f & f & 0 \\ b & b & 0 \end{bmatrix} = \frac{\hat{b}}{\hat{e}\hat{f}}$$

$$\begin{bmatrix} e & 0 & e \\ e & e & 0 \\ 0 & e & e \end{bmatrix} = \begin{bmatrix} e & e & 0 \\ 0 & e & e \\ e & 0 & e \end{bmatrix} = (-)^{2e}\frac{1}{\hat{e}\hat{e}}$$

$$\begin{bmatrix} e & e & 0 \\ e & 0 & e \\ 0 & e & e \end{bmatrix} = \begin{bmatrix} 0 & e & e \\ e & e & 0 \\ e & 0 & e \end{bmatrix} = \frac{1}{\hat{e}\hat{e}}$$

The reduction of two spherical harmonics into one is also needed,

$$\left[Y^{[k]}(\hat{\boldsymbol{r}})Y^{[l]}(\hat{\boldsymbol{r}})\right]^{[L]}_{M} = Q^{L}_{kl}Y^{[L]}_{M}(\hat{\boldsymbol{r}}).$$

It is represented by the diagram of fig. 4.6. The coefficient Q, i.e., the box in the figure is associated with a Wigner 3-j symbol, eq. (4.16),

$$Q^{L}_{kl} = \frac{1}{\sqrt{4\pi}}(-)^{(k+l+L)/2}\hat{k}\hat{l}\begin{pmatrix} k & l & L \\ 0 & 0 & 0 \end{pmatrix} = \frac{1}{\hat{L}}[k|l|L] \geqslant 0.$$

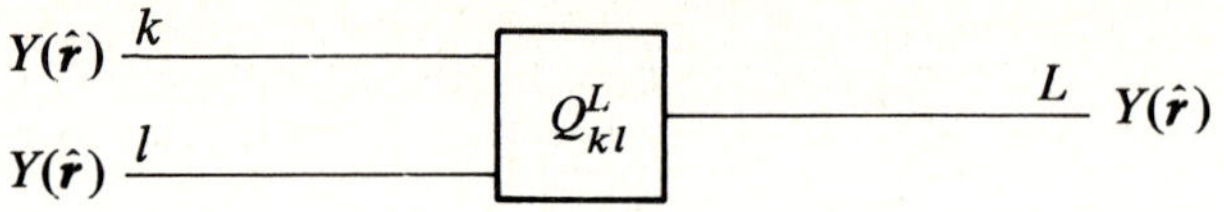

Fig. 4.6.

Finally, it is convenient to introduce a compact notation for vector coupling when working both in angular momentum and isospin spaces. For example, combined wave functions in space and isospace will be denoted as

$$\phi^{[jt]}_{m\tau} = \psi^{[j]}_{m}\eta^{[t]}_{\tau} \tag{4.27}$$

and we write for a coupled state

$$[\phi^{[j_1t_1]}\phi^{[j_2t_2]}]^{[JT]}_{M_JM_T} = \sum (j_1j_2m_1m_2|JM_J)(t_1t_2\tau_1\tau_2|TM_T)\phi^{[j_1t_1]}_{m_1\tau_1}\phi^{[j_2t_2]}_{m_2\tau_2}.$$

The basic recoupling of the four tensors both in space and isospace,

$$[[\varphi^{[j_1t_1]}\varphi^{[j_2t_2]}]^{[J_1T_1]}[\varphi^{[j_3t_3]}\varphi^{[j_4t_4]}]^{[J_2T_2]}]^{[JT]}$$

$$= \sum \begin{bmatrix} j_1 & j_2 & J_1 \\ j_3 & j_4 & J_2 \\ J_3 & J_4 & J \end{bmatrix} \begin{bmatrix} t_1 & t_2 & T_1 \\ t_3 & t_4 & T_2 \\ T_3 & T_4 & T \end{bmatrix} [[\varphi^{[j_1t_1]}\varphi^{[j_3t_3]}]^{[J_3T_3]}[\varphi^{[j_2t_2]}\varphi^{[j_4t_4]}]^{[J_4T_4]}]^{[JT]},$$

is then represented by a single recoupling graph, fig. 4.7, with a hatched top to indicate recoupling in both angular momentum and isospin spaces. Likewise, for a combined change of coupling order both in space and isospace the phase is given by a box, fig. 4.8, as in fig. 4.5, but with a hatched top.

Sometimes it will be convenient to use double indices even if a quantity is a tensor only in one space. For example, we may write

$$\psi^{[l0]}\eta^{[0t]} = \psi^{[l]}\eta^{[t]}. \tag{4.28}$$

Here in the notation of the left-hand side the position of the index indicates the space in which the quantity has a tensorial character.

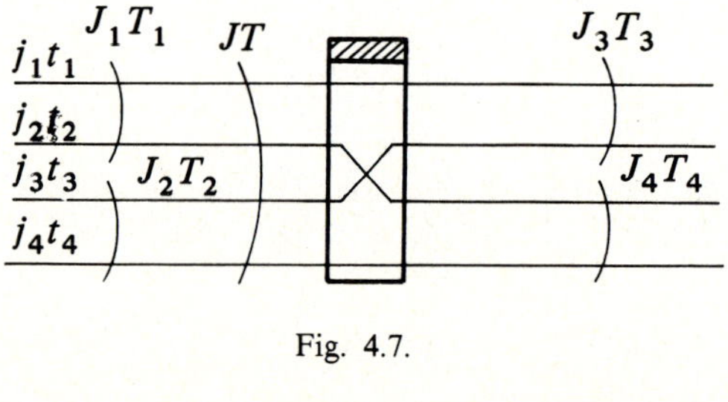

Fig. 4.7.

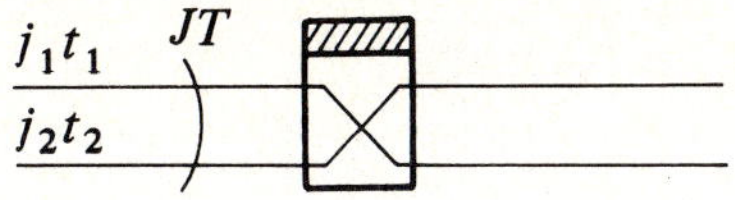

Fig. 4.8.

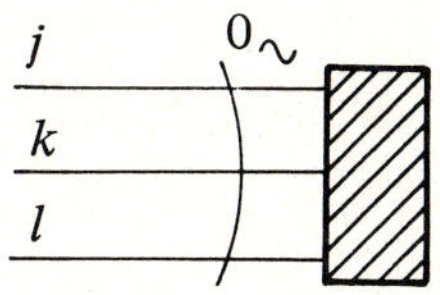

$$= [\psi_{\mathrm{f}}^{[j]}|\Omega^{[k]}|\psi_{\mathrm{i}}^{[l]}]$$

Fig. 4.9.

4.3.2. Graphs for invariant matrix elements

Invariant matrix elements, defined in sect. 4.2.4, are represented by hatched boxes in which terminate the ingoing tensor lines associated with the wave functions and operators of the matrix element as shown by fig. 4.9 for the expression

$$M = \left[\psi_{\mathrm{f}}^{[j]}|\Omega^{[k]}|\psi_{\mathrm{i}}^{[l]}\right].$$

The incoming tensor lines in that graph are of course coupled to zero, owing to the definition of the invariant matrix element. Furthermore, in this diagram the upper line corresponds to the bra function, indicated by the tilde. This ~ sign sometimes will be omitted from the diagram with the understanding that a bra line must always enter an end box at the top.

The special case with $\Omega^{[k]} = 1^{[0]}$ is the overlap matrix element, i.e., the normalization already encountered in eqs. (4.8) and (4.13)

$$\left[\psi^{[j]}|\psi^{[j]}\right] = \not{\sum}\left[\tilde{\psi}^{[j]}\psi^{[j]}\right]^{[0]} = \hat{j}.$$

It is represented by the hatched box of fig. 4.10. Note that the invariant norm is equal to $\hat{j}$ and not to 1.

As for coupled tensors, sometimes we use for compactness invariant matrix elements defined both in spin-orbit and isospin spaces. For example,

$$M = \left[\psi_{\mathrm{f}}^{[jt]}|\Omega^{[JT]}|\psi_{\mathrm{i}}^{[j't']}\right]$$

or

$$\left[\psi^{[jt]}|\psi^{[jt]}\right] = \hat{j}\hat{t}.$$

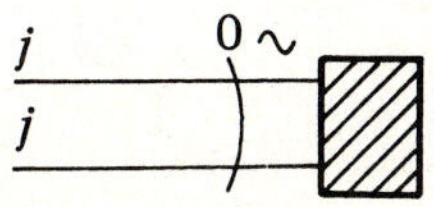

$$= [\tilde{\psi}^{[j]}|\psi^{[j]}] = \hat{j}$$

Fig. 4.10.

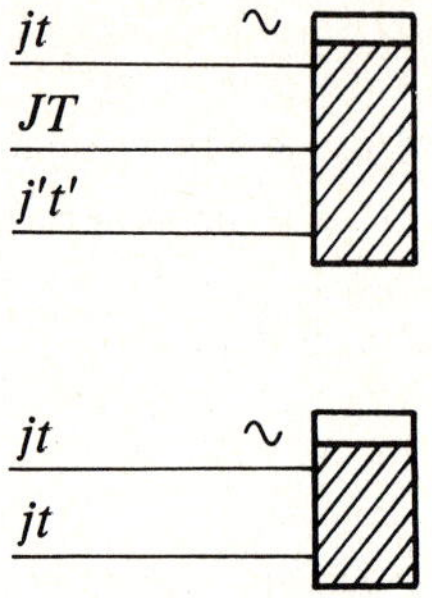

$$= [\psi_f^{[jt]} \, |\Omega^{[JT]}| \, \psi_i^{[j't']}]$$

$$= [\psi^{[jt]} \, | \, \psi^{[jt]}] = \hat{j}\,\hat{t}$$

Fig. 4.11.

They are represented by a hatched box diagram, now with white tops, as shown in fig. 4.11. The bracket indicating the overall coupling to zero is omitted as this coupling is implied by the invariant end boxes.

Finally, the insertion of a complete orthonormal set of states (α denotes all quantum numbers besides j)

$$\delta^3(\boldsymbol{x}-\boldsymbol{y}) = \sum_{\alpha jm} \psi_{\alpha m}^{[j]+}(\boldsymbol{x})\psi_{\alpha m}^{[j]}(\boldsymbol{y}) = \sum_{\alpha j} \hat{j}\left[\tilde{\psi}_\alpha^{[j]}(\boldsymbol{x})\psi_\alpha^{[j]}(\boldsymbol{y})\right]^{[0]} \tag{4.29}$$

or if $\boldsymbol{q}$ is a continuous quantum number

$$\delta^3(\boldsymbol{x}-\boldsymbol{y}) = \sum_j \hat{j}\int \mathrm{d}^3q\left[\tilde{\psi}_{\boldsymbol{q}}^{[j]}(\boldsymbol{x})\psi_{\boldsymbol{q}}^{[j]}(\boldsymbol{y})\right]^{[0]},$$

is represented by a hatched box with two outgoing tensors coupled to zero, as shown in fig. 4.12. In this figure the sum over α, j is understood and the upper line is a bra, hence the tilde superscript. In the case of purely orbital functions the identity (4.29) reads, recalling that the contrastandard spherical harmonics are self-conjugate, sect. 4.2.1,

$$\delta^3(\boldsymbol{x}-\boldsymbol{y}) = \sum_{\alpha l} \hat{l}\left[Y^{[l]}(\hat{\boldsymbol{x}})Y^{[l]}(\hat{\boldsymbol{y}})\right]^{[0]} F_\alpha(x)F_\alpha(y), \tag{4.30}$$

and no tilde is required in fig. 4.12. This remark will be used later for simplifying the graphs: we usually shall omit the tilde on the bra lines when they represent self-conjugate tensors.

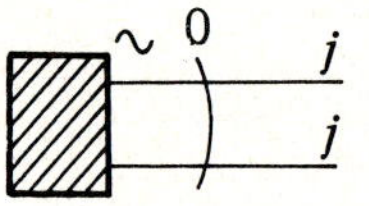

$$= \sum_{\alpha j} \hat{j}\,[\tilde{\psi}_\alpha^{[j]}(x)\,\psi_\alpha^{[j]}(y)]^{[0]}$$

Fig. 4.12.

4.3.3. *Application of the graphical method*

In order to show how tensorial calculations are done with the graphical method, we first treat three specific examples. They correspond to the following general cases: factorization of matrix elements, deriving of recoupling identities, use of the completeness relation of eq. (4.29). The general rules for the drawing and the evaluation of graphs which emerge from these examples will be given in the next section.

(i) Factorization of one- and two-body matrix elements. Consider the case of two systems (1) and (2), described by the wave functions $\Psi(1)$ and $\Phi(2)$, and on which act respectively the two operators $\Theta(1)$ and $\Omega(2)$. We evaluate the invariant matrix element

$$M = \left[[\Psi^{[I]}(1)\Phi^{[J]}(2)]^{[K]} | [\Theta^{[\lambda]}(1)\Omega^{[\mu]}(2)]^{[\nu]} | [\Psi^{[I']}(1)\Phi^{[J']}(2)]^{[K']} \right]. \quad (4.31)$$

The corresponding diagram, fig. 4.13, is obtained by drawing on the left of the page from top to bottom the entering horizontal lines associated with the tensors of M, read in their order from left to right in the expression (4.31). The initial couplings are represented by brackets joining the corresponding tensor lines. All the lines are of course coupled to a total angular momentum zero, since M is an invariant. The successive recoupling boxes are drawn from left to right so as to lead to the terminal hatched invariant boxes representing the separate invariant matrix elements for systems (1) and (2) respectively.

As indicated earlier, summations over all new quantum numbers are implied. In the present case these are the angular momenta X, Y, Z, V, W appearing at the exits of the recoupling boxes. However, we immediately see from the graph that the values zero are imposed on V and W by the final invariant matrix elements. We also see on the diagram several intermediate invariant triple products. Owing to the associative property of the triple product we see that the K and ν entering the first box must be coupled to K', hence $Z = K'$. The values for the other intermediate quantities are imposed by the invariant triple products of the hatched terminal boxes. Hence $X = I'$ and $Y = J'$. Thus supplying these values for the new angular momenta we

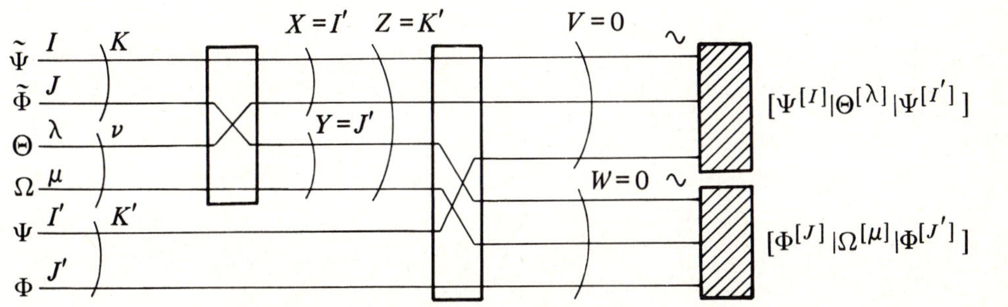

Fig. 4.13.

directly read from the graph 4.13

$$M=\begin{bmatrix} I & J & K \\ \lambda & \mu & \nu \\ I' & J' & K' \end{bmatrix}\begin{bmatrix} I' & J' & K' \\ I' & J' & K' \\ 0 & 0 & 0 \end{bmatrix}\left[\Psi^{[I]}|\Theta^{[\lambda]}|\Psi^{[I']}\right]\left[\Phi^{[J]}|\Omega^{[\mu]}|\Phi^{[J']}\right]$$

$$=\frac{\hat{K}'}{\hat{I}'\hat{J}'}\begin{bmatrix} I & J & K \\ \lambda & \mu & \nu \\ I' & J' & K' \end{bmatrix}\left[\Psi^{[I]}|\Theta^{[\lambda]}|\Psi^{[I']}\right]\left[\Phi^{[J]}|\Omega^{[\mu]}|\Phi^{[J']}\right],$$

where we have used for the second recoupling coefficient its value from table 4.2.

The special cases where the operator acts on one of the systems only, are obtained immediately by substituting in (4.31) the mock unit operator $1^{[0]}$ for one of the operators.

For example acting on system (1), the operator $\Theta^{[\lambda]}\Omega^{[\mu]}$ reads $\Theta^{[\lambda]}(1)1^{[0]}(2)$ and the corresponding mock tensor line of rank one is introduced in fig. 4.14, dashed line. Now one of the invariant matrix elements becomes a norm box. The result is, using the values of table 4.2,

$$M=\frac{\hat{K}'}{\hat{J}\hat{I}'}\begin{bmatrix} I & J & K \\ \lambda & 0 & \lambda \\ I' & J & K' \end{bmatrix}\left[\Psi^{[I]}(1)|\Theta^{[\lambda]}(1)|\Psi^{[I']}(1)\right]\left[\Phi^{[J]}(2)|\Phi^{[J']}(2)\right]$$

$$=\delta_{JJ'}\frac{\hat{K}'}{\hat{I}'}\begin{bmatrix} I & J & K \\ \lambda & 0 & \lambda \\ I' & J & K' \end{bmatrix}\left[\Psi^{[I]}|\Theta^{[\lambda]}|\Psi^{[I']}\right].$$

Acting on system (2) only, the operator becomes $1^{[0]}\Omega^{[\mu]}$ and yields from the diagram of fig. 4.15

$$M=\delta_{II'}\frac{\hat{K}'}{\hat{J}'}\begin{bmatrix} I & J & K \\ 0 & \mu & \mu \\ I & J' & K' \end{bmatrix}\left[\Phi^{[J]}(2)|\Omega^{[\mu]}(2)|\Phi^{[J']}(2)\right].$$

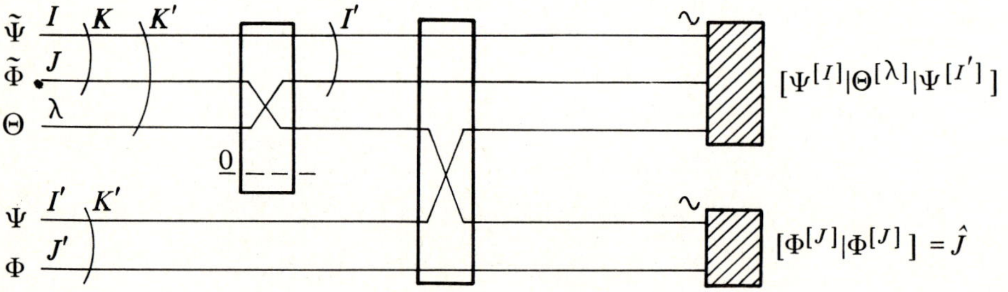

Fig. 4.14.

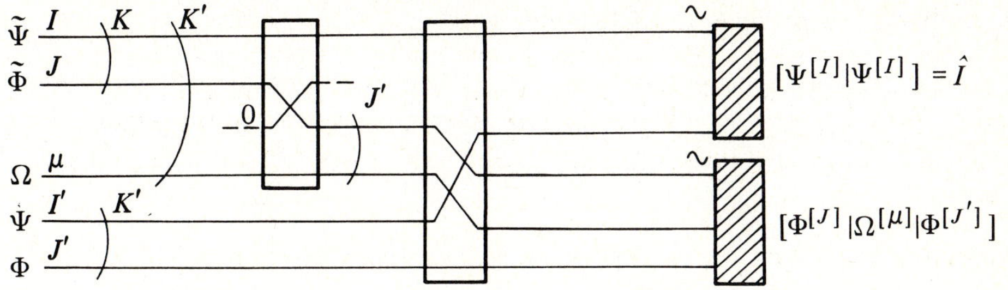

Fig. 4.15.

Similarly, the normalization of coupled states is given by fig. 4.16,

$$\left[\left[\Psi^{[I]}(1)\Phi^{[J]}(2)\right]^{[K]}\middle|\left[\Psi^{[I]}(1)\Phi^{[J]}(2)\right]^{[K]}\right] = \begin{bmatrix} I & J & K \\ 0 & 0 & 0 \\ I & J & K \end{bmatrix}\begin{bmatrix} I & J & K \\ I & J & K \\ 0 & 0 & 0 \end{bmatrix}\hat{I}\hat{J} = \hat{K}.$$

Of course this last simple result agrees with the definition of the invariant norm, eqs. (4.8) and (4.13).

Note that on all the diagrams the recouplings are done so that bra wave functions enter the invariant end boxes at the top. From now on, when there is no possible confusion, the tilde of the bra lines may be omitted in the diagrams.

(ii) Recoupling identities. In order to demonstrate recoupling identities we recognize that a given overall recoupling in general can be achieved by more than one recoupling graph. For example, we have from fig. 4.17 the orthogonality relation of the recoupling coefficients

$$\sum_{hi}\begin{bmatrix} a & b & c \\ d & e & f \\ h & i & g \end{bmatrix}\begin{bmatrix} a & d & h \\ b & e & i \\ c' & f' & g \end{bmatrix} = \delta_{cc'}\delta_{ff'}.$$

Likewise from fig. 4.18 we read the orthogonality relation for 6-*j* type coefficients

$$\sum_{\mu}\begin{bmatrix} j & k & J \\ l & 0 & l \\ \mu & k & I \end{bmatrix}\begin{bmatrix} j & l & \mu \\ k & 0 & k \\ J' & l & I \end{bmatrix} = \delta_{JJ'}.$$

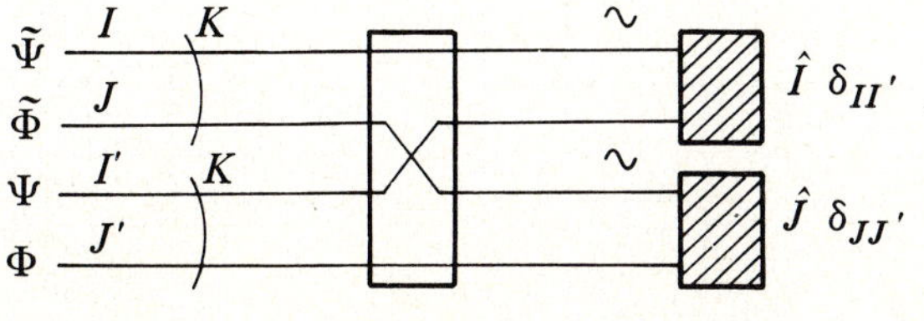

Fig. 4.16.

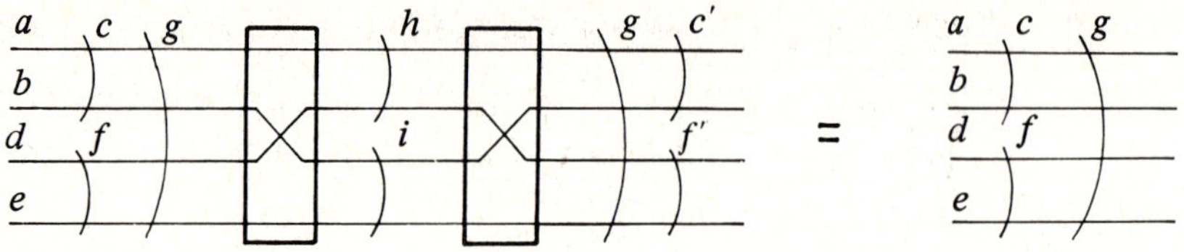

Fig. 4.17.

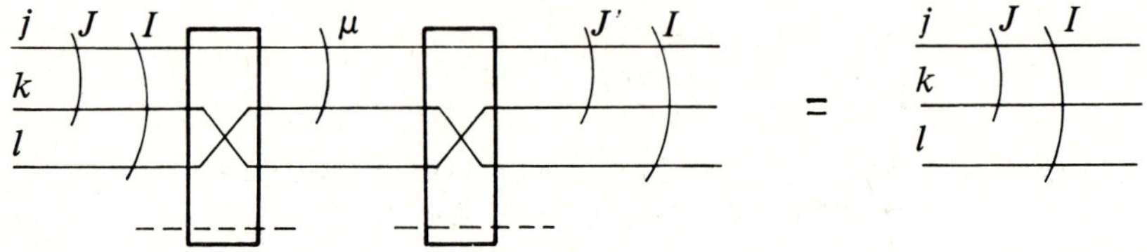

Fig. 4.18.

An example of a more complicated identity is given by the two graphs of fig. 4.19 which achieve the same recoupling transformation. Hence

$$\sum_{\mu}(-)^{j+l-\mu}\begin{bmatrix} j & k & J \\ l & 0 & l \\ \mu & k & I \end{bmatrix}\begin{bmatrix} l & j & \mu \\ k & 0 & k \\ J' & j & I \end{bmatrix} = (-)^{j+l-J-J'+2k}\begin{bmatrix} k & j & J \\ l & 0 & l \\ J' & j & I \end{bmatrix}.$$

Finally, a quite general situation is represented by the two equivalent recoupling diagrams of fig. 4.20. They give the identity

$$\sum_{PQ}\begin{bmatrix} A & B & J \\ C & D & K \\ P & Q & I \end{bmatrix}\begin{bmatrix} a & b & A \\ e & f & C \\ S & U & P \end{bmatrix}\begin{bmatrix} c & d & B \\ g & h & D \\ T & V & Q \end{bmatrix}\begin{bmatrix} S & U & P \\ T & V & Q \\ M & N & I \end{bmatrix}$$

$$= \sum_{\alpha\beta\gamma\delta}\begin{bmatrix} a & b & A \\ c & d & B \\ \alpha & \beta & J \end{bmatrix}\begin{bmatrix} e & f & C \\ g & h & D \\ \gamma & \delta & K \end{bmatrix}\begin{bmatrix} \alpha & \beta & J \\ \gamma & \delta & K \\ M & N & I \end{bmatrix}\begin{bmatrix} a & c & \alpha \\ e & g & \gamma \\ S & T & M \end{bmatrix}\begin{bmatrix} b & d & \beta \\ f & h & \delta \\ U & V & N \end{bmatrix}.$$

The first diagram has two fewer intermediate summations. A large number of other

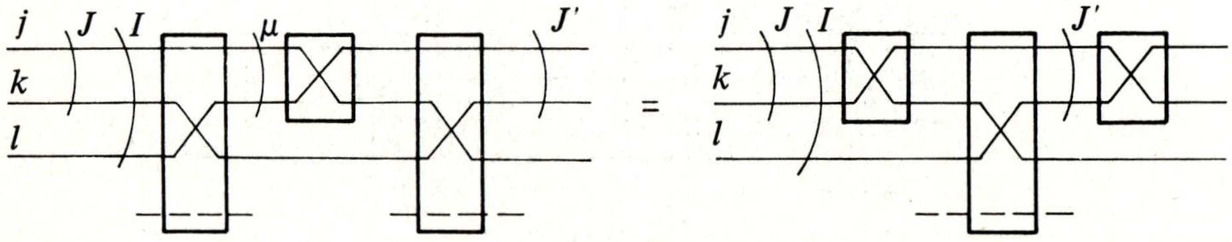

Fig. 4.19.

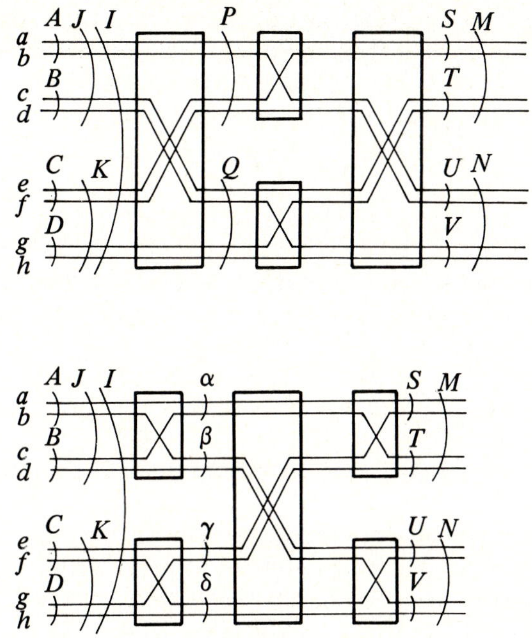

Fig. 4.20.

recoupling identities are contained in this expression. They are obtained by putting one or more of the input angular momenta to zero.

We see that in general many equivalent recoupling graphs can be drawn and that they may have different numbers of intermediate summations. By drawing several diagrams one easily can find the most economical recoupling transformation, before ever writing down the algebraic expressions.

(iii) Insertion of a complete set. The elementary diagram for inserting a complete set of states $\psi_\alpha^{[j]}$ was introduced in fig. 4.12. We show now that the diagram of fig. 4.21 drawn so that it has one incoming line representing $\psi_\alpha^{[j]}$ and one outgoing line representing $\psi_{\alpha'}^{[j']}$ indeed fulfill the identity

$$\psi_\alpha^{[j]} = \sum_{j'\alpha'} \delta_{jj'}\delta_{\alpha\alpha'}\psi_{\alpha'}^{[j']}.$$

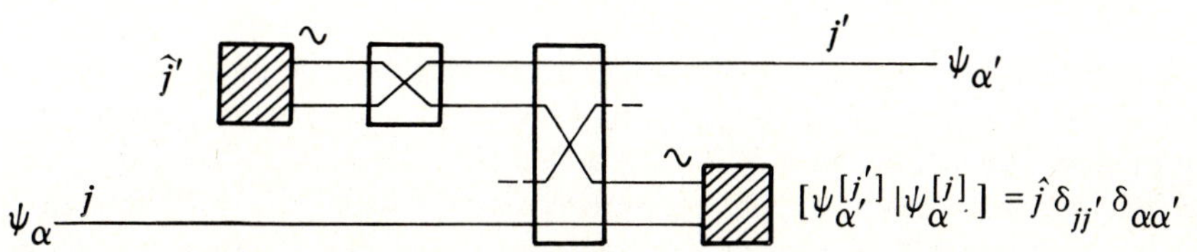

Fig. 4.21.

From the diagram 4.21 we read, eq. (4.29),

$$\psi_{\alpha}^{[j]} = \sum_{j'\alpha'} \hat{j}'(-)^{2j'} \begin{bmatrix} j' & j' & 0 \\ 0 & j & j \\ j' & 0 & j \end{bmatrix} [\psi_{\alpha'}^{[j']} | \psi_{\alpha}^{[j]}] \psi_{\alpha'}^{[j']}.$$

With the invariant norm value (4.13) and the value of the recoupling coefficient, table 4.2, we obtain indeed the required identity. For purely orbital functions the crossing box can be omitted in accordance with the remark after eq. (4.30). In that case the crossing sign is always positive (integer angular momenta) and the wave functions are self-conjugate.

4.3.4. *Recoupling graph rules*

In summary, the following rules must be obeyed when working with recoupling graphs:

(i) The diagram starts with lines drawn from top to bottom in the order in which the tensorial quantities appear from left to right in the initial algebraic expression.

(ii) Only two adjacent lines or two adjacent already coupled groups of lines can be coupled.

(iii) Lines can cross only within a recoupling box.

(iv) Only lines which are coupled can be recoupled. A group of lines coupled to a given angular momentum which remains intact in a recoupling behaves like a single line with that same angular momentum.

(v) The algebraic result is obtained by writing the product of all the values associated with the graph symbols of the diagram. All quantum numbers are to be summed over except those given initially or fixed by invariants.

This way, one first will draw a recoupling diagram. Then one will try to simplify it by choosing a different order of recoupling, until one is satisfied that no further improvements are possible. Only at this time is it necessary to supply the graph with labels, writing immediately all the special values for the quantum numbers imposed from the right by the invariant products and matrix elements. Now the final algebraic expression can be directly read off the graph. This expression is complete; no additional factors or phases have to be supplied.

Further rules will be required for dealing with non-commuting tensors such as creation and annihilation operators. They will be given in sect. 4.6.

4.3.5. *Extension of the graphical method to SU(n)*

All the techniques and results of the previous sections have been written for SU(2) tensors. They are, however, directly applicable to the case of SU(n) upon the following generalizations. (i) SU(2) has one Casimir operator while SU(n) has $n-1$ Casimir operators. Thus in SU(2) a complete classification of a one-body irreducible tensor requires a single quantum number, viz. J, while it requires a set of $n-1$ quantum numbers in SU(n). This set may be denoted as $a = (\alpha, \beta, \dots)$. (ii) In SU(2) the reduction of the direct product of two irreducible tensors J_1 and J_2 into a sum of irreducible product tensors J has only unit multiplicities. That means that each possible value of J appears once and only once. In SU(n) in the same reduction multiplicities higher than unity arise. Hence a complete classification into irreducible tensors with quantum numbers $a=(\alpha,\beta,\dots)$ requires one additional quantum number, ν, the multiplicity quantum number, to distinguish between these different product tensors. We designate the individual tensors by the complete set

$$A=(a,\nu)=(\alpha,\beta\dots\nu).$$

The coupling of two tensors in the SU(n) case is written

$$\psi^{[C]}_{M_C}=\left[\phi^{[A]}\phi^{[B]}\right]^{[C]}_{M_C}=\sum(ABM_AM_B|CM_C)\phi^{[A]}_{M_A}\phi^{[B]}_{M_B},$$

with the SU(n) vector coupling coefficients. Here of course the M's stand for the sets of projection quantum numbers. For example, in SU(3) a possible choice for the M's is T, T_z, Y, i.e., isospin, its projection, hypercharge.

Note that the invariant product of two tensors, i.e. the coupling to the singlet $J=0$ in SU(2) or to $a=(0,0,\dots)$ in SU(n), has always multiplicity unity. Also, there exists only one invariant triple product in SU(n). This allows the introduction of invariant matrix elements for the operators $\mathcal{O}^{[A]}$ of SU(n), in complete analogy with SU(2),

$$M=\left[\phi^{[A]}|\mathcal{O}^{[A]}|\phi^{[B]}\right].$$

This way we see that the recoupling expressions and the corresponding recoupling graphs of SU(n) are identical to those of SU(2) with the replacement of (i) the SU(2) indices J by the SU(n) composite indices A, (ii) the SU(2) recoupling coefficients by the SU(n) recoupling coefficients.

4.4. STATES, OPERATORS, AND MATRICES

4.4.1. Invariant form of states and operators

When carrying out computations it is very useful to work with invariant forms for the state vectors and the operators. We now describe and construct these quantities.

We consider first the invariant form of a state. A polarized pure state is written as

$$\Psi=\sum_M W^{(I)}_M\Psi^{[I]}_M,$$

where $W^{(I)}_M$ are complex amplitudes which depend in particular on the preparation of the state. Written in coupled form this state is an invariant according to sect. 4.2.2,

$$\Psi=\hat{I}\left[W^{[I]}\Psi^{[I]}\right]^{[0]}. \tag{4.32}$$

The corresponding invariant bra is, eq. (4.4),

$$\Psi^{+} = \hat{I}[\tilde{W}^{[I]}\tilde{\Psi}^{[I]}]^{[0]}.$$

The amplitudes $W^{[I]}$ are in general not self-conjugate as will be discussed below and the tilde must be retained.

In order to obtain the correct normalization of the state

$$\langle \Psi | \Psi \rangle = 1,$$

the polarization amplitudes $W_M^{[I]}$ must be such that their invariant norm matrix element fulfills

$$[W^{[I]} | W^{[I]}] = \frac{1}{\hat{I}}. \tag{4.33}$$

Here again the tilde on the bra is omitted as it is redundant with the bra notation. This result arises from the evaluation of the graph of fig. 4.22 which yields

$$1 = \langle \Psi | \Psi \rangle = \hat{I}^2 \not{\Sigma} [\tilde{W}^{[I]}\tilde{\Psi}^{[I]}]^{[0]}[W^{[I]}\Psi^{[I]}]^{[0]}$$

$$= \hat{I}^2 \begin{bmatrix} I & I & 0 \\ I & I & 0 \\ 0 & 0 & 0 \end{bmatrix} [W^{[I]} | W^{[I]}][\Psi^{[I]} | \Psi^{[I]}].$$

Let us note that the invariant amplitude overlap $[W^{[I]} | W^{[I]}]$ is represented in fig. 4.22 by a cross-hatched box to distinguish it from the single-hatched box which is associated with the wave function overlap $[\Psi^{[I]} | \Psi^{[I]}]$.

As an example, the invariant pure state Ψ which has good magnetic quantum number M along the axis Oz is described by the set of polarization amplitudes which obey the condition

$$|W_{M'}^{(I)}| = \delta_{MM'}.$$

This indeed fulfills the normalization, eq. (4.33). However the phase of $W_M^{(I)}$ is still open. It will be discussed in detail in the next section.

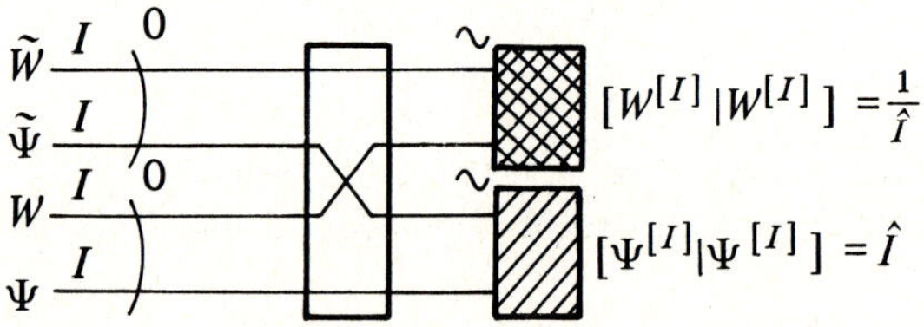

Fig. 4.22.

Likewise we define the invariant form Ω for a multipole operator $\Omega^{[\lambda]}$, introducing polarization amplitudes $\omega^{[\lambda]}$

$$\Omega = \hat{\lambda}[\omega^{[\lambda]}\Omega^{[\lambda]}]^{[0]}.$$

The amplitudes $\omega_\mu^{[\lambda]}$ are determined by the physical situation, e.g. by the orientation of the experimental apparatus which is described by the tensor operator $\Omega^{[\lambda]}$. Some examples of polarization amplitudes will arise in the context of the electromagnetic interaction discussed in Chapter 9. For a scalar operator they are unity, $\omega_0^{[0]} = 1$.

4.4.2. *Phases of the state amplitudes*

The phase freedom associated with the definition of invariant states will be used to achieve a real symmetric hamiltonian matrix. The amplitude phases needed for that result will be discussed in detail and given in Chapter 7 when constructing the energy matrix.

Here we only discuss how to achieve by a proper definition of the phases of the state amplitudes $W_M^{(I)}$ an invariant form of definite reality, i.e. a real or a pure imaginary form. The form being an invariant this reality character is maintained under rotation. The definition of invariant states of definite reality character can be helpful in handling transition or energy matrix element calculations.

Invariant forms of definite reality can be always achieved for non-relativistic states and for relativistic integer spin states. However, relativistic half-integer spin states have necessarily spinor components of different reality character, as discussed below.

We first show how the phases of the amplitudes $W_M^{(I)}$ must be chosen such that a "real" polarized state has an invariant form which is real. We consider three simple examples and then we give the general rule.

A real polarized orbital state with projection $m = 0$ along the Oz axis is phased according to

$$\psi \approx Y_{l0} \approx P_l(\cos\theta)$$

and the definition (4.3) for $Y_0^{[l]}$ entails for the invariant form to be real

$$\psi = \hat{l}[W^{[l]}Y^{[l]}] = \sum_m W_m^{(l)}\psi_m^{[l]},$$

$$W_m^{(l)} = i^l\delta_{m0}.$$

For a polarized spin-$\frac{1}{2}$ state with $m = \frac{1}{2}$ along the Oz axis,

$$\chi_{1/2}^{[1/2]} = \begin{pmatrix}1\\0\end{pmatrix},$$

we have the real form

$$\psi \approx \chi_{1/2}^{[1/2]} = W_{1/2}^{(1/2)}\chi_{1/2}^{[1/2]},$$

with

$$W_m^{(1/2)} = \delta_{m1/2}.$$

However, a polarized spin-1 state, with $m = 0$ along the Oz axis, to be defined real needs the phasing

$$\psi \approx \boldsymbol{e}_z = W_0^{(1)} e_0^{[1]},$$

with

$$W_m^{(1)} = i\delta_{m0}.$$

More generally, in order to yield real invariant forms defined according to eq. (4.32), all one-body integer angular momentum tensors of rank $2I + 1$ (orbital, spin or isospin wave functions) phased according to the relation (4.3) in order to be self-conjugate, require in the definition of their associated standard polarization amplitudes $W_M^{(I)}$ a phase i^I which is M-independent. Hence the contrastandard amplitudes $W_M^{[I]}$ in eq. (4.32) have the phase $(-i)^I$. In contrast, one-body half-integer tensors (i.e., half-integer spin or isospin wave functions) do not require that phase in their associated invariant form amplitudes. In the above examples, because of the M-independence of the phase, the invariant forms (4.32) of the wave functions ψ are respectively

$$\psi = \hat{l}\left[W^{[l]} Y^{[l]}\right]^{[0]}, \qquad \text{with } W_m^{[l]} = (-i)^l \delta_{m0},$$

$$\psi = \sqrt{2}\left[W^{[1/2]} \chi^{[1/2]}\right]^{[0]}, \qquad \text{with } W_m^{[1/2]} = \delta_{m1/2},$$

$$\psi = \hat{1}\left[W^{[1]} e^{[1]}\right]^{[0]}, \qquad \text{with } W_m^{[1]} = (-i)\delta_{m0}.$$

Although the reality of these forms, being invariants, is maintained under rotation of the coordinate system, the terms in the sum over the m's are not real individually.

For a coupled configuration ψ_α containing a product of orbital, spin, and isospin tensors

$$\psi_\alpha = \hat{I}\hat{T}\left[W_\alpha^{[IT]}\left[Y^{[l_1 0]} \ldots \chi^{[s_1 0]} \ldots \eta^{[0 t_1]} \ldots\right]_\alpha^{[IT]}\right]^{[00]},$$

the generalization of the above rule shows that in order to yield a real invariant form the contrastandard amplitudes $W_{\alpha M_I M_T}^{[IT]}$ must carry the phase

$$\xi_\alpha = (-i)^{\delta_\alpha}$$

with

$$\delta_\alpha = l_1 + \cdots + s_1 + \cdots + t_1 + \cdots,$$

where only integer spins s_i and integer isospins t_i contribute.

Likewise one may define the phases ξ_α such that for example the invariant forms of the state vectors are pure imaginary. This is achieved by multiplying the above phases by a factor i.

Several general remarks are in order:

(i) Not all state vector invariant forms can be made of definite reality. This is the case for relativistic half-integer spin state vectors since their upper and lower spinor components have opposite reality character.

(ii) The phases ξ_α are independent of the coupling scheme of the state vectors ψ_α and in particular of I and T. Of course, their choice does not affect the normalization condition (4.33),

$$\left[W_\alpha^{[IT]} | W_\alpha^{[IT]}\right] = \frac{1}{\hat{I}\hat{T}} .$$

(iii) For a general pure state which is a linear superposition of configurational basis state vectors ψ_α

$$\Psi = \sum_\alpha X_\alpha \psi_\alpha = \hat{I}\hat{T} \sum_\alpha X_\alpha \left[W_\alpha^{[IT]} \psi_\alpha^{[IT]}\right]^{[00]},$$

the state amplitudes $W_\alpha^{[IT]}$ depend on a particular configuration α only via the phase ξ_α. Hence

$$\Psi = \hat{I}\hat{T} \left[W^{[IT]} \sum_\alpha X_\alpha \xi_\alpha \psi_\alpha^{[IT]}\right]^{[00]} .$$

An off-diagonal matrix element therefore will contain a phase originating from

$$\left[W_\alpha^{[IT]} | W_\beta^{[IT]}\right] = \xi_\alpha^* \xi_\beta \frac{1}{\hat{I}\hat{T}} = i^{\delta_\alpha - \delta_\beta} \frac{1}{\hat{I}\hat{T}} .$$

We will use the following notation

$$\eta_{\alpha\beta} = i^{\delta_\alpha - \delta_\beta}$$

and retain these phases explicitly in all formal expressions for the matrix elements. Particular choices of the phases η which yield a real symmetric hamiltonian matrix are discussed in sect. 7.2.2.

4.4.3. Matrix elements

When writing the matrix element of an operator between two pure states in terms of their invariant forms defined above, one obtains from fig. 4.23

$$M = \langle \Psi | \Omega | \Phi \rangle = \hat{I}\hat{\lambda}\hat{J} \left[\left[\tilde{W}^{[I]} \Psi^{[I]} \right]^{[0]} \left| \left[\omega^{[\lambda]} \Omega^{[\lambda]} \right]^{[0]} \right| \left[V^{[J]} \Phi^{[J]} \right]^{[0]} \right]$$

$$= \hat{I}\hat{\lambda}\hat{J} \begin{bmatrix} I & I & 0 \\ \lambda & \lambda & 0 \\ J & J & 0 \end{bmatrix} \begin{bmatrix} J & J & 0 \\ J & J & 0 \\ 0 & 0 & 0 \end{bmatrix} \left[\Psi^{[I]} | \Omega^{[\lambda]} | \Phi^{[J]} \right] \left[\tilde{W}^{[I]} \omega^{[\lambda]} V^{[J]} \right]^{[0]}$$

$$= \left[\Psi^{[I]} | \Omega^{[\lambda]} | \Phi^{[J]} \right] \left[\tilde{W}^{[I]} \omega^{[\lambda]} V^{[J]} \right]^{[0]} .$$

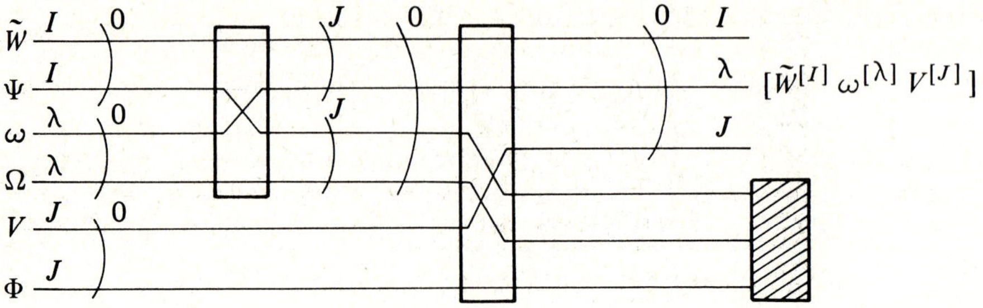

Fig. 4.23.

Thus a general matrix element is the product of two invariants. In this expression, the dynamics is contained in the invariant matrix element of the operator. The information on the angular distributions, polarizations, etc., which depends on the experimental set-up is contained in the invariant product of the three amplitudes $[\tilde{W}^{[I]}\omega^{[\lambda]}V^{[J]}]^{[0]}$.

We shall not be concerned with the case of impure states which requires the introduction of density matrices.

The simplest case is the hamiltonian operator H, a scalar, with $\omega_0^{[0]}=1$. Thus the diagonal energy of an invariant pure state $\mathbf{\Psi}$, is given in terms of the invariant matrix element of H by, eq. (4.33),

$$\langle\Psi|H|\Psi\rangle=\left[W^{[I]}|W^{[I]}\right]\left[\Psi^{[I]}|H|\Psi^{[I]}\right]=\frac{1}{\hat{I}}\left[\Psi^{[I]}|H|\Psi^{[I]}\right],$$

while for an off-diagonal matrix element we have

$$\langle\Psi|H|\Phi\rangle=\left[W^{[I]}|V^{[I]}\right]\left[\Psi^{[I]}|H|\Phi^{[I]}\right].$$

Finally, the matrix elements of an hermitian operator fulfill

$$\langle\psi_\beta|\mathbb{O}|\psi_\alpha\rangle=\langle\psi_\alpha|\mathbb{O}|\psi_\beta\rangle^*.$$

Thus they are of the general form

$$\langle\psi_\alpha|\mathbb{O}|\psi_\beta\rangle=\langle\psi_\alpha|S|\psi_\beta\rangle+i\langle\psi_\alpha|A|\psi_\beta\rangle,$$

where S is a real symmetric matrix and A a real antisymmetric matrix. Thus substituting for the invariant states ψ_α and ψ_β their definitions, sect. 4.4.2,

$$\begin{aligned}\langle\psi_\alpha|\mathbb{O}|\psi_\beta\rangle&=\left[W_\alpha^{[I]}|W_\beta^{[I]}\right]\left[\psi_\alpha^{[I]}|\mathbb{O}|\psi_\beta^{[I]}\right]\\&=\eta_{\alpha\beta}\frac{1}{\hat{I}}\left[\psi_\alpha^{[I]}|\mathbb{O}|\psi_\beta^{[I]}\right]=i^{\delta_\alpha-\delta_\beta}\frac{1}{\hat{I}}\left[\psi_\alpha^{[I]}|\mathbb{O}|\psi_\beta^{[I]}\right],\end{aligned}$$

we have the property of the invariant matrix element

$$\left[\psi_\beta^{[I]}|S|\psi_\alpha^{[I]}\right]=(-)^{\delta_\alpha-\delta_\beta}\left[\psi_\alpha^{[I]}|S|\psi_\beta^{[I]}\right],$$

$$\left[\psi_\beta^{[I]}|A|\psi_\alpha^{[I]}\right]=-(-)^{\delta_\alpha-\delta_\beta}\left[\psi_\alpha^{[I]}|A|\psi_\beta^{[I]}\right].$$

4.5. SPECIAL APPLICATIONS

4.5.1. Vector algebra

Here we collect the invariant forms of vector algebra. They are based on the representation of a "vector" $\boldsymbol{A}$ constructed with the unit vectors $\boldsymbol{e}_i$,

$$\boldsymbol{A}=A_x\boldsymbol{e}_x+A_y\boldsymbol{e}_y+A_z\boldsymbol{e}_z=\sum_m A_m^{(1)}e_m^{[1]}=\sum_m(-)^{1+m}A_{-m}^{[1]}e_m^{[1]}$$

$$=\hat{1}[A^{[1]}e^{[1]}]^{[0]}=\hat{1}[e^{[1]}A^{[1]}]^{[0]}. \tag{4.34}$$

Note that $\boldsymbol{A}$ is an invariant as is evident from the last two forms. The set of contrastandard self-conjugate unit vectors $e_m^{[1]}$ has been already introduced in sect. 4.2.2 in the context of the spin-1 wave function.

We already have calculated directly in sect. 4.2.3 the scalar product and the vector product of two vectors $\boldsymbol{A}$ and $\boldsymbol{B}$. These results can be obtained using their invariant form by means of recoupling. The scalar product of two vectors is according to fig. 4.24a

$$\boldsymbol{A}\cdot\boldsymbol{B}=\hat{1}[A^{[1]}e^{[1]}]^{[0]}\cdot\hat{1}[B^{[1]}e^{[1]}]^{[0]}$$

$$=\hat{1}^2\begin{bmatrix}1&1&0\\1&1&0\\0&0&0\end{bmatrix}[e^{[1]}|e^{[1]}][A^{[1]}B^{[1]}]^{[0]}=\hat{1}[A^{[1]}B^{[1]}]^{[0]},$$

which is the result of eq. (4.9). Here use has been made of the value of the invariant overlap

$$[e^{[1]}|e^{[1]}]=\hat{1}.$$

More generally, let us consider the invariant form of a vector function $\boldsymbol{F}$, i.e. a function containing a unit vector

$$\boldsymbol{F}=[\ldots e^{[1]}\ldots]^{[0]}.$$

An example is given by the vector spherical harmonics, eq. (4.6), the invariant form of which is

$$\boldsymbol{F}=\hat{J}[W^{[J]}e^{[1]}Y^{[l]}]^{[0]}.$$

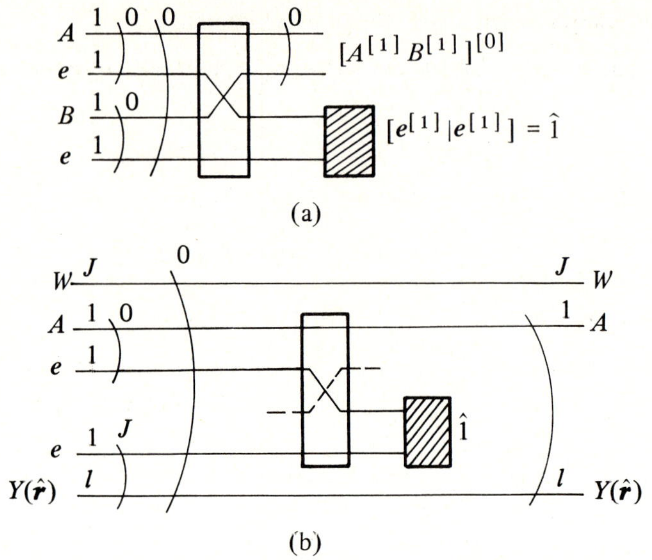

Fig. 4.24.

The scalar product of $\boldsymbol{F}$ with $\boldsymbol{A}$ is, generalizing the result of fig. 4.24b drawn for the above example,

$$\boldsymbol{A}\cdot\boldsymbol{F}=\hat{1}[A^{[1]}e^{[1]}]^{[0]}\cdot[\ldots e^{[1]}\ldots]^{[0]}=[\ldots A^{[1]}\ldots]^{[0]}. \tag{4.35}$$

The invariant $\boldsymbol{A}$, eq. (4.34), having angular momentum zero, can be inserted any place in the diagram. In words, the scalar product $\boldsymbol{A}\cdot\boldsymbol{F}$ is obtained by replacing the unit vector $e^{[1]}$ in $\boldsymbol{F}$ by the vector tensor $A^{[1]}$ without altering the coupling scheme in $\boldsymbol{F}$.

The invariant form of the vector product, the components of which have been given in eq. (4.10), is

$$\boldsymbol{C}=\boldsymbol{A}\times\boldsymbol{B}=\hat{1}[C^{[1]}e^{[1]}]^{[0]}=\sqrt{6}\,[A^{[1]}B^{[1]}e^{[1]}]^{[0]}. \tag{4.36}$$

Likewise the invariant form of the vector product of a vector $\boldsymbol{A}$ with the vector function $\boldsymbol{F}$ is

$$\boldsymbol{C}=\boldsymbol{A}\times\boldsymbol{F}=\sqrt{2}\,[\ldots[A^{[1]}e^{[1]}]^{[1]}\ldots]^{[0]}. \tag{4.37}$$

In words, the vector product is obtained by replacing in the vector function the unit vector $\boldsymbol{e}$ by the vector product $\boldsymbol{A}\times\boldsymbol{e}$. Note the difference between the numerical coefficients of (4.36) and (4.37).

Finally, the invariant form of the pseudo-scalar product of three vectors is

$$\boldsymbol{A}\times\boldsymbol{B}\cdot\boldsymbol{C}=\sqrt{6}\,[A^{[1]}B^{[1]}C^{[1]}]^{[0]},$$

which is obtained by applying (4.35) to (4.36).

4.5.2. Vector analysis

The formulae of the previous section can be applied to vector analysis by using

$$\nabla = \nabla_x e_x + \nabla_y e_y + \nabla_z e_z = \hat{1}[\nabla^{[1]} e^{[1]}]^{[0]}.$$

The divergence of a vector field $\boldsymbol{A}$ is in this way

$$\operatorname{div} \boldsymbol{A} = \nabla \cdot \boldsymbol{A} = \hat{1}[\nabla^{[1]} A^{[1]}]^{[0]},$$

or, more generally, for the vector function $\boldsymbol{F}$ defined above

$$\operatorname{div} \boldsymbol{F} = \nabla \cdot \boldsymbol{F} = [\ldots \nabla^{[1]} \ldots]^{[0]}.$$

Here $\nabla^{[1]}$ is at the position of, and replaces, the unit vector $e^{[1]}$ in $\boldsymbol{F}$ without alteration of the coupling scheme. Of course, before carrying out the divergence, $\nabla^{[1]}$ has to be positioned next to the operand by appropriate recoupling.

Next, using the result (4.36) for the vector product, the rotational of a field $\boldsymbol{B}$ is in invariant form

$$\operatorname{rot} \boldsymbol{B} = \nabla \times \boldsymbol{B} = \sqrt{6}[\nabla^{[1]} B^{[1]} e^{[1]}]^{[0]},$$

or, for the vector function $\boldsymbol{F}$

$$\operatorname{rot} \boldsymbol{F} = \sqrt{2}[\ldots [e^{[1]} \nabla^{[1]}]^{[1]} \ldots]^{[0]}.$$

Note here the inverted order of $e^{[1]}$ and $\nabla^{[1]}$, and again the absence of the normalizing factor $\hat{1}$ as in eq. (4.37). As above, the rot operator of the right-hand side must then be positioned next to the operand by appropriate recoupling.

Finally the invariant form of the gradient operation is written as

$$\operatorname{grad} f = \nabla f = \hat{1}[\nabla^{[1]} e^{[1]}]^{[0]} f = \hat{1}[\nabla^{[1]} f e^{[1]}]^{[0]}.$$

We now apply these formulae to spherical vector multipoles defined in contra-standard form as

$$\psi_M^{[J]} = \sqrt{\frac{2}{\pi}}\, j_\lambda(pr) \boldsymbol{Y}_{\lambda M}^{[J]} = \left[e^{[1]} \phi_p^{[\lambda]}\right]_M^{[J]},$$

where, see eq. (4.6),

$$\boldsymbol{Y}_{\lambda M}^{[J]} = [e^{[1]} Y^{[\lambda]}]_M^{[J]}$$

and where the normalized orbital functions $\phi_p^{[\lambda]}$ are

$$\phi_{pm}^{[\lambda]} = \sqrt{\frac{2}{\pi}}\, j_\lambda(pr) Y_m^{[\lambda]}(\hat{\boldsymbol{r}}).$$

The invariant form of these functions is constructed with the amplitudes $W_M^{[J]}$

$$\psi = \hat{J}\left[W^{[J]} e^{[1]} \phi_p^{[\lambda]}\right]^{[0]}.$$

This case is particularly important for handling the calculation of the matrix elements of the spin-1 fields. In all the following use will be made of eqs. (4.19) through (4.24) when evaluating the action of the gradient operator.

The divergence of the vector spherical function ψ is, in invariant form

$$\operatorname{div} \psi = \hat{J}\left[W^{[J]}\, \nabla^{[1]} \phi_p^{[\lambda]}\right]^{[0]},$$

which is represented by fig. 4.25. This diagram yields, with eq. (4.30) and fig. 4.24,

$$\operatorname{div} \psi = \sum_{JJ'} \hat{J}\hat{J}' \begin{bmatrix} J' & J' & 0 \\ 0 & J & J \\ J' & 0 & J \end{bmatrix} \int q^2 \,\mathrm{d}q \left[\phi_q^{[J]} |\, \nabla^{[1]} | \phi_p^{[\lambda]}\right]^{[0]} \left[W^{[J]} \phi_q^{[J]}\right]^{[0]}$$

$$= \delta_{JJ'} \alpha_{J\lambda}\, p \left[W^{[J]} \phi_p^{[J]}\right]^{[0]}. \tag{4.38}$$

Here the factor $\alpha_{J\lambda}$ comes from the invariant gradient matrix element, see eqs. (4.23) and (4.24). Explicitly,

$$\operatorname{div} \psi = \begin{cases} \sqrt{J+1}\, p \left[W^{[J]} \phi_p^{[J]}\right]^{[0]} & \text{for } \lambda = J+1, \\ -\sqrt{J}\, p \left[W^{[J]} \phi_p^{[J]}\right]^{[0]} & \text{for } \lambda = J-1. \end{cases}$$

The rotational of the vector spherical function ψ is

$$\operatorname{rot} \psi = \sqrt{2}\, \hat{J}\left[W^{[J]} \left[e^{[1]}\, \nabla^{[1]}\right]^{[1]} \phi_p^{[\lambda]}\right]^{[0]}.$$

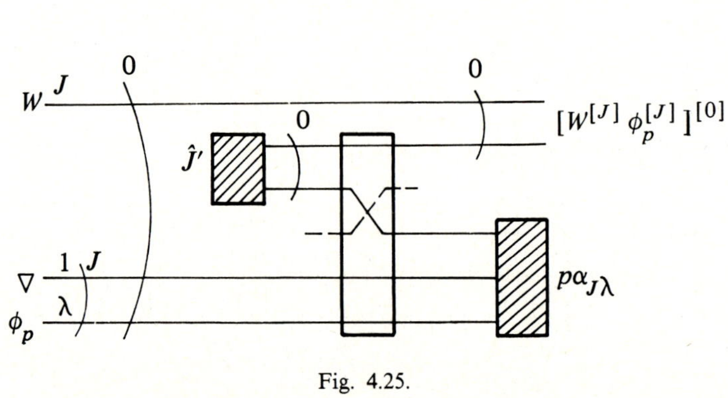

Fig. 4.25.

It is evaluated with the diagram of fig. 4.26 which yields

$$\operatorname{rot}\psi=\sqrt{2}\,\hat{J}\sum_{\nu}\begin{bmatrix}1&1&1\\0&\lambda&\lambda\\1&\nu&J\end{bmatrix}\hat{\nu}\begin{bmatrix}\nu&\nu&0\\0&\nu&\nu\\\nu&0&\nu\end{bmatrix}p\alpha_{\nu\lambda}\left[W^{[J]}e^{[1]}\phi_p^{[\nu]}\right]^{[0]}$$

$$=-p\begin{cases}\sqrt{J}\left[W^{[J]}e^{[1]}\phi_p^{[J]}\right]^{[0]} & \text{for } \lambda=J+1\\ \left(\sqrt{J}\left[W^{[J]}e^{[1]}\phi_p^{[J+1]}\right]^{[0]}+\sqrt{J+1}\left[W^{[J]}e^{[1]}\phi_p^{[J-1]}\right]^{[0]}\right) & \text{for } \lambda=J\\ \sqrt{J+1}\left[W^{[J]}e^{[1]}\phi_p^{[J]}\right]^{[0]} & \text{for } \lambda=J-1\,.\end{cases}\tag{4.39}$$

Here use has been made of the values of the recoupling coefficients

$$\begin{bmatrix}1&1&1\\0&\lambda&\lambda\\1&\nu&J\end{bmatrix}=\begin{cases}(-)^{J+\lambda}\dfrac{\hat{\nu}\sqrt{J}}{\hat{J}\sqrt{J+1}\,\sqrt{2}}\,, & \text{for } \lambda=J+1\,,\ \nu=J \text{ or } \lambda=J\,,\ \nu=J+1\\ (-)^{J+\lambda+1}\dfrac{\hat{\nu}\sqrt{J+1}}{\hat{J}\sqrt{J}\,\sqrt{2}}\,, & \text{for } \lambda=J\,,\ \nu=J-1 \text{ or } \lambda=J-1\,,\ \nu=J\,.\end{cases}$$

In a similar way, one can check from the diagram of fig. 4.27 that

$$\operatorname{rot}\operatorname{rot}\psi=2\hat{J}\left[W^{[J]}\left[\left[\left[e^{[1]}\,\nabla^{[1]}\right]^{[1]}\nabla^{[1]}\right]^{[1]}\phi_p^{[\lambda]}\right]^{[J]}\right]^{[0]}$$

$$=2\hat{J}\sum_{\nu\mu}\begin{bmatrix}1&1&1\\0&\lambda&\lambda\\1&\nu&J\end{bmatrix}\hat{\nu}\begin{bmatrix}\nu&\nu&0\\0&\nu&\nu\\\nu&0&\nu\end{bmatrix}p\alpha_{\nu\lambda}\begin{bmatrix}1&1&1\\0&\nu&\nu\\1&\mu&J\end{bmatrix}\hat{\mu}$$

$$\times\begin{bmatrix}\mu&\mu&0\\0&\mu&\mu\\\mu&0&\mu\end{bmatrix}p\alpha_{\mu\nu}\left[W^{[J]}e^{[1]}\phi_p^{[\mu]}\right]^{[0]}.$$

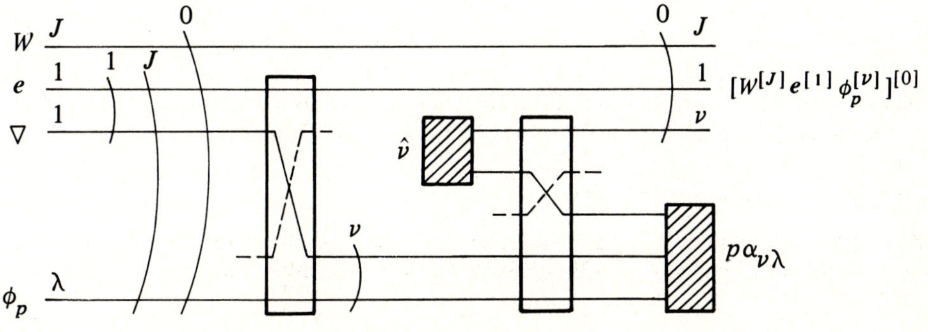

Fig. 4.26.

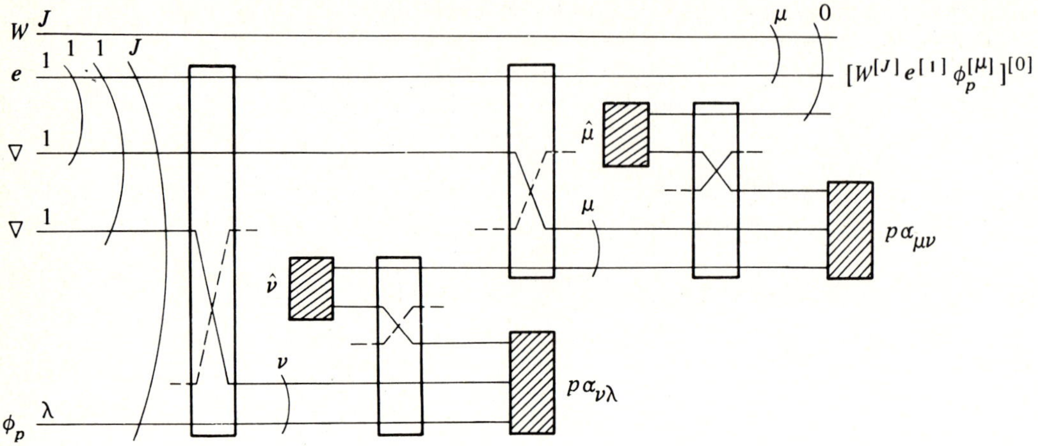

Fig. 4.27.

After simplification this expression becomes

$$\operatorname{rot}\operatorname{rot}\psi=\frac{p^2}{\hat{J}}\begin{cases}\sqrt{J}\left(\sqrt{J}\left[W^{[J]}e^{[1]}\phi_p^{[J+1]}\right]^{[0]}+\sqrt{J+1}\left[W^{[J]}e^{[1]}\phi_p^{[J-1]}\right]^{[0]}\right), & \text{if } \lambda=J+1\\ (2J+1)\left[W^{[J]}e^{[1]}\phi_p^{[J]}\right]^{[0]}, & \text{if } \lambda=J\\ \sqrt{J+1}\left(\sqrt{J}\left[W^{[J]}e^{[1]}\phi_p^{[J+1]}\right]^{[0]}+\sqrt{J+1}\left[W^{[J]}e^{[1]}\phi_p^{[J-1]}\right]^{[0]}\right), & \text{if } \lambda=J-1.\end{cases}$$

Finally, the gradient of the orbital function

$$\phi=\hat{J}\left[W^{[J]}\phi_p^{[J]}\right]^{[0]}$$

is in invariant form

$$\begin{aligned}\operatorname{grad}\phi&=\sum_{\lambda=J\pm1}p\alpha_{\lambda J}\left[W^{[J]}e^{[1]}\phi_p^{[\lambda]}\right]^{[0]}\\&=p\left(-\sqrt{J+1}\left[W^{[J]}e^{[1]}\phi_p^{[J+1]}\right]^{[0]}+\sqrt{J}\left[W^{[J]}e^{[1]}\phi_p^{[J-1]}\right]^{[0]}\right).\end{aligned}\tag{4.40}$$

Recall that all these formulae are valid only for the free wave multipoles, i.e., for the case where the radial function is a spherical Bessel function.

4.5.3. Spin-momentum coupling

An important example is the evaluation of the action of the operator $\boldsymbol{\sigma}\cdot\boldsymbol{p}$ on a spinorial wave function $\psi_{plm}^{[j]}$ defined as in (4.5). Here $s=\frac{1}{2}$,

$$\psi_{plm}^{[j]}(\boldsymbol{r})=\sqrt{\frac{2}{\pi}}\,j_l(pr)\left[\chi^{[s]}\hat{r}^{[l]}\right]_m^{[j]}=\left[\chi^{[s]}\phi_p^{[l]}\right]_m^{[j]},$$

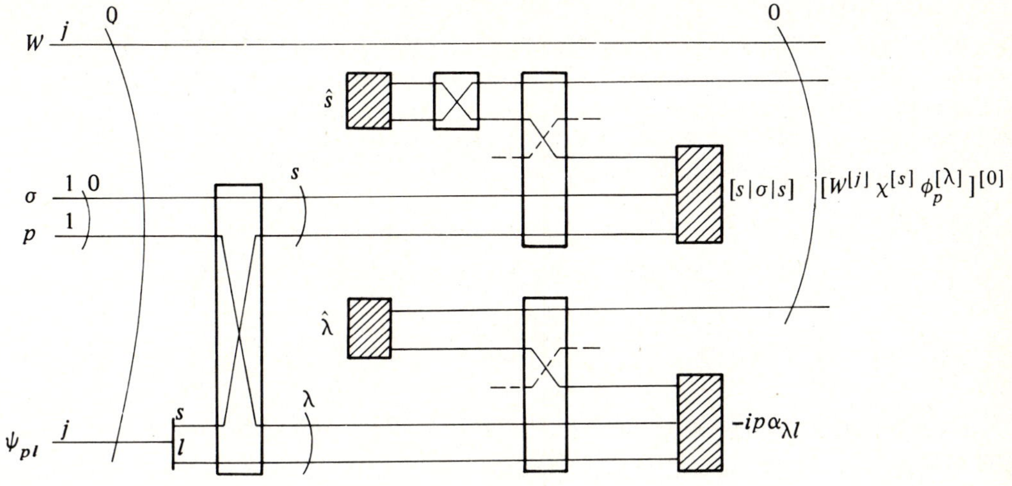

Fig. 4.28.

or in invariant form

$$\psi = \hat{j}\left[W^{[j]}\psi_{pl}^{[j]}\right]^{[0]}.$$

We have to calculate

$$\boldsymbol{\sigma}\cdot\boldsymbol{p}\psi = \hat{1}\left[\sigma^{[1]}p^{[1]}\right]^{[0]}\hat{j}\left[W^{[j]}\psi_{pl}^{[j]}\right]^{[0]}.$$

The recoupling diagram is given in fig. 4.28. Note that the lines representing the scalar $\boldsymbol{\sigma}\cdot\boldsymbol{p}$ can be inserted at liberty. The evaluation of the diagram, as in the cases of figs. 4.25 and 4.26, requires the insertion of a complete set both for the spin and for the orbital functions. As shown in fig. 4.21 a crossing box is associated with the spin completeness insertion, but it may be omitted for the orbital insertion.

The evaluation of the graph 4.28 yields

$$\boldsymbol{\sigma}\cdot\boldsymbol{p}\psi = \sum_{\lambda}\hat{1}\hat{j}\begin{bmatrix}1 & 1 & 0\\ s & l & j\\ s & \lambda & j\end{bmatrix}\hat{s}(-)^{2s}\begin{bmatrix}s & s & 0\\ 0 & s & s\\ s & 0 & s\end{bmatrix}[s|\sigma|s]\hat{\lambda}\begin{bmatrix}\lambda & \lambda & 0\\ 0 & \lambda & \lambda\\ \lambda & 0 & \lambda\end{bmatrix}$$

$$\times(-i\alpha_{\lambda l})p\left[W^{[j]}\chi^{[s]}\phi_p^{[\lambda]}\right]^{[0]}$$

$$= \frac{\hat{1}\hat{j}}{\hat{s}\hat{\lambda}}\begin{bmatrix}1 & 1 & 0\\ s & l & j\\ s & \lambda & j\end{bmatrix}\sqrt{6}\,\alpha_{\lambda l}p\left[W^{[j]}\psi_{p\lambda}^{[j]}\right]^{[0]},$$

where we have used the compact notation $[s|\sigma|s]$ for the matrix element of eq. (4.14). Here

$$\lambda = l + 1 \quad \text{for } j = l + \tfrac{1}{2} \quad \text{and} \quad \lambda = l - 1 \quad \text{for } j = l - \tfrac{1}{2},$$

as imposed by the triangular conditions contained in the recoupling coefficients. Introducing their numerical values

$$\begin{bmatrix} 1 & 1 & 0 \\ s & l & j \\ s & \lambda & j \end{bmatrix} = \begin{cases} \dfrac{1}{\sqrt{6}} \dfrac{1}{\sqrt{\lambda}} \dfrac{\hat{s}\hat{\lambda}}{\hat{1}} & \text{if } \lambda = l + 1 \\ -\dfrac{1}{\sqrt{6}} \dfrac{1}{\sqrt{l}} \dfrac{\hat{s}\hat{\lambda}}{\hat{1}} & \text{if } \lambda = l - 1, \end{cases}$$

and from (4.24) we get for both $\lambda = l + 1$ and $\lambda = l - 1$ the final simple expression

$$\boldsymbol{\sigma} \cdot \boldsymbol{p}\left[W^{[j]}\psi_{pl}^{[j]}\right]^{[0]} = -p\left[W^{[j]}\psi_{p\lambda}^{[j]}\right]^{[0]}.$$

This result will of course be of great use for calculating relativistic half-integer spin wave functions.

4.6. THE OCCUPATION NUMBER REPRESENTATION

4.6.1. *Standard and contrastandard creation and annihilation operators*

(i) Definitions. The quantization of a field is achieved by ascribing to the amplitudes $a_m^{(j)}$ of the classical solution the meaning of annihilation operators. Thus the annihilation part of a field in invariant form is written

$$\Phi^{(-)} = \sum_m a_m^{(j)}\phi_m^{[j]} = \hat{j}\left[a^{[j]}\phi^{[j]}\right]^{[0]}, \tag{4.41}$$

where the annihilation operators $a_m^{(j)}$ have the transformation properties of a standard tensorial set. The relation (see table 4.1)

$$a_m^{(j)} = (-)^{j+m} a_{-m}^{[j]}$$

has been used to write the right-hand side of eq. (4.41) with only contrastandard quantities coupled into an invariant product, as in eq. (4.7).

The creation operators are defined as the hermitian conjugates of the standard annihilation operators. They have contrastandard transformation properties as can be seen from table 4.1

$$a_m^{(j)+} = (-)^{2j} \tilde{a}_m^{[j]}. \tag{4.42}$$

The creation part of the field is then in invariant form

$$\Phi^{(+)} = \Phi^{(-)+} = \sum_m a_m^{(j)+}\phi_m^{[j]+} = (-)^{2j}\sum_m \tilde{a}_m^{[j]}\tilde{\phi}_m^{(j)} = \hat{j}\left[\tilde{a}^{[j]}\tilde{\phi}^{[j]}\right]^{[0]}, \tag{4.43}$$

recalling that, table 4.1

$$\phi_m^{[j]+} = (-)^{j-m}\phi_{-m}^{(j)+} = \tilde{\phi}_m^{(j)}.$$

Again the right-hand side of (4.43) is written only with contrastandard quantities.

(ii) Matrix representation. According to eq. (4.42) the tilde changes an annihilation into a creation operator. Indeed, when writing the Fock operators in a matrix representation the tilde denotes the transpose operation. For example for n bosons in a given state, $|n\rangle$, the action of the annihilation operator is

$$a|n\rangle = \sqrt{n}\,|n-1\rangle.$$

In matrix representation this relation reads

$$\begin{pmatrix} 0 & \sqrt{1} & & & & & & \\ & 0 & \sqrt{2} & & & & & \\ & & 0 & \cdot & & & & \\ & & & \cdot & \cdot & & & \\ & & & & 0 & \sqrt{n} & & \\ & & & & & 0 & \sqrt{n+1} & \\ & & & & & & \cdot & \cdot \end{pmatrix} \begin{pmatrix} 0 \\ 0 \\ \cdot \\ \cdot \\ 0 \\ 1 \\ \cdot \end{pmatrix} = \sqrt{n} \begin{pmatrix} 0 \\ 0 \\ \cdot \\ \cdot \\ 1 \\ 0 \\ \cdot \end{pmatrix}.$$

The hermitian conjugate operator creates a particle according to

$$a^+|n\rangle = \sqrt{n+1}\,|n+1\rangle,$$

which in matrix representation is given by the transposed operator matrix

$$\begin{pmatrix} 0 & & & & & & \\ \sqrt{1} & 0 & & & & & \\ & \sqrt{2} & \cdot & & & & \\ & \cdot & & \cdot & & & \\ & & \cdot & & 0 & & \\ & & & & \sqrt{n} & 0 & \\ & & & & & \sqrt{n+1} & \cdot \\ & & & & & & \cdot \end{pmatrix} \begin{pmatrix} 0 \\ 0 \\ \cdot \\ \cdot \\ \cdot \\ 1 \\ 0 \\ \cdot \end{pmatrix} = \sqrt{n+1} \begin{pmatrix} 0 \\ 0 \\ \cdot \\ \cdot \\ \cdot \\ 0 \\ 1 \\ \cdot \end{pmatrix}.$$

Similarly for fermions, a given state is occupied by at most one particle and the action of an annihilation operator b is

$$b|1\rangle = |0\rangle,$$

which in matrix form reads

$$\begin{pmatrix} 0 & 1 \\ 0 & 0 \end{pmatrix}\begin{pmatrix} 0 \\ 1 \end{pmatrix} = \begin{pmatrix} 1 \\ 0 \end{pmatrix},$$

while

$$b^+|0\rangle = |1\rangle$$

involves the transposed operator matrix

$$\begin{pmatrix} 0 & 0 \\ 1 & 0 \end{pmatrix}\begin{pmatrix} 1 \\ 0 \end{pmatrix} = \begin{pmatrix} 0 \\ 1 \end{pmatrix}.$$

We re-emphasize that the tilde always indicates a creation operator, be it in the standard or contrastandard representation, as can be seen from the relations given in table 4.1:

$$\tilde{a}_m^{(j)} = (-)^{j-m} a_{-m}^{(j)+},$$

$$\tilde{a}_m^{[j]} = (-)^{2j} a_m^{(j)+}.$$

(iii) Angular momentum coupled commutation relations. For quantization the creation and annihilation operators in their standard form are defined to fulfill the boson commutation or fermion anticommutation relations

$$\left[a_m^{(j)}, a_{m'}^{(j')+}\right]_\mp = \delta_{jj'}\delta_{mm'}.$$

In contrastandard form, these relations become

$$(-)^{j+m+2j'}\left(a_{-m}^{[j]}\tilde{a}_{m'}^{[j']} \mp \tilde{a}_{m'}^{[j']}a_{-m}^{[j]}\right) = \delta_{jj'}\delta_{mm'}.$$

Multiplying with the appropriate vector coupling coefficients and summing over the magnetic quantum numbers one obtains the coupled form of the commutation (anticommutation) relations

$$[a^{[j]}\tilde{a}^{[j']}]_M^{[I]} = \varepsilon(-)^{j+j'-I}[\tilde{a}^{[j']}a^{[j]}]_M^{[I]} + (-)^{2j}\hat{j}\delta_{jj'}\delta_{I0} \tag{4.44}$$

with

$$\varepsilon = 1 \text{ for boson operators},$$

$$\varepsilon = -1 \text{ for fermion operators}.$$

The graphical representation of the commutation (anticommutation) relations (4.44) is given in fig. 4.29. In this figure the two parts on the right correspond to the two terms of the right-hand side of eq. (4.44). We note that the crossing box indicating the sign change $(-)^{j+j'-I}$ associated with the permutation of the two coupled operators has a circle at the crossing point for the additional phase $\varepsilon = +1$ for commuting operators and $\varepsilon = -1$ for anticommuting operators. The second term of the right-hand side is indicated by a circled end box in fig. 4.29. This box is hatched according to our convention for invariant end boxes and it implies a Kronecker factor δ_{I0}. Note also that in contrast to the overlap or matrix element hatched end boxes, the tilde line enters this end box at the bottom. This is in conformity with the order of the operators in a non-vanishing vacuum expectation value.

Thus the contractions of two operators in coupled form are

$$\langle 0|[a^{[j]}\tilde{a}^{[j]}]^{[0]}|0\rangle = (-)^{2j}\hat{j},$$

$$\langle 0|[\tilde{a}^{[j]}\tilde{a}^{[j]}]^{[0]}|0\rangle = \langle 0|[a^{[j]}a^{[j]}]^{[0]}|0\rangle$$

$$= \langle 0|[\tilde{a}^{[j]}a^{[j]}]^{[0]}|0\rangle = 0. \tag{4.45}$$

The non-vanishing vacuum expectation value (4.45) has the graphical representation of fig. 4.30, which is the end diagram of the commutation relation, fig. 4.29. Note however that it has a different meaning. Fig. 4.29 is an operator equation, while fig. 4.30 represents an expectation value. Also note that the relation (4.45) is consistent with the normalisation of the vector $a_m^{(j)+}|0\rangle$

$$\langle 0|a_m^{(j)}a_m^{(j)+}|0\rangle = 1.$$

The crossing of fermion operator lines may occur not only in crossing boxes but also in recoupling boxes. To account for the resulting additional phase ε we shall also modify the fermion operator recoupling graphs by adding a circle at the crossing point as shown in fig. 4.31. Since a line in a crossing or recoupling box may represent several coupled individual fermion operators as in fig. 4.32, the circle

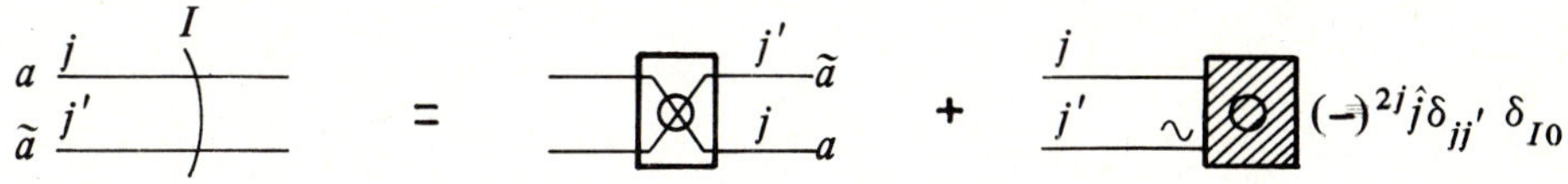

Fig. 4.29.

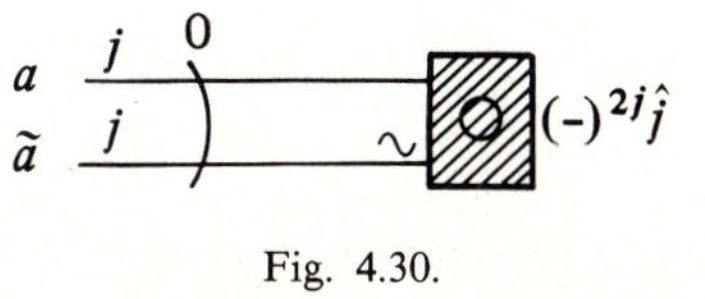

Fig. 4.30.

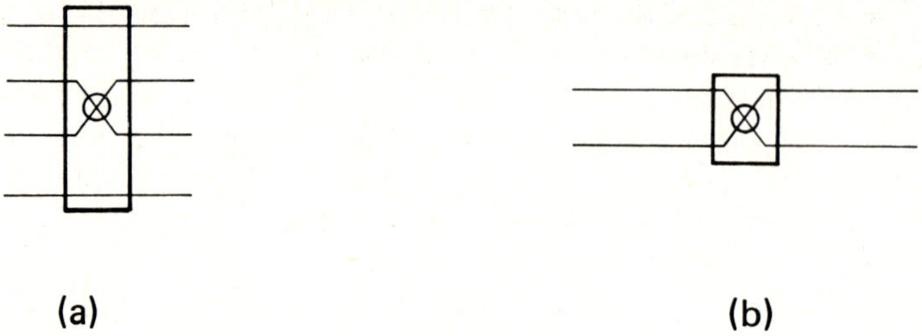

Fig. 4.31.

represents the phase

$$\varepsilon = (-)^{n_1 \times n_2},$$

where n_1 and n_2 are the number of fermion operators in the two crossing lines.

(iv) Single-particle state vectors. The single-particle state vectors are defined in terms of the creation operators as

$$|jm\rangle = a_m^{(j)+}|0\rangle = (-)^{2j}\tilde{a}_m^{[j]}|0\rangle .$$

They generate contrastandard single-particle states from the fields (4.41) and (4.43). Together with the commutation relations (4.44) we have indeed

$$\langle 0|(\Phi^{(-)} + \Phi^{(+)})|jm\rangle = \phi_m^{[j]} .$$

Likewise

$$\langle jm| = \langle 0|a_m^{(j)} = (-)^{j+m}\langle 0|a_{-m}^{[j]},$$

$$\langle jm|(\Phi^{(-)} + \Phi^{(+)})|0\rangle = \tilde{\phi}_m^{(j)} = (-)^{j+m}\tilde{\phi}_{-m}^{[j]} .$$

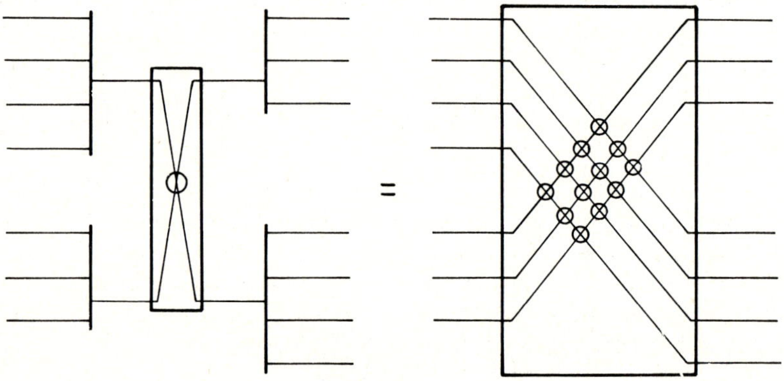

Fig. 4.32.

As usual the tilde on a wave function denotes the transpose of any spin function it may contain. The state vectors are normalized to

$$\langle jm|j'm'\rangle = \delta_{jj'}\delta_{mm'}.$$

Note that both $\tilde{a}_m^{[j]}$ and $\tilde{a}_m^{(j)}$ create particles, the former in a contrastandard state and the latter in a standard state according to

$$\langle 0|(\Phi^{(-)}+\Phi^{(+)})\tilde{a}_m^{[j]}|0\rangle = (-)^{2j}\phi_m^{[j]},$$

$$\langle 0|(\Phi^{(-)}+\Phi^{(+)})\tilde{a}_m^{(j)}|0\rangle = (-)^{2j}\phi_m^{(j)}.$$

Likewise for the annihilation operators $a_m^{[j]}$ and $a_m^{(j)}$

$$\langle 0|a_m^{[j]}(\Phi^{(-)}+\Phi^{(+)})|0\rangle = \tilde{\phi}_m^{[j]},$$

$$\langle 0|a_m^{(j)}(\Phi^{(-)}+\Phi^{(+)})|0\rangle = \tilde{\phi}_m^{(j)}.$$

Finally we introduce the invariant form of the single-particle state vectors with amplitudes $w_m^{[j]}$

$$|j\rangle = (-)^{2j}\hat{j}[w^{[j]}\tilde{a}^{[j]}]^{[0]}|0\rangle,$$

$$\langle j| = \langle 0|\hat{j}[\tilde{w}^{[j]}a^{[j]}]^{[0]}.$$

The amplitudes are normalized according to (4.33). Thus these invariant forms also fulfill

$$\langle j|j\rangle = 1.$$

4.6.2. Coupled basis vectors

Consider a system consisting of n identical particles, either fermions or bosons, distributed in several orbitals $j, k, l, \ldots$ Let an orthonormal set of Fock-space basis state vectors be denoted by $|\alpha I\rangle$, where I denotes the total angular momentum, and α all the other quantum numbers. Its invariant form is defined in terms of the creation operators as

$$|\alpha I\rangle = (-)^{2I}\hat{I}\left[W_\alpha^{[I]}\tilde{\mathscr{A}}_\alpha^{[I]}\right]^{[0]}|0\rangle,$$

where the $W_\alpha^{[I]}$ are complex amplitudes as discussed in sect. 4.4.

The operators $\tilde{\mathscr{A}}_\alpha^{[I]}$ are constructed with products of n creation operators. They are distributed in a given partition $n_j + n_k + n_l + \cdots = n$ among the states $j, k, l, \ldots$ This partition is specified in the overall quantum number α. The state vectors in the Fock representation are given in terms of the operators $a_m^{(j)+}$, which in the contrastandard representation are according to (4.42)

$$a_m^{(j)+} = (-)^{2j}\tilde{a}_m^{[j]}.$$

Therefore we have

$$\mathscr{a}^{(I)+}_{\alpha M} = (-)^{2I}\tilde{\mathscr{a}}^{[I]}_{\alpha M} = \sum_{\gamma} g_{\alpha\gamma}(-)^{2n_j j + 2n_k k + \cdots}\left[\tilde{a}^{[j]}\tilde{a}^{[k]}\ldots\right]^{[I]}_{\gamma M}. \tag{4.46}$$

The sum over the index γ is over all the coupling schemes within the given partition of α with the real weights $g_{\alpha\gamma}$, e.g., fractional parentage coefficients needed to achieve an orthonormal set of basis state vectors $|\alpha I\rangle$. Note that

$$(-)^{2I} = (-)^{2n_j j + 2n_k k + \cdots}.$$

With this choice of the phases the basis state creation operators $\tilde{\mathscr{a}}^{[I]}_{\alpha M}$ generate the basis wave functions $\psi^{[I]}_{\alpha M}$ from the fields Φ, eq. (4.41), according to

$$\psi^{[I]}_{\alpha M} = \langle 0|\Phi\Phi\Phi\cdots\tilde{\mathscr{a}}^{[I]}_{\alpha M}|0\rangle = (-)^{2I}\sum_{\gamma} g_{\alpha\gamma}\left[\phi^{[j]}\phi^{[k]}\phi^{[l]}\ldots\right]^{[I]}_{\gamma M}, \tag{4.47}$$

where the single-particle wave functions are coupled in the same way as the creation operators in eq. (4.46) and where the contractions have been given by (4.45).

The hermitian conjugate of a basis state vector is

$$\langle\alpha I| = \langle 0|\hat{I}\left[\tilde{W}^{[I]}_{\alpha}\mathscr{a}^{[I]}_{\alpha}\right]^{[0]}.$$

In this expression

$$\mathscr{a}^{[I]}_{\alpha M} = (-)^{I+M}\tilde{\mathscr{a}}^{[I]+}_{\alpha -M} = \zeta\sum_{\gamma} g_{\alpha\gamma}\left[a^{[j]}a^{[k]}\ldots\right]^{[I]}_{\gamma M}.$$

Here the annihilation operators have been re-ordered after conjugation to remain in the same lexicographic order as in the ket, with the same coupling γ. In addition there is a phase ζ associated with this reordering of the operators

$$\zeta = 1 \text{ for bosons},$$

$$\zeta = (-)^{\langle n/2\rangle} \text{ for fermions},$$

where $\langle\frac{1}{2}n\rangle$ is the integer part of $\frac{1}{2}n$ (the integer below or equal $\frac{1}{2}n$).

The vacuum expectation values of the basis state operators thus are

$$\langle 0|\mathscr{a}^{(I)}_{\alpha M}\mathscr{a}^{(I)+}_{\alpha M}|0\rangle = (-)^{I-M}\langle 0|\mathscr{a}^{[I]}_{\alpha -M}\tilde{\mathscr{a}}^{[I]}_{\alpha M}|0\rangle = 1,$$

or in coupled form

$$\langle 0|\left[\mathscr{a}^{[I]}_{\alpha}\tilde{\mathscr{a}}^{[I]}_{\alpha}\right]^{[0]}|0\rangle = (-)^{2I}\hat{I}. \tag{4.48}$$

Hence when working with orthonormalized bra and ket vectors in Fock space, the vacuum expectation value of their creation and annihilation operators $\tilde{\mathscr{A}}$, $\mathscr{A}$ is again represented by the graph of fig. 4.30, the value of which is given by eq. (4.48) in full analogy to eq. (4.45).

Herewith we see that the coupled Fock-space basis state vectors in their invariant form fulfill orthonormality

$$\langle \alpha I | \alpha' I' \rangle = \delta_{\alpha\alpha'} \delta_{II'} .$$

These results are directly applicable to basis vectors made of different types of particles,

$$|\alpha I\rangle = (-)^{2I} \hat{I} \left[W_{\alpha}^{[I]} [\tilde{\mathscr{A}}^{[K]} \tilde{\mathscr{B}}^{[L]}]_{\alpha}^{[I]} \right]^{[0]} |0\rangle ,$$

where $\tilde{\mathscr{A}}^{[K]}$ and $\tilde{\mathscr{B}}^{[L]}$ are coupled products of n_a and n_b creation operators $\tilde{a}^{[\,]}$ and $\tilde{b}^{[\,]}$. When taking the hermitian conjugate one obtains directly the product of the bras of the individual groups with an additional phase ξ in addition to the phases ζ contained in $\mathscr{A}$ and $\mathscr{B}$, resulting from commuting through the operators of $\mathscr{A}$ and $\mathscr{B}$,

$$\langle \alpha I | = \xi \langle 0 | \hat{I} \left[\tilde{W}_{\alpha}^{[I]} [\mathscr{A}^{[K]} \mathscr{B}^{[L]}]_{\alpha}^{[I]} \right]^{[0]}$$

with

$$\xi = +1 \qquad \text{if at least one of the particle types are bosons},$$

$$\xi = (-)^{n_a \times n_b} \qquad \text{if both types are fermions}.$$

It is easily checked from fig. 4.33 that this vector is normalized. The phase ξ cancels exactly the phase in the recoupling graph of the norm.

4.6.3. *Fractional parentage coefficients*

Let us consider a sub-system of n equivalent particles, i.e., n identical particles in a same shell. The particle indices are denoted $1, 2, \ldots, n$ and a symmetrized or antisymmetrized but not normalized wave function is represented by a single

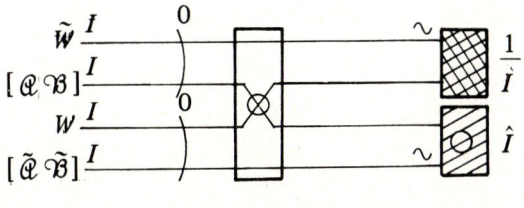

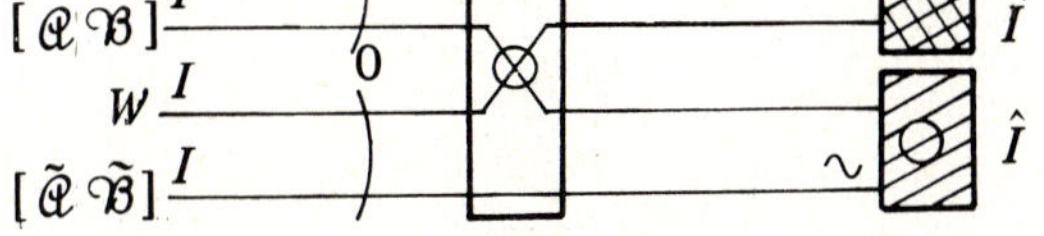

Fig. 4.33.

underlined ket

$$|\underline{12\dots n}\rangle = \sum_P (-)^{\varepsilon_P} |ij\dots x\rangle,$$

where the sum is over all the permutations

$$P = \begin{pmatrix} ij\dots x \\ 12\dots n \end{pmatrix}$$

of the particle indices. For fermions ε_P is the parity of the permutation. For bosons $\varepsilon_P = 0$.

The normalized symmetrized or antisymmetrized wave function is denoted by a double underlined ket

$$|\underline{\underline{12\dots n}}\rangle = \frac{1}{\sqrt{\mathfrak{N}}} |\underline{12\dots n}\rangle.$$

The one- and two-particle fractional parentage coefficients (CFP's) are defined respectively, for a shell j, by

$$|\underline{\underline{12\dots n}}(\alpha IM)\rangle = \sum_{\beta K} (j^{n-1}\beta K;\ j|\}j^n\alpha I)\left[|\underline{\underline{12\dots n-1}}(\beta K)\rangle |nj\rangle\right]^{[I]}_M, \quad (4.49)$$

$$|\underline{\underline{12\dots n}}(\alpha IM)\rangle = \sum_{\beta KL} (j^{n-2}\beta K;\ j^2L||\}j^n\alpha I)$$

$$\times \left[|\underline{\underline{12\dots n-2}}(\beta K)\rangle |\underline{\underline{n-1\,n}}(L)\rangle\right]^{[I]}_M. \quad (4.50)$$

Here the normalized symmetrized or antisymmetrized kets are coupled to total angular momenta $I, K, L\dots$ The quantum numbers $\alpha, \beta\dots$ stand for all the other quantum numbers needed to specify the states uniquely. For the two-particle ket with angular momentum L, the state $|\underline{\underline{n-1\,n}}(L)\rangle$ is unique with L even. Note in the right-hand side expansions the coupling order of the kets and also that the parent states are normalized and properly symmetrized (or antisymmetrized).

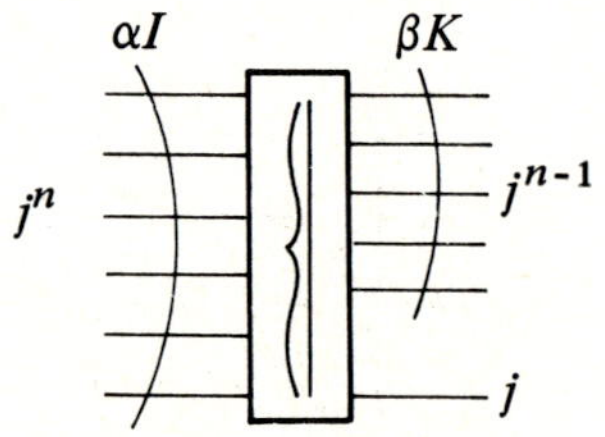

Fig. 4.34.

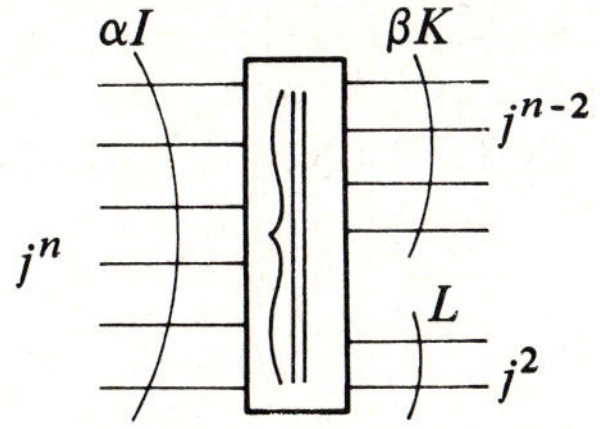

Fig. 4.35.

The graphical symbols for the one- and two-body fractional parentage coefficients which will be used in the recoupling diagrams are given in figs. 4.34 and 4.35 respectively.

In the Fock representation of sect. 4.6.2, coupled basis vectors are written as

$$\underline{\underline{|\,12\ldots n\,(\alpha IM)\rangle}} = (-)^{2I}\tilde{\mathscr{A}}^{[I]}_{n\alpha M}|0\rangle\,,$$

where the state creation operator $\tilde{\mathscr{A}}^{[I]}_{n\alpha M}$ is a product of n particle creation operators $\tilde{a}^{[j]}$ defined recursively by the expansions with the CFP's of eqs. (4.49) and (4.50)

$$\tilde{\mathscr{A}}^{[I]}_{n\alpha M} = \frac{1}{\sqrt{n}} \sum_{\beta K} (j^{n-1}\beta K;\, j|\}j^n\alpha I)\big[\tilde{\mathscr{A}}^{[K]}_{n-1\beta}\tilde{a}^{[j]}\big]^{[I]}_M, \tag{4.51}$$

$$\tilde{\mathscr{A}}^{[I]}_{n\alpha M} = \sqrt{\frac{2}{n(n-1)}} \sum_{\beta KL} (j^{n-2}\beta K;\, j^2 L|\}j^n\alpha I)\big[\tilde{\mathscr{A}}^{[K]}_{n-2\beta}\tilde{\mathscr{A}}^{[L]}_2\big]^{[I]}_M. \tag{4.52}$$

The square root factors in these equations arise from the fact that the creation operator products $\tilde{\mathscr{A}}_{n-1}\tilde{a}$ and $\tilde{\mathscr{A}}_{n-2}\tilde{\mathscr{A}}_2$ generate only symmetrized (or antisymmetrized) but not normalized states. In contrast the definitions of the CFP's, eqs. (4.49) and (4.50), are in terms of fully normalized states.

The corresponding expressions for the annihilation operators are obtained from

$$\mathscr{A}^{[I]}_{n\alpha M} = (-)^{I+M}\tilde{\mathscr{A}}^{[I]+}_{n\alpha -M},$$

which yields

$$\mathscr{A}^{[I]}_{n\alpha M} = \frac{\varepsilon}{\sqrt{n}} \sum_{\beta K} (j^{n-1}\beta K;\, j|\}j^n\alpha I)\big[\mathscr{A}^{[K]}_{n-1\beta}a^{[j]}\big]^{[I]}_M,$$

$$\mathscr{A}^{[I]}_{n\alpha M} = \sqrt{\frac{2}{n(n-1)}} \sum_{\beta KL} (j^{n-2}\beta K;\, j^2 L|\}j^n\alpha I)\big[\mathscr{A}^{[K]}_{n-2\beta}\mathscr{A}^{[L]}_2\big]^{[I]}_M.$$

Here in the first expression the phase ε is that part of the crossing phase (arising from the hermitian conjugation) which is not contained in the two operators of the right-hand side. It arises from crossing these two operators, hence

$$\varepsilon = (-)^{n-1}, \qquad \text{for fermions},$$

$$\varepsilon = 1, \qquad \text{for bosons},$$

or, in general,

$$\varepsilon = (-)^{2K}.$$

Since in these expressions the operator products on the right-hand side have the correct overall permutation symmetry, the CFP transformations are unitary and can be inverted. Thus we have, for example,

$$\left[\mathscr{A}^{[K]}_{n-1\,\beta} a^{[j]}\right]^{[I]}_{M} = \varepsilon\sqrt{n}\sum_{\alpha}\left(j^{n-1}\beta K;\, j|\}j^{n}\alpha I\right)\mathscr{A}^{[I]}_{n\alpha M}.$$

Herewith with expressions of this type, together with the normalization relation (4.48) we get

$$\langle 0|\left[\left[\mathscr{A}^{[K]}_{n-1\,\beta} a^{[j]}\right]^{[I]} \tilde{\mathscr{A}}^{[I]}_{n\alpha}\right]^{[0]}|0\rangle = \varepsilon\langle 0|\left[\mathscr{A}^{[I]}_{n\alpha}\tilde{\mathscr{A}}^{[K]}_{n-1\,\beta}\tilde{a}^{[j]}\right]^{[0]}|0\rangle$$

$$= \varepsilon(-)^{2I}\hat{I}\sqrt{n}\left(j^{n-1}\beta K;\, j|\}j^{n}\alpha I\right), \qquad (4.53)$$

and

$$\langle 0|\left[\left[\mathscr{A}^{[K]}_{n-2\,\beta}\left[a^{[j]}a^{[j]}\right]^{[L]}\right]^{[I]}\tilde{\mathscr{A}}^{[I]}_{n\alpha}\right]^{[0]}|0\rangle = \langle 0|\left[\mathscr{A}^{[I]}_{n\alpha}\tilde{\mathscr{A}}^{[K]}_{n-2\beta}\left[\tilde{a}^{[j]}\tilde{a}^{[j]}\right]^{[L]}\right]^{[0]}|0\rangle$$

$$= (-)^{2I}\hat{I}\sqrt{n(n-1)}\left(j^{n-2}\beta K;\, j^{2}L||\}j^{n}\alpha I\right).$$

(4.54)

These results will be useful for computing the vacuum expectation values arising in the many-body matrix elements of one- and two-body operators.

4.6.4. CFP's for the $1s^n$ pion cloud

The special case of n identical pions in the shell 1s is very simple but may play an important role in view of the high multiplicities associated with the stimulated emission of pions in the lowest energy orbital.

For the basis states $1s^n$, only the isospin part needs symmetrization. Now, for isospin-1 particles, all possible states are uniquely specified by the particle number n and the total isospin quantum number T. This happens because each state $|1s^n T\rangle$ is generated at most from two states, viz., $|1s^{n-1}T-1\rangle$ and $|1s^{n-1}T+1\rangle$. The state $|1s^{n-1}T\rangle$ would not be symmetric. Since the corresponding two amplitudes must fulfill the normalization condition, this yields a single possible state for each set of values of n and T.

We introduce a shortened notation for the CFP's of the $1s^n$ pion cloud, which will be used throughout,

$$\mathrm{CFP}_1^n(T, R) = \left(1s^{n-1}R; 1s|\}1s^n T\right), \tag{4.55}$$

$$\mathrm{CFP}_2^n(T, R, t) = \left(1s^{n-2}R; 1s^2 t\|\}1s^n T\right). \tag{4.56}$$

Furthermore, in the following equations the 1s pion creation operator is denoted $\tilde{a}^{[1]}$ and the creation operator for a normalized state $1s^n$ is denoted $\tilde{\mathscr{Q}}_n^{[T]}$.

Thus in order to generate the one-particle CFP's of the successive states $|1s^n T\rangle$ which by definition verify

$$\tilde{\mathscr{Q}}_n^{[T]} = \frac{1}{\sqrt{n}}\left\{\mathrm{CFP}_1^n(T, T-1)\left[\tilde{\mathscr{Q}}_{n-1}^{[T-1]}\tilde{a}^{[1]}\right]^{[T]} + \mathrm{CFP}_1^n(T, T+1)\left[\tilde{\mathscr{Q}}_{n-1}^{[T+1]}\tilde{a}^{[1]}\right]^{[T]}\right\},$$

let us consider the recursion

$$\begin{aligned}\tilde{\mathscr{Q}}_{n+1}^{[T]} = \frac{1}{\sqrt{n+1}\sqrt{n}}\Big\{&\mathrm{CFP}_1^{n+1}(T, T-1)\Big(\mathrm{CFP}_1^n(T-1, T-2)\Big[\Big[\tilde{\mathscr{Q}}_{n-1}^{[T-2]}\tilde{a}^{[1]}\Big]^{[T-1]}\tilde{a}^{[1]}\Big]^{[T]} \\ &+ \mathrm{CFP}_1^n(T-1, T)\Big[\Big[\tilde{\mathscr{Q}}_{n-1}^{[T]}\tilde{a}^{[1]}\Big]^{[T-1]}\tilde{a}^{[1]}\Big]^{[T]}\Big) \\ &+ \mathrm{CFP}_1^{n+1}(T, T+1)\Big(\mathrm{CFP}_1^n(T+1, T)\Big[\Big[\tilde{\mathscr{Q}}_{n-1}^{[T]}\tilde{a}^{[1]}\Big]^{[T+1]}\tilde{a}^{[1]}\Big]^{[T]} \\ &+ \mathrm{CFP}_1^n(T+1, T+2)\times\Big[\Big[\tilde{\mathscr{Q}}_{n-1}^{[T+2]}\tilde{a}^{[1]}\Big]^{[T+1]}\tilde{a}^{[1]}\Big]^{[T]}\Big)\Big\}.\end{aligned}$$

We recouple the terms of the right-hand side into the coupling scheme $\tilde{\mathscr{Q}}_{n-1}[\tilde{a}\tilde{a}]^{[t]}$. Here the terms with $t = 1$ must cancel in order that the state be symmetric, which

TABLE 4.3
One-particle fractional parentage coefficients $CFP_1^n(T, R)$ for $1s^n$ pions

$n = 3$

T \ R	0	2
1	0.745356	0.666667
3		1.

$n = 4$

T \ R	1	3
0	1.	
2	0.836660	0.547723
4		1.

$n = 5$

T \ R	0	2	4
1	0.683130	0.730297	
3		0.878310	0.478091
5			1.

$n = 6$

T \ R	1	3	5
0	1.		
2	0.774597	0.632456	
4		0.902671	0.430331
6			1.

yields for the CFP_1's the recursion relation

$$CFP_1^{n+1}(T, T-1)CFP_1^n(T-1, T)\begin{bmatrix} T & 1 & T-1 \\ 0 & 1 & 1 \\ T & 1 & T \end{bmatrix}$$

$$+CFP_1^{n+1}(T, T+1)CFP_1^n(T+1, T)\begin{bmatrix} T & 1 & T+1 \\ 0 & 1 & 1 \\ T & 1 & T \end{bmatrix} = 0 .$$

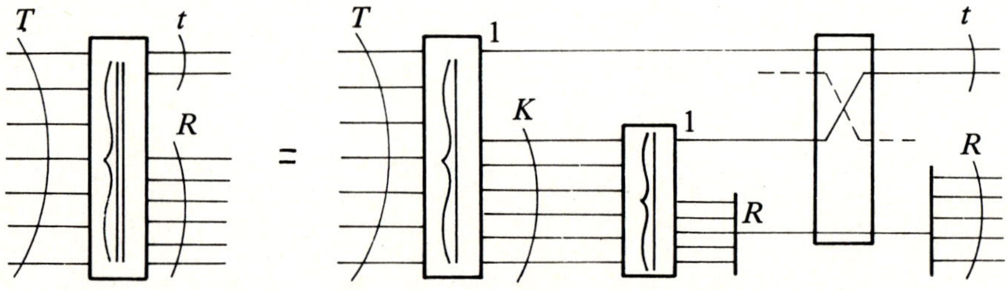

Fig. 4.36.

TABLE 4.4
Two-particle fractional parentage coefficients $CFP_2^n(T, R, t)$ for $1s^n$ pions

$n = 3$

R / T	1	
	$t = 0$	$t = 2$
1	0.745356	0.666667
3		1.

$n = 4$

R / T	0		2	
	$t = 0$	$t = 2$	$t = 0$	$t = 2$
0	0.745356			0.666667
2		0.623610	0.623610	0.471405
4				1.

$n = 5$

R / T	1		3	
	$t = 0$	$t = 2$	$t = 0$	$t = 2$
1	0.683130	0.611010		0.4
3		0.734847	0.547723	0.4
5				1.

$n = 6$

R / T	0		2		4	
	$t = 0$	$t = 2$	$t = 0$	$t = 2$	$t = 0$	$t = 2$
0	0.683130			0.730230		
2		0.529150	0.632456	0.478091		0.302372
4				0.792825	0.494413	0.356348
6						1.

Together with the normalization condition

$$\mathrm{CFP}_1^n(T, T-1)^2 + \mathrm{CFP}_1^n(T, T+1)^2 = 1\,,$$

and the values $\mathrm{CFP}_1^2(2,1) = 1, \mathrm{CFP}_1^2(0,1) = 1$, the CFP_1^n can thus be calculated. Their numerical values up to $n = 6$ are given in table 4.3.

The two-particle CFP's can now be obtained in terms of two successive one-particle CFP's according to the recoupling diagram of fig. 4.36 which yields

$$\mathrm{CFP}_2^n(T, R, t) = \sum_K \begin{bmatrix} 1 & 0 & 1 \\ 1 & R & K \\ t & R & T \end{bmatrix} \mathrm{CFP}_1^n(T, K)\mathrm{CFP}_1^{n-1}(K, R)\,.$$

The coefficients CFP_2^n calculated up to $n = 6$ are given in table 4.4.

CHAPTER 5

Discretized fields and configuration spaces

5.1. INTRODUCTION

The expressions for the quantized relativistic free fields for particles of spins 0, $\frac{1}{2}$, 1, and $\frac{3}{2}$ are given in sect. 5.2 in full detail as an expansion on a discretized basis in good angular momentum and parity quantum numbers. This basis consists of harmonic oscillator functions in momentum space constructed according to the method of Chapter 3. The forms of the associated relativistic single particle basis wave functions are given in sect. 5.3.

In the remaining chapters of this book we shall show how to use these fields to construct the interaction matrix elements for many-body systems. To that end we shall treat explicitly the different examples taken from a selection of particular configuration spaces. We re-emphasize that these examples are given only to demonstrate the application of the present methods and are not meant to describe actual physical systems. Thus tables of many-particle (fermion and boson) configurations constructed with these discretized fields are given in sect. 5.4 for the simple many-body systems which have the quantum numbers of the deuteron, the delta, the nucleon, the pion and the vacuum. These basis vector states will be used in the following chapters to calculate our examples of energy, center-of-mass and electromagnetic transition matrix elements.

5.2. INVARIANT DISCRETIZED FIELD EXPANSIONS

5.2.1. *Discretized spin-0 fields*

The discretization of the spin-0 field has already been obtained in sect. 3.2. This discretized multipole expansion will now be written in the contrastandard notations of Chapter 4 as sums of invariants.

The plane wave solution of the Klein-Gordon equation for particles of spin 0 and isospin t, with $t = 1$ for the pion and $t = 0$ for the hypothetical σ-meson has been given in sect. 2.3.1 in the Heisenberg picture. We now rewrite it using only contrastandard quantities in invariant form as in eqs. (4.41) and (4.43). We begin with

$$\varphi(\boldsymbol{r}) = \left(\frac{1}{2\pi}\right)^{3/2} \int \mathrm{d}^3 p \, \frac{1}{\sqrt{2E}} \hat{t}\left(\left[a_{\boldsymbol{p}}^{[t]} \eta^{[t]}\right]^{[0]} \mathrm{e}^{i\boldsymbol{p}\cdot\boldsymbol{r}} + \left[\tilde{a}_{\boldsymbol{p}}^{[t]} \eta^{[t]}\right]^{[0]} \mathrm{e}^{-i\boldsymbol{p}\cdot\boldsymbol{r}}\right).$$

Here the field is written in the Schrödinger picture by putting in eq. (2.8) the time to zero. The $\tilde{a}_{\boldsymbol{p}}^{[t]}$ denotes the creation operator for a particle with momentum $\boldsymbol{p}$ and isospin t, as defined in sect. 4.6. The coupling to an invariant refers to isospin only and the isospin wave functions $\eta_\tau^{[t]}$ are normalized according to eq. (4.8):

$$[\eta^{[t]}|\eta^{[t']}] = \hat{t}\delta_{tt'}.$$

Recall that as discussed in sect. 4.2.1 the isospin wave functions for integer isospin are self-conjugate and thus they do not require a tilde in the second term of the field expansion. The commutation relations (2.9) read in coupled form, as shown in sect. 4.6.1, eq. (4.44),

$$[a_{\boldsymbol{p}}^{[t]}\tilde{a}_{\boldsymbol{p}'}^{[t']}]_M^{[T]} = (-)^{t+t'-T}[\tilde{a}_{\boldsymbol{p}'}^{[t']}a_{\boldsymbol{p}}^{[t]}]_M^{[T]} + (-)^{2t}\hat{t}\delta_{tt'}\delta_{T0}\delta^3(\boldsymbol{p}-\boldsymbol{p}').$$

Now, the invariant form of the plane wave multipole expansion is

$$\begin{aligned} \mathrm{e}^{i\boldsymbol{p}\cdot\boldsymbol{r}} &= 4\pi\sum_{lm} i^l j_l(pr)Y_m^{(l)}(\hat{\boldsymbol{p}})Y_m^{[l]}(\hat{\boldsymbol{r}}) \\ &= 4\pi\sum_l i^l j_l(pr)\hat{l}[\hat{p}^{[l]}\hat{r}^{[l]}]^{[0]}, \end{aligned} \qquad (5.1)$$

using the short-hand notation

$$\hat{p}_m^{(l)} = Y_m^{(l)}(\hat{\boldsymbol{p}}) \qquad \text{and} \qquad \hat{r}_m^{[l]} = Y_m^{[l]}(\hat{\boldsymbol{r}}).$$

Discretization of the field is now carried out according to the method explained in sect. 3.2. Multipole creation and annihilation operators are introduced with the unitary transformation

$$a_{pm\tau}^{[lt]} = \int \mathrm{d}^2 p\, \hat{p}_m^{[l]} a_{\boldsymbol{p}\tau}^{[t]}. \qquad (5.2)$$

The double superscript indices of $a_{pm\tau}^{[lt]}$ denote that the operator behaves as a contrastandard tensor component both in ordinary and isospin spaces according to the convention adopted in eqs. (4.27) and (4.28). The plane wave multipole expansion is thus in coupled form both in ordinary and isospin spaces

$$\varphi(\boldsymbol{r}) = \sqrt{\frac{2}{\pi}}\int p^2 \mathrm{d}p\frac{1}{\sqrt{2E}}\sum_l i^l j_l(pr)\hat{l}\hat{t}\left([a_p^{[lt]}\hat{r}^{[l]}\eta^{[t]}]^{[00]} + (-)^l[\tilde{a}_p^{[lt]}\hat{r}^{[l]}\eta^{[t]}]^{[00]}\right).$$

The radial unitary transformation introduced in eq. (3.3) yields the discretized multipole annihilation operators

$$a_{\nu m\tau}^{[lt]} = \int p^2 \mathrm{d}p\, f_{\nu l}(p) a_{pm\tau}^{[lt]}. \qquad (5.3)$$

Here the functions $f_{\nu l}(p)$ are the complete orthonormal set given by the harmonic oscillator functions, eq. (3.8). Combining the two transformations (5.2) and (5.3) yields

$$a^{[lt]}_{\nu m \tau} = \int \mathrm{d}^3 p\, f_{\nu l}(p)\, \hat{p}^{[l]}_m a^{[t]}_{p\tau}.$$

The hermitian conjugation of these operators in the discretized representation reads according to table 4.1

$$a^{[lt]+}_{\nu m \tau} = (-)^{l+m+t+\tau}\, \tilde{a}^{[lt]}_{\nu-m-\tau},$$

where $\tilde{a}^{[lt]}_{\nu-m-\tau}$ are the contrastandard creation operators of a particle in the state ν, l, m, t, τ. Their commutation relations are in invariant form

$$\left[a^{[lt]}_{\nu}\tilde{a}^{[l't']}_{\nu'}\right]^{[IT]}_{M_I M_T} = (-)^{l+l'+t+t'-I-T}\left[\tilde{a}^{[l't']}_{\nu'}a^{[lt]}_{\nu}\right]^{[IT]}_{M_I M_T} + (-)^{2t}\hat{l}\hat{t}\,\delta_{\nu\nu'}\delta_{ll'}\delta_{tt'}\delta_{I0}\delta_{T0},$$

where $I=0$, $T=0$ of course imply $M_I=0$, $M_T=0$ in the last right-hand term.

After substitution and integration over the linear momentum p we obtain the proper discretized invariant form of the spin-0 free field

$$\varphi(\boldsymbol{r}) = \sqrt{\tfrac{1}{2}}\sum_{\nu l} i^l g_{\nu l}(r)\hat{l}\hat{t}\left(\left[a^{[lt]}_{\nu}\hat{r}^{[l]}\eta^{[t]}\right]^{[00]} + (-)^l\left[\tilde{a}^{[lt]}_{\nu}\hat{r}^{[l]}\eta^{[t]}\right]^{[00]}\right). \tag{5.4}$$

The functions $g_{\nu l}(r)$ represent the wave packets over the energy variable and they have been defined in eq. (3.4). The double coupling notation of (4.27) and (4.28) is used: the contrastandard operator $a^{[lt]}_{\nu m \tau}$ is coupled into an invariant product with the contrastandard spherical harmonic $\hat{r}^{[l]}_m$ in ordinary space and with the contrastandard wave function $\eta^{[t]}_{\tau}$ in isospin space.

Likewise the canonical conjugate field defined by eq. (2.8) is given by

$$\pi(\boldsymbol{r}) = -i\sqrt{\tfrac{1}{2}}\sum_{\nu l} i^l h_{\nu l}(r)\hat{l}\hat{t}\left(\left[a^{[lt]}_{\nu}\hat{r}^{[l]}\eta^{[t]}\right]^{[00]} - (-)^l\left[\tilde{a}^{[lt]}_{\nu}\hat{r}^{[l]}\eta^{[t]}\right]^{[00]}\right), \tag{5.5}$$

where the functions $h_{\nu l}(r)$ are defined in eq. (3.6).

The equal time commutation relations between the field $\varphi(\boldsymbol{r})$ and its conjugate $\pi(\boldsymbol{r})$ are preserved by the unitary transformations employed in discretizing the fields, eq. (2.10),

$$[\varphi(\boldsymbol{r}), \pi(\boldsymbol{r}')]_- = i(2t+1)\delta^3(\boldsymbol{r}-\boldsymbol{r}'),$$

where the factor $(2t+1)$ arises from the summation over the different charge states.

We now redefine the above field expansions with phases more suited for the description of stationary states. The phases i^l which appear in (5.4) and (5.5) arise from the plane wave nature of the solutions. However, when setting up the basis states each term of these expansions will provide a single-particle basis state. In that context they can be phased independently. This freedom of choice is now used to eliminate the relative phases i^l between the different multipoles. Herewith the form of the basic fields is chosen phased according to

$$\varphi(\boldsymbol{r})=\sqrt{\tfrac{1}{2}}\sum_{\nu l} g_{\nu l}(r)\hat{l}\hat{t}\left(\left[a_\nu^{[lt]}\hat{\boldsymbol{r}}^{[l]}\eta^{[t]}\right]^{[00]}+\left[\tilde{a}_\nu^{[lt]}\hat{\boldsymbol{r}}^{[l]}\eta^{[t]}\right]^{[00]}\right), \tag{5.6}$$

$$\pi(\boldsymbol{r})=-i\sqrt{\tfrac{1}{2}}\sum_{\nu l} h_{\nu l}(r)\hat{l}\hat{t}\left(\left[a_\nu^{[lt]}\hat{\boldsymbol{r}}^{[l]}\eta^{[t]}\right]^{[00]}-\left[\tilde{a}_\nu^{[lt]}\hat{\boldsymbol{r}}^{[l]}\eta^{[t]}\right]^{[00]}\right). \tag{5.7}$$

5.2.2. *Discretized spin-$\frac{1}{2}$ fields*

We now turn to the spin-$\frac{1}{2}$ field obtained in sect. 2.3.2, eq. (2.13). Its discretization proceeds along the same lines as that of the spin-0 field, except for an additional simple coupling in the multipole expansion so as to introduce the total angular momentum $\boldsymbol{j}$ of the upper and lower components of the spinors and the need to evaluate the operator $\boldsymbol{\sigma}\cdot\boldsymbol{p}$ contained in the definition of the spinors, eqs. (2.11) and (2.12).

To that end the expansion (5.1) of the plane wave is introduced. In its uncoupled form, the field (2.13) is rewritten as an invariant according to eqs. (4.41) and (4.43) with isospin t and with spin $s=\frac{1}{2}$,

$$\psi(\boldsymbol{r})=\sqrt{\frac{2}{\pi}}\int \mathrm{d}^3p\sqrt{\frac{M}{E}}\sum_{\substack{lm\\ \sigma\tau}} i^l\Big(b_{\boldsymbol{p}\sigma\tau}^{(st)}\mathcal{U}_\sigma^{[s]}(\boldsymbol{p})\eta_\tau^{[t]}\hat{p}_m^{(l)}\hat{r}_m^{[l]}$$

$$+(-)^l\tilde{d}_{\boldsymbol{p}\sigma\tau}^{[st]}\mathcal{V}_\sigma^{(s)}(\boldsymbol{p})\eta_\tau^{(t)}\hat{p}_m^{[l]}\hat{r}_m^{(l)}\Big)j_l(pr).$$

Let us note that in the second term the standard form of the spinor is required in order to describe the proper spin projection of the antiparticles, in view of the right-hand side of eq. (2.12). This is associated with the fact that an antiparticle of spin projection σ is described by the absence of a negative energy solution with a spin projection $-\sigma$. Note also that the spin and isospin functions for the antiparticle are non-transposed. The transposed functions are associated with the hermitian conjugate field ψ^+. In its coupled form expressed in contrastandard quantities the field ψ is

$$\psi(\boldsymbol{r})=\sqrt{\frac{2}{\pi}}\int \mathrm{d}^3p\sqrt{\frac{M}{E}}\sum_{l} i^l\hat{l}\hat{s}\hat{t}\Big(\left[b_{\boldsymbol{p}}^{[st]}\mathcal{U}^{[s]}(\boldsymbol{p})\eta^{[t]}\right]^{[00]}$$

$$+(-)^l\left[\tilde{d}_{\boldsymbol{p}}^{[st]}\mathcal{V}^{[s]}(\boldsymbol{p})\eta^{[t]}\right]^{[00]}\Big)\left[\hat{\boldsymbol{p}}^{[l]}\hat{\boldsymbol{r}}^{[l]}\right]^{[0]}j_l(pr).$$

The spinors $\mathcal{U}$ and $\mathcal{V}$ are operators in position space acting on the functions

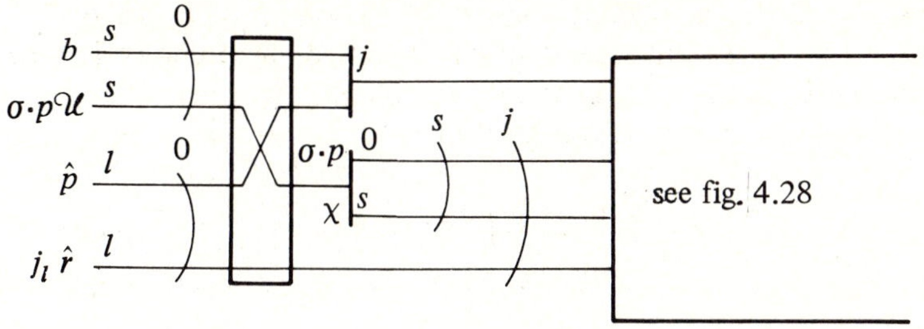

Fig. 5.1.

$\hat{r}^{[l]}_m j_l(pr)$. We separate the upper and lower components, and we evaluate the action of the operator $\boldsymbol{\sigma}\cdot\boldsymbol{p}$ in the lower components of the spinors according to sect. 4.5.3 and as shown in fig. 4.28. A simple recoupling is left, shown in fig. 5.1, to yield the field expansion in invariant form in terms of the spin spherical functions $Y^{[j]}_l(\hat{\boldsymbol{r}})$ of eq. (4.5)

$$\psi(\boldsymbol{r}) = \sqrt{\frac{2}{\pi}}\int \mathrm{d}^3p\,\frac{1}{\sqrt{2E}}\sum_{lj} i^l\hat{j}\hat{t}\left(\left[\left[b_p^{[st]}\hat{p}^{[l]}\eta^{[t]}\right]^{[j0]}\begin{pmatrix}\sqrt{E+M}\,Y^{[j]}_l(\hat{\boldsymbol{r}})\,j_l(pr)\\ -\sqrt{E-M}\,Y^{[j]}_\lambda(\hat{\boldsymbol{r}})\,j_\lambda(pr)\end{pmatrix}\right]^{[00]}\right.$$

$$\left.+(-)^l\left[\left[\tilde{d}_p^{[st]}\hat{p}^{[l]}\eta^{[t]}\right]^{[j0]}\begin{pmatrix}-\sqrt{E-M}\,Y^{[j]}_\lambda(\hat{\boldsymbol{r}})\,j_\lambda(pr)\\ \sqrt{E+M}\,Y^{[j]}_l(\hat{\boldsymbol{r}})\,j_l(pr)\end{pmatrix}\right]^{[00]}\right)$$

with $\lambda = l+1$ if $j = l+\frac{1}{2}$ and $\lambda = l-1$ if $j = l-\frac{1}{2}$, i.e. $\lambda = 2j-l$.

Discretized multipole annihilation operators are defined by a similar unitary transformation as that used for the spin-0 operators

$$b^{[jt]}_{\nu l m\tau} = \int \mathrm{d}^3p\, f_{\nu l}(p)\left[b^{[st]}_{p\tau}\hat{p}^{[l]}\right]^{[jt]}_{m\tau}.$$

The anti-commutation relations are preserved, and in invariant form as in eq. (4.44) we have

$$\left[b^{[jt]}_{\nu l}\tilde{b}^{[j't']}_{\nu' l'}\right]^{[IT]}_{M_I M_T} = -(-)^{j+j'+t+t'-I-T}\left[\tilde{b}^{[j't']}_{\nu' l'}b^{[jt]}_{\nu l}\right]^{[IT]}_{M_I M_T}$$

$$+(-)^{2j+2t}\,\hat{j}\hat{t}\,\delta_{\nu\nu'}\delta_{jj'}\delta_{ll'}\delta_{tt'}\delta_{I0}\delta_{T0}.$$

The same is done for the antiparticle operators.

Substituting the new operators in the plane wave expansion, and eliminating the plane wave phases i^l as explained above for the spin-0 field yields the desired discretized expansion for the spin-$\frac{1}{2}$ field,

$$\psi(\boldsymbol{r})=\sqrt{\tfrac{1}{2}}\sum_{\nu lj}\hat{j}\hat{t}\left(\left[b_{\nu l}^{[jt]}\begin{pmatrix} Y_l^{[j]}(\hat{\boldsymbol{r}})u_{\nu l}(r)\\ -Y_\lambda^{[j]}(\hat{\boldsymbol{r}})v_{\nu l\lambda}(r)\end{pmatrix}\eta^{[t]}\right]^{[00]}\right.$$

$$\left.+\left[\tilde{d}_{\nu l}^{[jt]}\begin{pmatrix} -Y_\lambda^{[j]}(\hat{\boldsymbol{r}})v_{\nu l\lambda}(r)\\ Y_l^{[j]}(\hat{\boldsymbol{r}})u_{\nu l}(r)\end{pmatrix}\eta^{[t]}\right]^{[00]}\right) \tag{5.8}$$

with $\lambda=l+1$ if $j=l+\frac{1}{2}$ and $\lambda=l-1$ if $j=l-\frac{1}{2}$. The hermitian conjugate field is

$$\psi(\boldsymbol{r})^{+}=\sqrt{\tfrac{1}{2}}\sum_{\nu lj}\hat{j}\hat{t}\left(\left[\tilde{b}_{\nu l}^{[jt]}\begin{pmatrix} \tilde{Y}_l^{[j]}(\hat{\boldsymbol{r}})u_{\nu l}(r)\\ -\tilde{Y}_\lambda^{[j]}(\hat{\boldsymbol{r}})v_{\nu l\lambda}(r)\end{pmatrix}^{\mathrm{T}}\tilde{\eta}^{[t]}\right]^{[00]}\right.$$

$$\left.+\left[d_{\nu l}^{[jt]}\begin{pmatrix} -\tilde{Y}_\lambda^{[j]}(\hat{\boldsymbol{r}})v_{\nu l\lambda}(r)\\ \tilde{Y}_l^{[j]}(\hat{\boldsymbol{r}})u_{\nu l}(r)\end{pmatrix}^{\mathrm{T}}\tilde{\eta}^{[t]}\right]^{[00]}\right).$$

The radial functions are, together with the factor $\sqrt{\frac{1}{2}}$ in (5.8),

$$u_{\nu l}(r)=\sqrt{\frac{2}{\pi}}\int p^2\,\mathrm{d}p\sqrt{\frac{E+M}{E}}\,f_{\nu l}(p)\,j_l(pr), \tag{5.9}$$

$$v_{\nu l\lambda}(r)=\sqrt{\frac{2}{\pi}}\int p^2\,\mathrm{d}p\sqrt{\frac{E-M}{E}}\,f_{\nu l}(p)\,j_\lambda(pr). \tag{5.10}$$

They are normalized such that,

$$\sum_{\nu l\lambda}\left(u_{\nu l}(r)u_{\nu l}(r')+v_{\nu l\lambda}(r)v_{\nu l\lambda}(r')\right)=2\frac{\delta(r-r')}{r^2}.$$

It may be verified on the discretized form (5.8) together with the similar expression for the conjugate field $\pi=i\psi^{+}$ that the anticommutation relations (2.14) between the free fields $\psi(\boldsymbol{r})$ and $\pi(\boldsymbol{r})$ are indeed fulfilled.

5.2.3. Discretized spin-1 fields

The vector field is treated in a same way as the spin-$\frac{1}{2}$ field with respect to coupling to good total angular momentum $\boldsymbol{J}=\boldsymbol{s}+\boldsymbol{l}$. However, in analogy to the usual expansions of the electromagnetic field, one more unitary transformation is introduced in order to define magnetic ($\mathfrak{M}$), electric ($\mathcal{E}$) and longitudinal ($\mathcal{L}$) multipolarities so as to separate the transverse and longitudinal parts, and to achieve states of good parity.

The spin-1 field is a four-vector field and as discussed in sect. 2.3.3 is of the form

$$\phi(\boldsymbol{r})=\begin{pmatrix}\boldsymbol{\phi}(\boldsymbol{r})\\ \phi_4(\boldsymbol{r})\end{pmatrix}.$$

The components $\boldsymbol{\phi}(\boldsymbol{r})$ represent the space-like part of the field with polarization unit vector $\boldsymbol{e}$ and they transform like the components of a spin-1 vector field under a space rotation. The component $\phi_4(\boldsymbol{r})$ represents the time-like part of the field and it transforms under a space rotation like a scalar field. This time-like field is fixed by a gauge condition, the Lorentz gauge given by eq. (2.17),

$$\dot{\phi}_4(\boldsymbol{r})=-i\,\mathrm{div}\,\boldsymbol{\phi}(\boldsymbol{r}).$$

We consider first the space-like components $\boldsymbol{\phi}(\boldsymbol{r})$ of the field. Their plane wave form is given in (2.21). Putting the time to zero, we go to the Schrödinger picture. Introducing the plane wave expansion (5.1) and using the form eq. (4.34) for the polarization vector $\boldsymbol{d}$, the field is in invariant form ($s=1$)

$$\boldsymbol{\phi}(\boldsymbol{r})=\sqrt{\frac{2}{\pi}}\int \mathrm{d}^3p\,\frac{N}{\sqrt{2E}}\hat{s}\hat{t}\sum_l i^l j_l(pr)\hat{l}\left[\hat{p}^{[l]}\hat{r}^{[l]}\right]^{[0]}$$
$$\times\left(\left[A_p^{[st]}e^{[s]}\eta^{[t]}\right]^{[00]}+(-)^l\left[\tilde{A}_p^{[st]}e^{[s]}\eta^{[t]}\right]^{[00]}\right).$$

Here the spin-1 function $e^{[s]}$ is the unit vector of sect. 4.2.2 and N is a normalization given below. Recoupling now the spin function $e^{[s]}$ and the orbital functions $\hat{r}^{[l]}$ into vector spherical harmonics with total angular momentum $\boldsymbol{J}$ defined as in eq. (4.6)

$$\boldsymbol{Y}_{lM}^{[J]}(\boldsymbol{r})=\left[e^{[s]}\hat{r}^{[l]}\right]_M^{[J]},$$

we get

$$\boldsymbol{\phi}(\boldsymbol{r})=\sqrt{\frac{2}{\pi}}\int \mathrm{d}^3p\,\frac{N}{\sqrt{2E}}\sum_{Jl} i^l j_l(pr)\hat{J}\hat{t}\left(\left[\left[A_p^{[st]}\hat{p}^{[l]}\eta^{[t]}\right]^{[J0]}\boldsymbol{Y}_l^{[J]}(\hat{\boldsymbol{r}})\right]^{[0]}\right.$$
$$\left.+(-)^l\left[\left[\tilde{A}_p^{[st]}\hat{p}^{[l]}\eta^{[t]}\right]^{[J0]}\boldsymbol{Y}_l^{[J]}(\hat{\boldsymbol{r}})\right]^{[0]}\right).$$

Now the discretized multipole annihilation operators are introduced as before

$$A_{\nu lM_\tau}^{[Jt]}=\int \mathrm{d}^3p\,f_{\nu J}(p)\left[A_{p\tau}^{[st]}\hat{p}^{[l]}\right]_{M_\tau}^{[Jt]}.$$

A new and final unitary transformation over the quantum numbers $l=J+n$, $n=-1,0,1$ (J fixed) is now applied to the multipole operators to achieve new operators $A_{\kappa\nu M_\tau}^{[Jt]}$ of a given multipole J and with a specified multipolarity κ. Namely, we define, the magnetic multipolarity operators, $\kappa = \mathfrak{M}$ and $l = J$

$$A^{[Jt]}_{\mathfrak{M}\nu M_T} = A^{[JT]}_{\nu J M_T},$$

the electric multipolarity operators, $\kappa = \mathfrak{E}$ and $l = J \pm 1$

$$A^{[Jt]}_{\mathfrak{E}\nu M_T} = \frac{\sqrt{J+1}}{\sqrt{2J+1}} A^{[Jt]}_{\nu(J-1)M_T} - \frac{\sqrt{J}}{\sqrt{2J+1}} A^{[Jt]}_{\nu(J+1)M_T},$$

and the longitudinal multipolarity operators, $\kappa = \mathfrak{L}$ and $l = J \pm 1$

$$A^{[Jt]}_{\mathfrak{L}\nu M_T} = \frac{\sqrt{J}}{\sqrt{2J+1}} A^{[Jt]}_{\nu(J-1)M_T} + \frac{\sqrt{J+1}}{\sqrt{2J+1}} A^{[Jt]}_{\nu(J+1)M_T}.$$

This well-known transformation of electrodynamics is obviously unitary and generates states of good parity.

The new operators still obey the commutation relations (4.44)

$$\left[A^{[Jt]}_{\kappa\nu} \tilde{A}^{[J't']}_{\kappa'\nu'}\right]^{[IT]}_{M_I M_T} = (-)^{J+J'+t+t'-I-T}\left[\tilde{A}^{[J't']}_{\kappa'\nu'} A^{[Jt]}_{\kappa\nu}\right]^{[IT]}_{M_I M_T}$$

$$+(-)^{2t}\hat{J}\hat{t}\delta_{\kappa\kappa'}\delta_{\nu\nu'}\delta_{JJ'}\delta_{tt'}\delta_{I0}\delta_{T0}.$$

Substitution of these definitions into the plane wave expansion of the space-like part of the field separates ϕ into three mutually orthogonal terms

$$\phi = \phi_{\mathfrak{M}} + \phi_{\mathfrak{E}} + \phi_{\mathfrak{L}}.$$

Keeping for simplicity only the annihilation part $\phi^{(-)}$ of the field, the fields for the three multipolarities are, after eliminating the plane wave phases i^l as explained for

TABLE 5.1
Lowest multipolarities for the spin-1 field

κ	ν	J	P	N_i
$\mathfrak{L}$	1	0	+	0
$\mathfrak{M}$	1	1	+	1
$\mathfrak{E}$	1	1	−	1
$\mathfrak{L}$	1	1	−	1
$\mathfrak{M}$	1	2	−	2
$\mathfrak{E}$	1	2	+	2
$\mathfrak{L}$	2	0	+	2
$\mathfrak{L}$	1	2	+	2

the spin-0 field in sect. 5.2.1, namely, i^J in (5.11) and i^{J-1} in (5.12) and (5.13):

$$\boldsymbol{\phi}^{(-)}_{\mathfrak{M}}(\boldsymbol{r}) = \sum_{\nu J} \hat{J}\hat{t}\left[A^{[Jt]}_{\mathfrak{M}\nu}\boldsymbol{\Omega}^{[J]}_{\mathfrak{M}\nu(0)}(\boldsymbol{r})\eta^{[t]}\right]^{[00]}, \tag{5.11}$$

$$\boldsymbol{\phi}^{(-)}_{\mathfrak{E}}(\boldsymbol{r}) = \sum_{\nu J} \hat{J}\hat{t}\left[A^{[Jt]}_{\mathfrak{E}\nu}\left(\frac{\sqrt{J+1}}{\sqrt{2J+1}}\boldsymbol{\Omega}^{[J]}_{\mathfrak{E}\nu(-1)}(\boldsymbol{r}) + \frac{\sqrt{J}}{\sqrt{2J+1}}\boldsymbol{\Omega}^{[J]}_{\mathfrak{E}\nu(+1)}(\boldsymbol{r})\right)\eta^{[t]}\right]^{[00]}, \tag{5.12}$$

$$\boldsymbol{\phi}^{(-)}_{\mathfrak{L}}(\boldsymbol{r}) = \sum_{\nu J} \hat{J}\hat{t}\left[A^{[Jt]}_{\mathfrak{L}\nu}\left(\frac{\sqrt{J}}{\sqrt{2J+1}}\boldsymbol{\Omega}^{[J]}_{\mathfrak{L}\nu(-1)}(\boldsymbol{r}) - \frac{\sqrt{J+1}}{\sqrt{2J+1}}\boldsymbol{\Omega}^{[J]}_{\mathfrak{L}\nu(+1)}(\boldsymbol{r})\right)\eta^{[t]}\right]^{[00]}. \tag{5.13}$$

We have introduced the abbreviation

$$\boldsymbol{\Omega}^{[J]}_{\kappa\nu(n)}(\boldsymbol{r}) = \mathcal{W}_{\kappa\nu Jn}(r)\boldsymbol{Y}^{[J]}_{l=J+n}(\hat{\boldsymbol{r}}) \tag{5.14}$$

with $n = -1, 0, +1$. The radial functions are

$$\mathcal{W}_{\kappa\nu Jn}(r) = \sqrt{\frac{2}{\pi}}\int p^2\,\mathrm{d}p\,N_\kappa\frac{1}{\sqrt{2E}}f_{\nu J}(p)\,j_{J+n}(pr), \tag{5.15}$$

We note that the wave packet form factor $f_{\nu J}(p)$ is indexed with the total angular momentum J while the plane wave radial multipole function $j_{J+n}(pr)$ is indexed with the orbital quantum number $l = J + n$. The normalization N_κ required by the commutation relations below, eq. (5.23), is

$$N_{\mathfrak{M}} = 1, \qquad N_{\mathfrak{E}} = 1, \qquad N_{\mathfrak{L}} = \frac{E}{M}, \tag{5.16}$$

where M is the mass of the particle. Note that the limit $M \to 0$ cannot be performed for the longitudinal multipolarity. This is associated with the transversality of the massless free fields, as discussed in sects. 2.3.3 and 2.7.

From the expression (5.11) through (5.15) one may check that the following relations hold:

$$\operatorname{div}\boldsymbol{\phi}_{\mathfrak{M}} = \operatorname{div}\boldsymbol{\phi}_{\mathfrak{E}} = 0,$$

$$\operatorname{rot}\boldsymbol{\phi}_{\mathfrak{L}} = 0.$$

The fourth component of the field, its time-like part, ϕ_4 is fixed in terms of the space-like components $\boldsymbol{\phi}$ by the Lorentz condition. Since the divergence of the transverse multipolarities vanishes, the Lorentz condition yields only a longitudinal contribution

$$\dot{\phi}_4 = -i \operatorname{div} \phi_{\mathfrak{L}}.$$

This expression is evaluated in the Heisenberg picture putting the time to zero at the end. The geometry is that of the calculation of the div in sect. 4.5.2, fig. 4.25. This yields with the phase choice (5.13) for the space part

$$\phi_4^{(-)}(\boldsymbol{r}) = -\sum_{\nu J} \hat{J}\hat{t}\mathfrak{X}_{\nu J}(r)\left[A_{\mathfrak{L}\nu}^{[Jt]}\hat{r}^{[J]}\eta^{[t]}\right]^{[00]} \tag{5.17}$$

with

$$\mathfrak{X}_{\nu J}(r) = \sqrt{\frac{2}{\pi}} \int p^2 \mathrm{d}p \frac{p}{M} \frac{1}{\sqrt{2E}} f_{\nu J}(p) j_J(pr). \tag{5.18}$$

The complete modes are the four-vector fields, again writing only the annihilation part

$$\phi_{\mathfrak{M}}^{(-)} = \begin{pmatrix} \phi_{\mathfrak{M}}^{(-)} \\ 0 \end{pmatrix}, \qquad \phi_{\mathfrak{E}}^{(-)} = \begin{pmatrix} \phi_{\mathfrak{E}}^{(-)} \\ 0 \end{pmatrix}, \qquad \phi_{\mathfrak{L}}^{(-)} = \begin{pmatrix} \phi_{\mathfrak{L}}^{(-)} \\ \phi_4^{(-)} \end{pmatrix},$$

and

$$\phi = \phi^{(-)} + \phi^{(+)}.$$

In our metric the two parts of the longitudinal multipolarity fulfill

$$\phi_{\mathfrak{L}}^{(+)} = \phi_{\mathfrak{L}}^{(-)+},$$

$$\phi_4^{(+)} = -\phi_4^{(-)+}.$$

The quantum numbers of the lowest multipolarities are given in table 5.1. The parity P includes the odd parity of the spin-1 wave functions $e^{[1]}$ for the space parts. Here the excitation quantum number N_i classifies the free energies as will be shown in sect. 6.2.3.

The canonical conjugate fields are defined in sect. 2.3.3 in the Heisenberg picture, eq. (2.19)

$$\bar{\pi} = \dot{\phi} - i \operatorname{grad} \phi_4,$$

$$\bar{\pi}_4 = \dot{\phi}_4 + i \operatorname{div} \phi = 0.$$

The last equation is simply the Lorentz condition. Thus the conjugate fields are only space-like and again performing the calculation in the Heisenberg picture and putting the time to zero at the end, their creation parts are

$$\bar{\boldsymbol{\pi}}_{\mathcal{M}}^{(+)}(\boldsymbol{r}) = i\sum_{\nu J} \hat{J}\hat{t}\left[\tilde{A}_{\mathcal{M}\nu}^{[Jt]}\boldsymbol{\Theta}_{\mathcal{M}\nu(0)}^{[J]}(\boldsymbol{r})\eta^{[t]}\right]^{[00]}, \tag{5.19}$$

$$\bar{\boldsymbol{\pi}}_{\mathcal{E}}^{(+)}(\boldsymbol{r}) = i\sum_{\nu J} \hat{J}\hat{t}\left[\tilde{A}_{\mathcal{E}\nu}^{[Jt]}\left(\frac{\sqrt{J+1}}{\sqrt{2J+1}}\boldsymbol{\Theta}_{\mathcal{E}\nu(-1)}^{[J]}(\boldsymbol{r}) + \frac{\sqrt{J}}{\sqrt{2J+1}}\boldsymbol{\Theta}_{\mathcal{E}\nu(+1)}^{[J]}(\boldsymbol{r})\right)\eta^{[t]}\right]^{[00]}, \tag{5.20}$$

$$\bar{\boldsymbol{\pi}}_{\mathcal{L}}^{(+)}(\boldsymbol{r}) = i\sum_{\nu J} \hat{J}\hat{t}\left[\tilde{A}_{\mathcal{L}\nu}^{[Jt]}\left(\frac{\sqrt{J}}{\sqrt{2J+1}}\boldsymbol{\Theta}_{\mathcal{L}\nu(-1)}^{[J]}(\boldsymbol{r}) - \frac{\sqrt{J+1}}{\sqrt{2J+1}}\boldsymbol{\Theta}_{\mathcal{L}\nu(+1)}^{[J]}(\boldsymbol{r})\right)\eta^{[t]}\right]^{[00]}. \tag{5.21}$$

Here, we have introduced the functions

$$\boldsymbol{\Theta}_{\kappa\nu(n)}^{[J]}(\boldsymbol{r}) = \mathcal{Y}_{\kappa\nu Jn}(r)\boldsymbol{Y}_{l=J+n}^{[J]}(\hat{\boldsymbol{r}})$$

with the radial wave packets, for $\kappa = \mathcal{M}$ and $\kappa = \mathcal{E}$,

$$\mathcal{Y}_{\kappa\nu Jn}(r) = \sqrt{\frac{2}{\pi}}\int p^2\,\mathrm{d}p\sqrt{\tfrac{1}{2}E}\, f_{\nu J}(p)\, j_{J+n}(pr), \tag{5.22}$$

and for $\kappa = \mathcal{L}$

$$\mathcal{Y}_{\mathcal{L}\nu Jn}(r) = \sqrt{\frac{2}{\pi}}\int p^2\,\mathrm{d}pM\sqrt{\frac{1}{2E}}\, f_{\nu J}(p)\, j_{J+n}(pr).$$

The creation part $\phi^{(+)}$ of the fields and their canonical conjugates $\pi^{(-)}$ are obtained according to sect. 4.6.1. For example, for the magnetic part we have

$$\boldsymbol{\phi}_{\mathcal{M}}^{(+)}(\boldsymbol{r}) = \sum_{\nu J} \hat{J}\hat{t}\left[\tilde{A}_{\mathcal{M}\nu}^{[Jt]}\boldsymbol{\Omega}_{\mathcal{M}\nu(0)}^{[J]}(\boldsymbol{r})\eta^{[t]}\right]^{[00]},$$

$$\bar{\boldsymbol{\pi}}_{\mathcal{M}}^{(-)}(\boldsymbol{r}) = -i\sum_{\nu J} \hat{J}\hat{t}\left[A_{\mathcal{M}\nu}^{[Jt]}\boldsymbol{\Theta}_{\mathcal{M}\nu(0)}^{[J]}(\boldsymbol{r})\eta^{[t]}\right]^{[00]}.$$

The values given in (5.16) for the normalization factors N_κ of the radial functions are determined by the normalization for given κ, J, n

$$\sum_{\nu} \mathcal{W}_{\kappa\nu Jn}(r)\mathcal{Y}_{\kappa\nu Jn}(r') = \frac{1}{2}\frac{\delta(r-r')}{r^2}.$$

This way the commutation relations of the fields are valid for each multipolarity separately

$$[\phi_\kappa(\boldsymbol{r}), \pi_{\kappa'}(\boldsymbol{r}')]_- = i(2t+1)\delta_{\kappa\kappa'}\delta^3(\boldsymbol{r}-\boldsymbol{r}'). \tag{5.23}$$

For completeness we now give the electric and magnetic fields $\boldsymbol{E}$ and $\boldsymbol{B}$ respectively, defined by eq. (2.18) in sect. 2.3.3,

$$\boldsymbol{E} = -\dot{\boldsymbol{\phi}}(\boldsymbol{r}) + i\,\text{grad}\,\phi_4(\boldsymbol{r}) = -\bar{\boldsymbol{\pi}},$$

$$\boldsymbol{B} = \text{rot}\,\boldsymbol{\phi}(r).$$

Thus the electric field $\boldsymbol{E}$ is already given by eqs. (5.19) through (5.21). For the magnetic field we have from the general expression for the rotation, eq. (4.39),

$$\boldsymbol{B}^{(+)}_{\mathfrak{M}}(\boldsymbol{r}) = -\sum_{\nu J}\hat{J}\hat{t}\left[\tilde{A}^{[Jt]}_{\mathfrak{M}\nu}\left(\frac{\sqrt{J+1}}{\sqrt{2J+1}}\boldsymbol{\omega}^{[J]}_{\mathfrak{M}\nu(-1)}(\boldsymbol{r}) + \frac{\sqrt{J}}{\sqrt{2J+1}}\boldsymbol{\omega}^{[J]}_{\mathfrak{M}\nu(+1)}(\boldsymbol{r})\right)\eta^{[t]}\right]^{[00]}, \tag{5.24}$$

$$\boldsymbol{B}^{(+)}_{\mathfrak{E}}(\boldsymbol{r}) = -\sum_{\nu J}\hat{J}\hat{t}\left[\tilde{A}^{[Jt]}_{\mathfrak{E}\nu}\boldsymbol{\omega}^{[J]}_{\mathfrak{E}\nu(0)}(\boldsymbol{r})\eta^{[t]}\right]^{[00]}, \tag{5.25}$$

$$\boldsymbol{B}^{(+)}_{\mathfrak{L}}(\boldsymbol{r}) = 0,$$

with

$$\boldsymbol{\omega}^{[J]}_{\kappa\nu(n)}(\boldsymbol{r}) = \mathcal{Z}_{\kappa\nu Jn}(r)\boldsymbol{Y}^{[J]}_{l=J+n}(\hat{\boldsymbol{r}}),$$

$$\mathcal{Z}_{\kappa\nu Jn}(r) = \sqrt{\frac{2}{\pi}}\int p^2\,\mathrm{d}p\frac{1}{\sqrt{2E}}pf_{\nu J}(p)j_{J+n}(pr). \tag{5.26}$$

A compact form for the fields can be achieved by introducing the coefficients $C_\kappa(n, J)$ of a unitary transformation

$$C_{\mathfrak{M}}(0, J) = 1,$$

$$C_{\mathfrak{E}}(-1, J) = \frac{\sqrt{J+1}}{\sqrt{2J+1}}, \qquad C_{\mathfrak{E}}(1, J) = \frac{\sqrt{J}}{\sqrt{2J+1}},$$

$$C_{\mathfrak{L}}(-1, J) = \frac{\sqrt{J}}{\sqrt{2J+1}}, \qquad C_{\mathfrak{L}}(1, J) = -\frac{\sqrt{J+1}}{\sqrt{2J+1}}. \tag{5.27}$$

Herewith the space-like parts $\boldsymbol{\phi}, \boldsymbol{\pi}$, eqs. (5.11) through (5.13) and (5.19) through

(5.21) write

$$\phi_\kappa^{(-)}(\boldsymbol{r}) = \sum_{\nu J n} \hat{J}\hat{t} C_\kappa(n, J) \mathcal{W}_{\kappa\nu J n}(r) \left[A_{\kappa\nu}^{[Jt]} \mathbf{Y}_{J+n}^{[J]}(\hat{\boldsymbol{r}}) \eta^{[t]} \right]^{[00]}, \tag{5.28}$$

$$\bar{\pi}_\kappa^{(+)}(\boldsymbol{r}) = i \sum_{\nu J n} \hat{J}\hat{t} C_\kappa(n, J) \mathcal{Y}_{\kappa\nu J n}(r) \left[\tilde{A}_{\kappa\nu}^{[Jt]} \mathbf{Y}_{J+n}^{[J]}(\hat{\boldsymbol{r}}) \eta^{[t]} \right]^{[00]}. \tag{5.29}$$

The time-like part $\phi_4^{(-)}$ is still given by eq. (5.17).

5.2.4. Discretized spin-$\frac{3}{2}$ fields

We now describe the discretization of the spin-$\frac{3}{2}$ fields. As explained in sect. 2.6, the potential field Φ contains in the transverse gauge spinors of the form, eq. (2.65), setting $S = \frac{3}{2}$,

$$\mathcal{U}_m^{[S]} = \sqrt{\frac{E+M}{2M}} \begin{pmatrix} \chi_m^{[S]} \\ 0 \\ \dfrac{-\hat{1}\left[\sigma^{[1]} p^{[1]}\right]^{[0]}}{E+M} \chi_m^{[S]} \\ 0 \end{pmatrix}$$

for the particles and, eq. (2.66),

$$\mathcal{V}_m^{(S)} = \sqrt{\frac{E+M}{2M}} \begin{pmatrix} \dfrac{-\hat{1}\left[\sigma^{[1]} p^{[1]}\right]^{[0]}}{E+M} \chi_m^{(S)} \\ 0 \\ \chi_m^{(S)} \\ 0 \end{pmatrix}$$

for the antiparticles. Note the standard form of the antiparticles and the minus sign in front of the small components. The spin-$\frac{3}{2}$ matrices, here in contrastandard form, are given by eq. (2.64)

$$\sigma_0^{[1]} = -i\sigma_z, \text{ etc}\ldots$$

The spin $S = \frac{3}{2}$ wave functions have the normalization

$$\left[\chi^{[S]} | \chi^{[S]}\right] = \hat{S} = 2\,.$$

The transverse gauge used here simplifies the spinors, since in that gauge the components with spin $S-1$ are absent in the potential field Φ. In other words, the 2-component spin-$\frac{1}{2}$ wave functions, u and v of eq. (2.51), equal zero in the above expressions. On the other hand, in the Θ field, eq. (2.45), which has also to be introduced in the formalism and which plays the role of the strength field, the spin $(S-1) = \frac{1}{2}$ parts do not vanish in the spinors. This can be seen for example from the defining equation (2.45) since the γ matrices (2.53) contain elements off-diagonal in S. Hence the spinors for the Θ field have spin-$\frac{1}{2}$ wave functions $\chi_m{}^{[\frac{1}{2}]}$ in the position of the zeroes above, normalized as

$$[\chi^{[1/2]}|\chi^{[1/2]}] = \sqrt{2} .$$

However we shall avoid writing the explicit expression for the strength field Θ and use instead their defining equation (2.45) in terms of the potential field Φ.

Introducing the isospin t, the plane wave expansion (2.67) can now be written in invariant form

$$\Phi(\boldsymbol{r}) = \sqrt{\frac{2}{\pi}} \int \mathrm{d}^3 p \frac{1}{\sqrt{2E}} \sum_l i^l \hat{l}\hat{S}\hat{t}\Big(\big[B_p^{[St]}\mathcal{U}^{[S]}(\boldsymbol{p})\eta^{[t]}\big]^{[00]}$$

$$+ (-)^l \big[\tilde{D}_p^{[St]}\mathcal{V}^{[S]}(\boldsymbol{p})\eta^{[t]}\big]^{[00]}\Big)\big[\hat{p}^{[l]}\hat{r}^{[l]}\big]^{[0]} j_l(pr) . \qquad (5.31)$$

The action of all the 12×12 matrices introduced in sect. 2.6 is now defined by giving the invariant matrix elements of their constituent sub-matrices, as for example in (2.56). To that end we write as usual the contrastandard form of the 12×12 cartesian matrices γ_μ of sect. 2.6 as

$$\gamma_0^{[1]} = -i\gamma_z \text{ etc} \ldots ,$$

and likewise for the 6×6 matrices, eq. (2.54) and (2.55),

$$S_0^{[1]} = \tfrac{3}{2}\sigma_0^{[1]} = -iS_z , \text{ etc} \ldots ,$$

$$T_0^{[1]} = \tfrac{3}{2}\tau_0^{[1]} = -iT_z , \text{ etc} \ldots ,$$

where S_z and T_z are given for the spin-$\frac{3}{2}$ case by eqs. (2.62) and (2.63). Recall that the notation S_z and T_z refers to the diagonal and off-diagonal character in spin of the matrices. We have according to the Wigner-Eckart theorem, in the form (4.11), for example

$$\left\langle \chi_m^{[3/2]} \middle| S_\mu^{[1]} \middle| \chi_{m'}^{[3/2]} \right\rangle = (-)^{3/2+m}(-)^{3/2-1+3/2}\begin{pmatrix} \frac{3}{2} & 1 & \frac{3}{2} \\ -m & \mu & m' \end{pmatrix}\big[\tfrac{3}{2}|S^{[1]}|\tfrac{3}{2}\big] .$$

From the values given for S_z, eq. (2.62), we obtain, in agreement with eq. (4.15) for the spin operator $S^{[1]}$

$$\big[\tfrac{3}{2}|S^{[1]}|\tfrac{3}{2}\big] = i\sqrt{15} .$$

Likewise,

$$\big[\tfrac{1}{2}|S^{[1]}|\tfrac{1}{2}\big] = i\sqrt{\tfrac{3}{2}} ,$$

$$\big[\tfrac{1}{2}|S^{[1]}|\tfrac{3}{2}\big] = \big[\tfrac{3}{2}|S^{[1]}|\tfrac{1}{2}\big] = 0 ,$$

or, using the definition (2.54) for the spin-$\frac{3}{2}$ operator $\sigma^{[1]}$,

$$\left[\tfrac{3}{2}|\sigma^{[1]}|\tfrac{3}{2}\right] = i2\sqrt{\tfrac{5}{3}}\ ,$$

$$\left[\tfrac{1}{2}|\sigma^{[1]}|\tfrac{1}{2}\right] = i\sqrt{\tfrac{2}{3}}\ .$$

Similarly with the values (2.63) for T_z we obtain the invariants, noting that $\tilde{T}^{[1]} \neq T^{[1]}$ for spin changing operators,

$$\left[\tfrac{1}{2}|\tilde{T}^{[1]}|\tfrac{3}{2}\right] = \left[\tfrac{3}{2}|\tilde{T}^{[1]}|\tfrac{1}{2}\right]^* = i\sqrt{12}\ ,$$

$$\left[\tfrac{1}{2}|T^{[1]}|\tfrac{1}{2}\right] = \left[\tfrac{3}{2}|T^{[1]}|\tfrac{3}{2}\right] = 0,$$

or, from (2.55),

$$\left[\tfrac{1}{2}|\tilde{\tau}^{[1]}|\tfrac{3}{2}\right] = \left[\tfrac{3}{2}|\tilde{\tau}^{[1]}|\tfrac{1}{2}\right]^* = i4\sqrt{\tfrac{1}{3}}\ .$$

We now turn to the discretization of the fields. We proceed as usual by recoupling first the terms of the expansion (5.31) into multipoles of total angular momentum $\boldsymbol{j} = \boldsymbol{S} + \boldsymbol{l}$. For a given parity π and a given j, one has in general two possible values for l. These $\pi l j$ states are orthogonal and needed for completeness. Then, following the lines of sect. 4.5.3, the evaluation of the action of the operators $\boldsymbol{\sigma}\cdot\boldsymbol{p}$ in the spinors is carried out as in fig. 4.28 together with fig. 5.1. We thus obtain

$$\Phi(\boldsymbol{r}) = \sqrt{\frac{2}{\pi}}\int \mathrm{d}^3p\,\frac{1}{\sqrt{2E}}\,\frac{1}{\sqrt{2M}}\sum_{lj} i^l\hat{j}\hat{t}$$

$$\times\left(\left[\left[B_p^{[St]}\hat{p}^{[l]}\eta^{[t]}\right]^{[j0]}\begin{pmatrix}\sqrt{E+M}\,Y_{Sl}^{[j]}(\hat{\boldsymbol{r}})\,j_l(pr)\\ 0\\ \sum_\lambda\sqrt{E-M}\,Y_{S\lambda}^{[j]}(\hat{\boldsymbol{r}})\,j_\lambda(pr)\beta_{l\lambda j}\\ 0\end{pmatrix}\right]^{[00]}\right.$$

$$\left.+(-)^l\left[\left[\tilde{D}_p^{[St]}\hat{p}^{[l]}\eta^{[t]}\right]^{[j0]}\begin{pmatrix}\sum_\lambda\sqrt{E-M}\,Y_{S\lambda}^{[j]}(\hat{\boldsymbol{r}})\,j_\lambda(pr)\beta_{l\lambda j}\\ 0\\ \sqrt{E+M}\,Y_{Sl}^{[j]}(\hat{\boldsymbol{r}})\,j_l(pr)\\ 0\end{pmatrix}\right]^{[00]}\right).$$

The spin and isospin functions for the antiparticle creation part of the field are non-transposed, as explained in sect. 5.2.2 for the spin-$\frac{1}{2}$ field. Here the coefficients $\beta_{l\lambda j}$ are given by the graph of fig. 4.28 in terms of the above invariants and including the minus sign in front of the small components

$$\beta_{l\lambda j} = -\hat{1}\begin{bmatrix} 1 & 1 & 0 \\ S & l & j \\ S & \lambda & j \end{bmatrix}\hat{S}(-)^{2S}\begin{bmatrix} S & S & 0 \\ 0 & S & S \\ S & 0 & S \end{bmatrix}[S|\sigma|S]\hat{\lambda}\begin{bmatrix} \lambda & \lambda & 0 \\ 0 & \lambda & \lambda \\ \lambda & 0 & \lambda \end{bmatrix}(-i\alpha_{\lambda l}). \tag{5.32}$$

Before giving their values let us note that in the case of spin $\frac{1}{2}$, sect. 5.2.2, only one value λ arose in the small components for given j and l, namely $\lambda = l+1$ for $j = l+\frac{1}{2}$ and $\lambda = l-1$ for $j = l-\frac{1}{2}$. Here one value is possible for each of the cases $j = l \pm \frac{3}{2}$ and two values for each of the cases $j = l \pm \frac{1}{2}$, hence the summation over λ in the small components of the spinors. The possible values of λ are

$$\lambda = \begin{cases} l+1 & \text{if } j = l+\frac{3}{2} \\ l-1 & \text{if } j = l-\frac{3}{2} \\ l \pm 1 & \text{for both } j = l+\frac{1}{2} \text{ and } j = l-\frac{1}{2}. \end{cases}$$

Thus the non-vanishing values of $\beta_{l\lambda j}$ calculated with (5.32) are

$$\beta_{l,l+1,l+3/2} = \left(\frac{2l+5}{3(2l+3)}\right)^{1/2},$$

$$\beta_{l,l-1,l-3/2} = \left(\frac{2l-3}{3(2l-1)}\right)^{1/2},$$

$$\beta_{l,l+1,l+1/2} = \left(\frac{16l(l+2)}{9(2l+1)(2l+3)}\right)^{1/2},$$

$$\beta_{l,l-1,l+1/2} = \left(\frac{2l+3}{3(2l+1)}\right)^{1/2},$$

$$\beta_{l,l+1,l-1/2} = \left(\frac{2l-1}{3(2l+1)}\right)^{1/2},$$

$$\beta_{l,l-1,l-1/2} = \left(\frac{16(l+1)(l-1)}{9(2l-1)(2l+1)}\right)^{1/2}.$$

We finally introduce as before discretized multipole operators $B_{\nu l}^{[jt]}$ by the unitary transformation

$$B_{\nu l m_j m_t}^{[jt]} = \int d^3p\, f_{\nu l}(p)\left[B_{p m_t}^{[St]}\hat{p}^{[l]}\right]_{m_j m_t}^{[jt]}.$$

They conserve the anticommutation relations (2.68) which write in invariant coupled form

$$\left[B_{\nu l}^{[jt]}\tilde{B}_{\nu' l'}^{[j't']}\right]_{M_I M_T}^{[IT]} = -(-)^{j+j'+t+t'-I-T}\left[\tilde{B}_{\nu' l'}^{[j't']}B_{\nu l}^{[jt]}\right]_{M_I M_T}^{[IT]}$$
$$+(-)^{2j+2t}\,\hat{j}\hat{t}\,\delta_{\nu\nu'}\delta_{ll'}\delta_{jj'}\delta_{tt'}\delta_{I0}\delta_{T0}. \tag{5.33}$$

The same is done for the antiparticle operators.

Substituting the new operators and eliminating the plane wave phase i^l as usual, one obtains the discretized expansion for the spin-$\frac{3}{2}$ field

$$\Phi(\boldsymbol{r}) = \frac{1}{\sqrt{2}}\frac{1}{\sqrt{2M}}\sum_{\nu lj}\hat{j}\hat{t}\left(\left[B_{\nu l}^{[jt]}\begin{pmatrix} Y_{Sl}^{[j]}(\hat{\boldsymbol{r}})u_{\nu l}(r) \\ 0 \\ \sum_{\lambda} Y_{S\lambda}^{[j]}(\hat{\boldsymbol{r}})v_{\nu l\lambda}(r)\beta_{l\lambda j} \\ 0 \end{pmatrix}\eta^{[t]}\right]^{[00]}\right.$$
$$\left.+\left[\tilde{D}_{\nu l}^{[jt]}\begin{pmatrix} \sum_{\lambda} Y_{S\lambda}^{[j]}(\hat{\boldsymbol{r}})v_{\nu l\lambda}(r)\beta_{l\lambda j} \\ 0 \\ Y_{Sl}^{[j]}(\hat{\boldsymbol{r}})u_{\nu l}(r) \\ 0 \end{pmatrix}\eta^{[t]}\right]^{[00]}\right), \tag{5.34}$$

where the radial functions $u_{\nu l}(r)$ and $v_{\nu l\lambda}(r)$ are the same wave packet discretized forms as for the spin-$\frac{1}{2}$ case and they are given by eqs. (5.9) and (5.10).

5.3. THE SINGLE-PARTICLE BASIS FUNCTIONS

In a Fock representation the single-particle states for the spins 0, 1, $\frac{1}{2}$ and $\frac{3}{2}$ respectively are,

$$|\nu lm;\, t\tau\rangle = \tilde{a}_{\nu m\tau}^{[lt]}|0\rangle,$$

$$|\nu ljm;\, t\tau\rangle = \tilde{b}_{\nu lm\tau}^{[jt]}|0\rangle,$$

$$|\kappa\nu JM;\, t\tau\rangle = \tilde{A}_{\kappa\nu M\tau}^{[Jt]}|0\rangle,$$

$$|\nu ljm;\, t\tau\rangle = \tilde{B}_{\nu lm\tau}^{[jt]}|0\rangle.$$

The corresponding wave functions are obtained from the fields as the vacuum expectation value of, for example, for spin 0, eq. (5.6),

$$\varphi_{\nu m\tau}^{[lt]}(\boldsymbol{r}) = C\langle 0|\varphi(\boldsymbol{r})\tilde{a}_{\nu m\tau}^{[lt]}|0\rangle, \tag{5.35}$$

where C is a phase. Note that in setting up a linear eigenvalue problem, this phase C

can be freely chosen for each basis state. This freedom of choice has already been used to drop the phase i^l of the plane wave expansion. Furthermore the phases of the type $(-)^{2l+2t}$ as discussed in sect. 4.6.2, in particular eq. (4.47), are always $+1$ for all the fields considered here.

This way the single-particle basis states for spin 0 are

$$\varphi_{\nu m\tau}^{[lt]}(\boldsymbol{r}) = \sqrt{\tfrac{1}{2}}\, g_{\nu l}(r)\hat{r}_m^{[l]}\eta_\tau^{[t]}.$$

They are normalized according to the spin-0 boson density, the form of J_4 in eq. (2.31),

$$i\int \mathrm{d}^3r\, \varphi_{\nu m\tau}^{[lt]}(\boldsymbol{r})^* \frac{\overleftrightarrow{\partial}}{\partial t}\varphi_{\nu' m'\tau'}^{[l't']}(\boldsymbol{r}) = \delta_{\nu\nu'}\delta_{ll'}\delta_{mm'}\delta_{tt'}\delta_{\tau\tau'}.$$

Here $f\overleftrightarrow{\partial}g/\partial t = f\dot{g} - \dot{f}g$ and the time derivatives are evaluated by going back to the Heisenberg picture and putting the time equal to zero at the end. This normalization is needed to achieve the commutation relations for the fields.

For spin $\frac{1}{2}$ the basis single-particle wave functions are

$$\psi_{\nu lm\tau}^{[jt]}(\boldsymbol{r}) = \frac{1}{\sqrt{2}}\begin{pmatrix} u_{\nu l}(r)Y_{lm}^{[j]}(\hat{\boldsymbol{r}}) \\ -v_{\nu l\lambda}(r)Y_{\lambda m}^{[j]}(\hat{\boldsymbol{r}}) \end{pmatrix}\eta_\tau^{[t]},$$

normalized according to the spin-$\frac{1}{2}$ density, i.e. the current J_4 in (2.32),

$$\int \mathrm{d}^3r\, \psi_{\nu lm\tau}^{[jt]+}(\boldsymbol{r})\psi_{\nu' l'm'\tau'}^{[j't']}(\boldsymbol{r}) = \delta_{\nu\nu'}\delta_{jj'}\delta_{ll'}\delta_{mm'}\delta_{tt'}\delta_{\tau\tau'}.$$

For spin 1 the basis wave functions are, with the phase choice of eqs. (5.11), (5.12), (5.13), and (5.17)

$$\phi_{\kappa\nu M\tau}^{[Jt]}(\boldsymbol{r}) = \mathcal{F}_{\kappa\nu M}^{[J]}(\boldsymbol{r})\eta_\tau^{[t]},$$

where

$$\mathcal{F}_{\mathcal{M}\nu M}^{[J]}(\boldsymbol{r}) = \left(\boldsymbol{\Omega}_{\mathcal{M}\nu(0)M}^{[J]}(\boldsymbol{r}), 0\right),$$

$$\mathcal{F}_{\mathcal{E}\nu M}^{[J]}(\boldsymbol{r}) = \left(\frac{\sqrt{J+1}}{\sqrt{2J+1}}\boldsymbol{\Omega}_{\mathcal{E}\nu(-1)M}^{[J]}(\boldsymbol{r}) + \frac{\sqrt{J}}{\sqrt{2J+1}}\boldsymbol{\Omega}_{\mathcal{E}\nu(+1)M}^{[J]}(\boldsymbol{r}), 0\right),$$

$$\mathcal{F}_{\mathcal{L}\nu M}^{[J]}(\boldsymbol{r}) = \left(\frac{\sqrt{J}}{\sqrt{2J+1}}\boldsymbol{\Omega}_{\mathcal{L}\nu(-1)M}^{[J]}(\boldsymbol{r}) - \frac{\sqrt{J+1}}{\sqrt{2J+1}}\boldsymbol{\Omega}_{\mathcal{L}\nu(+1)M}^{[J]}(\boldsymbol{r}), -\mathcal{X}_{\nu J}(r)\hat{r}_M^{[J]}\right).$$

These spin-1 functions are four-vectors. The magnetic and electric multipolarities have only space-like components, which is expressed in the fact that the quantities $\mathbf{\Omega}$ contain vector spherical harmonics. The longitudinal multipolarity has in addition the scalar time-like (fourth) component, i.e. the last term. They are normalized according to the spin-1 boson density, related to the conserved current (2.33), which also underlies the commutation relations.

Finally, for spin $\frac{3}{2}$ the single-particle basis wave functions are, from eq. (5.34),

$$\Phi^{[jt]}_{\nu l m \tau}(\boldsymbol{r}) = \frac{1}{\sqrt{2}} \frac{1}{\sqrt{2M}} \begin{pmatrix} u_{\nu l}(r) Y^{[j]}_{Slm}(\hat{\boldsymbol{r}}) \\ 0 \\ \sum_{\lambda} \beta_{l\lambda j} v_{\nu l\lambda}(r) Y^{[j]}_{S\lambda m}(\hat{\boldsymbol{r}}) \\ 0 \end{pmatrix} \eta^{[t]}_{\tau}. \qquad (5.36)$$

5.4. EXAMPLES OF CONFIGURATION SPACES

The many-body problem for finite systems is constructed in terms of state vectors defined with respect to the above discretized fields. They are coupled to the quantum numbers of the system, i.e. its total spin I, parity P, isospin T, and, if applicable, $\mathcal{G}$-parity G. They are properly symmetrized and antisymmetrized in the bosons and fermions respectively. The construction of the configurations and of their matrix elements is the subject of the next chapters.

At this stage one is confronted with the task of deciding on an appropriate model for describing the desired physical system. All the well-known approximations of non-relativistic finite systems may be employed for relativistic systems. For example the interactions may be treated to second order to generate one-boson exchange interactions, as described in appendix A.3., or as elementary vertices in a truncated Hilbert space, choosing a set of particular configurations for setting up the secular problem. This includes the generalization of specific approximation schemes, such as the Hartree-Fock or BCS methods, the RPA, etc.

Of course the nature of the elementary fields depends on the physical systems and it is part of the model. In this book we generally have used the language of nuclear physics in that we call the fields "nucleons", "mesons"... In atomic physics the names would be "electrons", "photons", and in QCD "quarks", "gluons".

With this in mind we give here a few model Hilbert spaces for light hadrons in the picture of meson-dressed nucleons. This will provide us with a fairly complete set of examples to be treated in detail in the next chapters and covering most of the situations one may encounter.

In our examples the basic fields are the pion, the sigma meson, the vector meson, and the nucleon $s = \frac{1}{2}$, $t = \frac{1}{2}$. We will use the systems of table 5.2 as examples of composite systems made of these basic fields and list their configurations in some model spaces.

TABLE 5.2
Quantum numbers for the configuration spaces

	B	I	P	T	G
vacuum	0	0	+	0	+
pion	0	0	−	1	−
nucleon	1	$\frac{1}{2}$	+	$\frac{1}{2}$	
delta	1	$\frac{3}{2}$	+	$\frac{3}{2}$	
deuteron	2	1	+	0	

The configurations, denoted $\{\alpha\}$, can be grouped according to their mean energy numbers $\mathcal{E}_\alpha$ defined in (3.11). It is more convenient to replace $\mathcal{E}_\alpha$ by a discrete index N_α

$$\mathsf{N}_\alpha = \sum_i (m_i + N_i) - \mathsf{N}_g. \tag{5.37}$$

Here, the sum is over the particles i of the configuration as in (3.11). Adopting the standard convention $(\nu_i l_i) = 1\text{s}, 1\text{p}, 2\text{s}, 1\text{d}, \ldots, N_i$ is taken to be the excitation quantum number of sect. 3.3. The discrete index m_i is associated with the mass parameter M_i of the particles for example by the convention of table 5.3. N_g refers to the ground configuration.

This way the configuration index N_α can be used to classify the configurations in a hierarchy of states with increasing energy and it can be employed in the truncation criterion for the Hilbert space.

We now give the notation for the couplings of the pions, heavy bosons and fermions making up the configurations. This notation will be used in the tables below and throughout the next chapters.

TABLE 5.3
Effective mass parameter convention

	m_i
pion	1
sigma	3
rho	4
omega	4
nucleon	6

The configurations generally contain a pion cloud of $n_1, n_2 \ldots$ pions in states $\pi_1 = \nu_1 l_1$, $\pi_2 = \nu_2 l_2 \ldots$ In the tables at most three distinct pion states appear in one configuration. The pion cloud is thus defined by the distributions n_1, n_2, n_3 in states π_1, π_2, π_3 with the standard coupling scheme where *LP* and *TP* denote the total angular momentum and isospin respectively of the pion cloud

$$|[\text{pion cloud}]^{[LP\,TP]}\rangle = \Big|\Big[\Big[[(\pi_1)^{n_1}]^{[LP_1 TP_1]}[(\pi_2)^{n_2}]^{[LP_2 TP_2]}\Big]^{[LP_{12} TP_{12}]} \times [(\pi_3)^{n_3}]^{[LP_3 TP_3]}\Big]^{[LP\,TP]}\Big\rangle . \tag{5.38}$$

At most one heavy boson, σ, ρ or ω, is contained in the configurations of the tables. The quantum numbers of the heavy boson are denoted by κ, ν_h, J, t_h (h stands for "heavy boson"). In the tables the index κ takes on the explicit names $\mathfrak{M}$, $\mathfrak{E}$ and $\mathfrak{L}$ and we write $\kappa_{\nu J}$, for example $\mathfrak{M}_{11}$.

The fermion number of the systems of table 5.2 is 0, 1 or 2. The fermion single-particle states are denoted by the indices *N* and *M* with quantum numbers $\nu_N l_N j_N$ and $\nu_M l_M j_M$ respectively. They are coupled to total angular momentum and total isospin *JF* and *TF*. In the case of a single fermion $JF = j_N$ and $TF = t_N$ of course.

TABLE 5.4
Vacuum configurations, $I = 0$, $T = 0$

N_α	π_1	π_2	σ	*LP*	*TP*	LP_1	TP_1	LP_2	TP_2
0									
2	$1s^2$								
3			1s						
4	1s	2s							
	$1p^2$								
	$1s^4$								
5			2s						
	$1s^2$		1s	0	0				
6	1s	3s							
	$2s^2$								
	1p	2p							
	$1d^2$								
	$1s^3$	2s				0	1		
	$1s^2$	$1p^2$				0	0	0	0
	$1s^2$	$1p^2$				0	2	0	2
	$1s^6$								

A complete configuration with total angular momentum I and total isospin T has the standard coupling scheme

$$|\alpha^{[IT]}\rangle = \left|\left[\left[\left[\text{heavy boson}\right]^{[J\,t_h]}[\text{pions}]^{[LP\,TP]}\right]^{[JB\,TB]}[\text{fermions}]^{[JF\,TF]}\right]^{[IT]}\right\rangle, \quad (5.39)$$

with the notation JB, TB for the boson cloud and JF, TF for the fermion part.

TABLE 5.5
Pion configurations, $I=0$, $T=1$

N_α	π_1	π_2	π_3	σ	ρ	LP_1	TP_1	LP_2	TP_2	LP_{12}	TP_{12}
0	1s										
2	2s										
	$1s^3$										
3	1s			1s							
4	3s										
	$1s^2$	2s				0	0	0	1		
	$1s^2$	2s				0	2	0	1		
	1s	$1p^2$						0	0		
	1s	$1p^2$						0	2		
	$1s^5$										
	1s				$\mathcal{L}_{10}$						
5	1s			2s							
	2s			1s							
	1p			1p							
	$1s^3$			1s							
6	3s										
	$1s^2$	2s				0	0				
	$1s^2$	2s				0	2				
	1s	$2s^2$						0	0		
	1s	$2s^2$						0	2		
	1s	1p	2p							1	0
	1s	1p	2p							1	1
	1s	1p	2p							1	2
	1s	$1d^2$						0	0		
	1s	$1d^2$						0	2		
	$1p^2$	1d				2	0				
	$1p^2$	1d				2	2				
	$1p^2$	2s				0	0				
	$1p^2$	2s				0	2				
	2s				$\mathcal{L}_{10}$						
	1p				$\mathcal{E}_{11}$						
	1p				$\mathcal{L}_{11}$						
	1s				$\mathcal{L}_{20}$						

This coupling order and notation is kept throughout in the tables and in the examples treated in the following chapters.

Some configuration spaces are now given. The vacuum and pion configurations, tables 5.4 and 5.5, are few in number and we list all configurations up to $N_\alpha = 6$. Owing to selection rules, not all heavy vector mesons participate in this limited configuration space. For the other cases the number of possible configurations is much larger. Hence in tables 5.6, 5.7 and 5.8 we limit the lists to $N_\alpha = 4$ for the nucleon and the deuteron and $N_\alpha = 3$ for the delta. However, we list separately in tables 5.9, 5.10 and 5.11 the configurations containing one heavy boson for some higher values of N_α.

TABLE 5.6

Nucleon configurations, $I = \frac{1}{2}$, $T = \frac{1}{2}$

N_α	N	π_1	π_2	σ	ρ	ω	LP	TP	LP_1	TP_1
0	$1s_{1/2}$									
2	$2s_{1/2}$									
	$1p_{1/2}$	1s								
	$1s_{1/2}$	1p								
	$1s_{1/2}$	$1s^2$					0	0		
3	$1s_{1/2}$			1s						
4	$3s_{1/2}$									
	$1s_{1/2}$	2p								
	$1p_{1/2}$	2s								
	$1p_{3/2}$	1d								
	$2s_{1/2}$	1p								
	$1d_{3/2}$	1p								
	$2s_{1/2}$	$1s^2$					0	0		
	$1s_{1/2}$	1s	2s				0	0		
	$1s_{1/2}$	1s	2s				0	1		
	$1p_{1/2}$	1s	1p				1	0		
	$1p_{1/2}$	1s	1p				1	1		
	$1p_{3/2}$	1s	1p				1	0		
	$1p_{3/2}$	1s	1p				1	1		
	$1s_{1/2}$	$1p^2$					0	0		
	$1s_{1/2}$	$1p^2$					1	1		
	$1p_{1/2}$	$1s^3$					0	1		
	$1s_{1/2}$	$1s^2$	1p				1	1	0	0
	$1s_{1/2}$	$1s^2$	1p				1	1	0	2
	$1s_{1/2}$	$1s^4$					0	0		
	$1s_{1/2}$					$\mathcal{L}_{10}$				
	$1s_{1/2}$				$\mathcal{L}_{10}$					

TABLE 5.7
Delta configurations, $I=\frac{3}{2}$, $T=\frac{3}{2}$

N_α	N	π_1	π_2	σ	ρ	LP	TP	LP_1	TP_1	JB	TB
0	$1s_{1/2}$	1p									
	$1p_{3/2}$	1s									
2	$1s_{1/2}$	2p									
	$1p_{1/2}$	1d									
	$1p_{3/2}$	2s									
	$1p_{3/2}$	1d									
	$2s_{1/2}$	1p									
	$1d_{3/2}$	1p									
	$1d_{5/2}$	1p									
	$2p_{3/2}$	1s									
	$1s_{1/2}$	1s	1d			2	1				
	$1s_{1/2}$	1s	1d			2	2				
	$1p_{1/2}$	1s	1p			1	1				
	$1p_{1/2}$	1s	1p			1	2				
	$1p_{3/2}$	1s	1p			1	1				
	$1p_{3/2}$	1s	1p			1	2				
	$1s_{1/2}$	$1p^2$				2	2				
	$1s_{1/2}$	$1p^2$				1	1				
	$1p_{3/2}$	$1s^3$				0	1				
	$1s_{1/2}$	$1s^2$	1p			1	1	0	0		
	$1s_{1/2}$	$1s^2$	1p			1	1	0	2		
	$1s_{1/2}$	$1s^2$	1p			1	2	0	2		
3	$1s_{1/2}$				$\mathcal{M}_{11}$						
	$1s_{1/2}$	1p		1s						1	1
	$1p_{3/2}$	1s		1s						0	1
	$1s_{1/2}$	1s		1p						1	1

TABLE 5.8
Deuteron configurations, $I=1$, $T=0$

N_α	N	M	π_1	π_2	σ	ω	JF	TF	LP	TP	LP_1	TP_1
0	$1s_{1/2}$	$1s_{1/2}$										
2	$1s_{1/2}$	$2s_{1/2}$										
	$1s_{1/2}$	$1d_{3/2}$										
	$1p_{1/2}$	$1p_{1/2}$										
	$1p_{1/2}$	$1p_{3/2}$										
	$1p_{3/2}$	$1p_{3/2}$										
	$1s_{1/2}$	$1s_{1/2}$	1p				0	1				
	$1s_{1/2}$	$1p_{1/2}$	1s				1	1				
	$1s_{1/2}$	$1p_{3/2}$	1s				1	1				
	$1s_{1/2}$	$1s_{1/2}$	$1s^2$				1	0	0	0		
3	$1s_{1/2}$	$1s_{1/2}$			1s		1	0				
4	$1s_{1/2}$	$2d_{3/2}$										
	$1p_{1/2}$	$2p_{1/2}$										
	$1p_{3/2}$	$2p_{3/2}$										
	$1p_{1/2}$	$2p_{3/2}$										
	$1p_{3/2}$	$2p_{1/2}$										
	$1p_{3/2}$	$1f_{5/2}$										

TABLE 5.8 (continued)

N_α	N	M	π_1	π_2	σ	ω	JF	TF	LP	TP	LP_1	TP_1
	$2s_{1/2}$	$2s_{1/2}$										
	$2s_{1/2}$	$1d_{3/2}$										
	$1d_{3/2}$	$1d_{3/2}$										
	$1d_{3/2}$	$1d_{5/2}$										
	$1d_{5/2}$	$1d_{5/2}$										
	$1s_{1/2}$	$3s_{1/2}$										
	$1s_{1/2}$	$1s_{1/2}$	2p				0	1				
	$1s_{1/2}$	$2s_{1/2}$	1p				0	1				
	$1s_{1/2}$	$2s_{1/2}$	1p				1	1				
	$1s_{1/2}$	$1d_{3/2}$	1p				1	1				
	$1s_{1/2}$	$1d_{3/2}$	1p				2	1				
	$1s_{1/2}$	$1d_{5/2}$	1p				2	1				
	$1p_{1/2}$	$1p_{1/2}$	1p				0	1				
	$1p_{1/2}$	$1p_{1/2}$	1p				2	1				
	$1p_{3/2}$	$1p_{3/2}$	1p				0	1				
	$1p_{3/2}$	$1p_{3/2}$	1p				2	1				
	$1p_{1/2}$	$1p_{3/2}$	1p				1	1				
	$1p_{1/2}$	$1p_{3/2}$	1p				2	1				
	$1s_{1/2}$	$2p_{1/2}$	1s				1	1				
	$1s_{1/2}$	$2p_{3/2}$	1s				1	1				
	$1p_{1/2}$	$2s_{1/2}$	1s				1	1				
	$1p_{3/2}$	$2s_{1/2}$	1s				1	1				
	$1p_{1/2}$	$1d_{3/2}$	1s				1	1				
	$1p_{3/2}$	$1d_{3/2}$	1s				1	1				
	$1p_{3/2}$	$1d_{5/2}$	1s				1	1				
	$1s_{1/2}$	$1p_{1/2}$	2s				1	1				
	$1s_{1/2}$	$1p_{1/2}$	1d				1	1				
	$1s_{1/2}$	$1p_{3/2}$	2s				1	1				
	$1s_{1/2}$	$1p_{3/2}$	1d				1	1				
	$1s_{1/2}$	$1p_{3/2}$	1d				2	1				
	$1s_{1/2}$	$2s_{1/2}$	$1s^2$				1	0	0	0		
	$1s_{1/2}$	$1d_{3/2}$	$1s^2$				1	0	0	0		
	$1p_{1/2}$	$1p_{1/2}$	$1s^2$				1	0	0	0		
	$1p_{1/2}$	$1p_{3/2}$	$1s^2$				1	0	0	0		
	$1p_{3/2}$	$1p_{3/2}$	$1s^2$				1	0	0	0		
	$1s_{1/2}$	$1s_{1/2}$	1s	2s			1	0	0	0		
	$1s_{1/2}$	$1s_{1/2}$	1s	1d			1	0	2	0		
	$1s_{1/2}$	$1s_{1/2}$	$1p^2$				1	0	0	0		
	$1s_{1/2}$	$1s_{1/2}$	$1p^2$				0	1	1	1		
	$1s_{1/2}$	$1s_{1/2}$	$1p^2$				1	0	2	0		
	$1s_{1/2}$	$1p_{1/2}$	1s	1p			0	0	1	0		
	$1s_{1/2}$	$1p_{1/2}$	1s	1p			0	1	1	1		
	$1s_{1/2}$	$1p_{1/2}$	1s	1p			1	1	1	1		
	$1s_{1/2}$	$1p_{1/2}$	1s	1p			1	0	1	0		
	$1s_{1/2}$	$1p_{3/2}$	1s	1p			1	0	1	0		
	$1s_{1/2}$	$1p_{3/2}$	1s	1p			1	1	1	1		
	$1s_{1/2}$	$1p_{3/2}$	1s	1p			2	1	1	1		
	$1s_{1/2}$	$1p_{3/2}$	1s	1p			2	0	1	0		
	$1s_{1/2}$	$1s_{1/2}$	$1s^2$	1p			0	1	1	1	0	0
	$1s_{1/2}$	$1s_{1/2}$	$1s^2$	1p			0	1	1	1	0	2
	$1s_{1/2}$	$1p_{1/2}$	$1s^3$				1	1	0	1		
	$1s_{1/2}$	$1p_{3/2}$	$1s^3$				1	1	0	1		
	$1s_{1/2}$	$1s_{1/2}$	$1s^4$				1	0	0	0		
	$1s_{1/2}$	$1s_{1/2}$				$\mathfrak{L}_{10}$	1	0				

TABLE 5.9

Nucleon configurations with heavy bosons for $N_\alpha = 5$ and 6; $I = \frac{1}{2}$, $T = \frac{1}{2}$

N_α	N	π_1	σ	ρ	ω	LP	TP	JB	TB
5	$2s_{1/2}$		1s						
	$1p_{1/2}$		1p						
	$1p_{3/2}$		1p						
	$1s_{1/2}$		2s						
	$1s_{1/2}$				$\mathcal{M}_{11}$				
	$1s_{1/2}$			$\mathcal{M}_{11}$					
	$1s_{1/2}$	1p	1s						
	$1p_{1/2}$	1s	1s						
	$1s_{1/2}$	1s	1p						
	$1s_{1/2}$	$1s^2$	1s			0	0		
6	$2s_{1/2}$			$\mathcal{L}_{10}$					
	$1p_{1/2}$			$\mathcal{E}_{11}$					
	$1p_{3/2}$			$\mathcal{E}_{11}$					
	$1p_{1/2}$			$\mathcal{L}_{11}$					
	$1p_{3/2}$			$\mathcal{L}_{11}$					
	$2s_{1/2}$				$\mathcal{L}_{10}$				
	$1p_{1/2}$				$\mathcal{E}_{11}$				
	$1p_{3/2}$				$\mathcal{E}_{11}$				
	$1p_{1/2}$				$\mathcal{L}_{11}$				
	$1p_{3/2}$				$\mathcal{L}_{11}$				
	$1s_{1/2}$				$\mathcal{L}_{20}$				
	$1s_{1/2}$	1p		$\mathcal{L}_{10}$				1	0
	$1s_{1/2}$	1p		$\mathcal{L}_{10}$				1	1
	$1p_{1/2}$	1s		$\mathcal{L}_{10}$				0	0
	$1p_{1/2}$	1s		$\mathcal{L}_{10}$				0	1
	$1s_{1/2}$	1s		$\mathcal{E}_{11}$				1	0
	$1s_{1/2}$	1s		$\mathcal{E}_{11}$				1	1
	$1s_{1/2}$	1s		$\mathcal{L}_{11}$				1	0
	$1s_{1/2}$	1s		$\mathcal{L}_{11}$				1	1
	$1s_{1/2}$	1p			$\mathcal{L}_{10}$			1	1
	$1p_{1/2}$	1s			$\mathcal{L}_{10}$			0	1
	$1s_{1/2}$	1s			$\mathcal{E}_{11}$			1	1
	$1s_{1/2}$	1s			$\mathcal{L}_{11}$			1	1
	$1s_{1/2}$	$1s^2$		$\mathcal{L}_{10}$		0	0	0	1
	$1s_{1/2}$	$1s^2$		$\mathcal{L}_{10}$		0	2	0	1
	$1s_{1/2}$	$1s^2$			$\mathcal{L}_{10}$	0	0	0	0

CHAPTER 6

Elementary free energy and interaction operators

6.1. INTRODUCTION

The various parts of the hamiltonian field operators are given in the discretized representation of the free field solutions of Chapter 5. This is done by expressing the free energy and interaction operators in the occupation number representation. It is this occupation number representation which will be used in the next chapters to carry out the evaluation of the matrix elements of the hamiltonian for the system. In sect. 6.2 we derive the discretized expansions of the free hamiltonian H_0. The single-particle states defined in Chapter 5 are of course not eigenstates of H_0. Therefore non-diagonal terms of H_0 connecting different radial quantum numbers will arise. The final simple expressions for the different fields are collected in sect. 6.2.4.

In the next sections we treat the various terms of the interaction hamiltonian H_I given in Chapter 2. With the phase convention adopted for the basis state vectors, and which will be discussed in detail in Chapter 7, the hamiltonian matrix will be real, i.e. symmetric. Hence it is sufficient here to treat only one of the two processes which are related by time reversal and which occupy mirror positions in the hamiltonian matrix. However, the corresponding invariant matrix elements themselves, which appear as the coefficients of the interaction operator expansions in this chapter, are not necessarily symmetric, as shown in sect. 4.4.3. The choice of which of the two time-reversal related processes is treated will be indicated for each vertex. The real symmetric hamiltonian matrix itself will be constructed in Chapter 7.

The way the momentum conservation is realized at each vertex between the interacting particles represented by wave packets over momenta is demonstrated in the last section.

6.2. FREE ENERGIES

We first consider separately the free energy operators for the different spin fields and summarize the results at the end of the section.

6.2.1. Spin-0 free energy

The spin-0 free field hamiltonian from sect. 2.3.1, is written in terms of the discretized expansion (5.6) carrying out an integration by parts in the space derivatives

$$\int d^3r :H_0: = \tfrac{1}{2}\int d^3r :(-\partial_4\varphi\partial_4\varphi + \partial_i\varphi\partial_i\varphi + M^2\varphi^2):$$

$$= \frac{1}{2\pi}\int d^3r \sum_{\nu l}\sum_{\nu' l'} \hat{l}\hat{l}'\hat{t}^2 \int p^2 dp \int p'^2 dp' \frac{1}{\sqrt{EE'}} 2 j_l(pr) f_{\nu l}(p)$$

$$:\left[\tilde{a}_\nu^{[lt]} \hat{r}^{[l]} \eta^{[t]}\right]^{[00]} (-\overleftarrow{\partial}_4 \vec{\partial}_4 - \nabla^2 + M^2)$$

$$\times \left[a_{\nu'}^{[l't]} \hat{r}^{[l']} \eta^{[t]}\right]^{[00]} : j_{l'}(p'r) f_{\nu' l'}(p') .$$

The factor 2 arises from the two identical contributions of the normal products $:\tilde{a}a:$ and $:a\tilde{a}:$. Using the relation

$$(-\nabla^2 + M^2) j_l(pr) \hat{r}^{[l]} = (p^2 + M^2) j_l(pr) \hat{r}^{[l]},$$

we get

$$\int d^3r :H_0: = \sum_{\nu'\nu l} \hat{l}\hat{t} : \left[\tilde{a}_\nu^{[lt]} a_{\nu'}^{[lt]}\right]^{[00]} : \int p^2 dp\, f_{\nu l}(p) E f_{\nu' l}(p),$$

where in the last factor the operator E in momentum space stands for $E = \sqrt{p^2 + M^2}$. M is the mass parameter of the spin-0 field. The interpretation of the last factor is evident: it generalizes the mean energy of the discretized wave-packets to the terms off-diagonal in the radial quantum number ν.

6.2.2. Spin-$\frac{1}{2}$ free energy

The evaluation of the free energy of the spin-$\frac{1}{2}$ field is involved but will yield a very simple expression. In order to demonstrate the method of evaluation we shall compute this case in full detail. We will not repeat the calculation for the free energy of the other fields.

The $s = \frac{1}{2}$, $t = \frac{1}{2}$ free field hamiltonian from sect. 2.3.2 is calculated by introducing in it the expansion (5.8). Omitting the antiparticle part and defining for the transposed spinors the notation $\begin{pmatrix}\alpha\\ \beta\end{pmatrix}^{\mathrm{T}} = (\alpha\beta)$ we get

$$\int d^3r :H_0: = \int d^3r :(\bar{\psi}\gamma\cdot\nabla\psi + M\bar{\psi}\psi):$$

$$= \tfrac{1}{2} \sum_{\substack{\nu l \lambda j\\ \nu' l' \lambda' j'}} \hat{j}\hat{j}'\hat{t}^2 : \left[\tilde{b}_{\nu l}^{[jt]} \begin{pmatrix} \tilde{Y}_l^{[j]}(\hat{r}) u_{\nu l}(r) \\ -\tilde{Y}_\lambda^{[j]}(\hat{r}) v_{\nu l \lambda}(r) \end{pmatrix}^{\mathrm{T}} \tilde{\eta}^{[t]} \right]^{[00]}$$

$$\times\begin{pmatrix} M & \boldsymbol{\sigma}\cdot\boldsymbol{p} \\ \boldsymbol{\sigma}\cdot\boldsymbol{p} & -M \end{pmatrix}\left[b_{\nu' l'}^{[j't]} \begin{pmatrix} \tilde{Y}_{l'}^{[j']}(\hat{\boldsymbol{r}}) u_{\nu' l'}(r) \\ -\tilde{Y}_{\lambda'}^{[j']}(\hat{\boldsymbol{r}}) v_{\nu' l'\lambda'}(r) \end{pmatrix} \eta^{[t]} \right]^{[00]} :$$

$$= \sum_{\substack{\nu l \lambda j \\ \nu' l' \lambda' j'}} \hat{j}\hat{t} : \left[\tilde{b}_{\nu l}^{[jt]} b_{\nu' l'}^{[j't]} \right]^{[00]} : \Bigg\{ \tfrac{1}{2} \int r^2 \mathrm{d} r\, M \big(u_{\nu l}(r) u_{\nu' l'}(r) - v_{\nu l \lambda}(r) v_{\nu' l' \lambda'}(r) \big) \delta_{ll'} \delta_{\lambda\lambda'}$$

$$- \frac{1}{2} \frac{[s|\sigma|s]}{\hat{j}\hat{1}} \left(\begin{bmatrix} s & l & j \\ s & \lambda' & j \\ 1 & 1 & 0 \end{bmatrix} \left[u_{\nu l} \hat{r}^{[l]} | p^{[1]} | v_{\nu' l' \lambda'} \hat{r}^{[\lambda']} \right] \right.$$

$$\left. - \begin{bmatrix} s & \lambda & j \\ s & l' & j \\ 1 & 1 & 0 \end{bmatrix} \left[v_{\nu l \lambda} \hat{r}^{[\lambda]} | p^{[1]} | u_{\nu' l'} \hat{r}^{[l']} \right] \right) \Bigg\} \delta_{jj'} .$$

The geometry of the terms involving $\boldsymbol{\sigma}\cdot\boldsymbol{p}$ is similar to that treated in sect. 4.5.3. In order to obtain the factor $\delta_{ll'}$ in the first contribution we have used the fact that $\delta_{\lambda\lambda'}$ for a given $j = j'$ determines uniquely the l-value. The radial integrals of the first term with (5.9) and (5.10) are given by

$$\int r^2 \mathrm{d} r\, M \big(u_{\nu l}(r) u_{\nu' l'}(r) - v_{\nu l \lambda}(r) v_{\nu' l' \lambda'}(r) \big) = \int p^2 \mathrm{d} p \frac{2M^2}{E} f_{\nu l}(p) f_{\nu' l'}(p) \delta_{ll'} .$$

In the second term, $[s|\sigma|s]$ is given by (4.14). With (4.19) and (4.20) we have

$$\left[u_{\nu l} \hat{r}^{[l]} | p^{[1]} | v_{\nu' l' \lambda'} \hat{r}^{[\lambda']} \right]$$

$$= \begin{cases} -\mathrm{i}\sqrt{\lambda'+1} \int r^2 \mathrm{d} r\, u_{\nu l}(r) \left(\dfrac{\partial}{\partial r} - \dfrac{\lambda'}{r} \right) v_{\nu' l' \lambda'}(r) & \text{for } j = l - \tfrac{1}{2},\quad \lambda' + 1 = l = l' \\[2ex] -\mathrm{i}\sqrt{\lambda'} \int r^2 \mathrm{d} r\, u_{\nu l}(r) \left(\dfrac{\partial}{\partial r} + \dfrac{\lambda'+1}{r} \right) v_{\nu' l' \lambda'}(r) & \text{for } j = l + \tfrac{1}{2},\quad \lambda' - 1 = l = l' . \end{cases}$$

Likewise,

$$\left[v_{\nu l \lambda}(r) \hat{r}^{[\lambda]} | p^{[1]} | u_{\nu' l'}(r) \hat{r}^{[l']} \right]$$

$$= \begin{cases} -\mathrm{i}\sqrt{l'} \int r^2 \mathrm{d} r\, v_{\nu l \lambda}(r) \left(\dfrac{\partial}{\partial r} + \dfrac{l'+1}{r} \right) u_{\nu' l'}(r) & \text{for } j = l - \tfrac{1}{2},\quad \lambda + 1 = l = l' \\[2ex] -\mathrm{i}\sqrt{l'+1} \int r^2 \mathrm{d} r\, v_{\nu l \lambda}(r) \left(\dfrac{\partial}{\partial r} - \dfrac{l'}{r} \right) u_{\nu' l'}(r) & \text{for } j = l + \tfrac{1}{2},\quad \lambda - 1 = l = l' . \end{cases}$$

Utilizing the relations (4.21) and (4.22) we get

$$\int r^2 \mathrm{d}r\, u_{\nu l}(r)\left(\frac{\partial}{\partial r} - \frac{\lambda'}{r}\right) v_{\nu' l' \lambda'}(r) = -\int \mathrm{d}p \frac{p^4}{E} f_{\nu l}(p) f_{\nu' l'}(p) \delta_{ll'},$$

$$\int r^2 \mathrm{d}r\, u_{\nu l}(r)\left(\frac{\partial}{\partial r} + \frac{\lambda'+1}{r}\right) v_{\nu' l' \lambda'}(r) = \int \mathrm{d}p \frac{p^4}{E} f_{\nu l}(p) f_{\nu' l'}(p) \delta_{ll'},$$

$$\int r^2 \mathrm{d}r\, v_{\nu l \lambda}(r)\left(\frac{\partial}{\partial r} + \frac{l'+1}{r}\right) u_{\nu' l'}(r) = \int \mathrm{d}p \frac{p^4}{E} f_{\nu l}(p) f_{\nu' l'}(p) \delta_{ll'},$$

$$\int r^2 \mathrm{d}r\, v_{\nu l \lambda}(r)\left(\frac{\partial}{\partial r} - \frac{l'}{r}\right) u_{\nu' l'}(r) = -\int \mathrm{d}p \frac{p^4}{E} f_{\nu l}(p) f_{\nu' l'}(p) \delta_{ll'}.$$

Furthermore, with the values

$$\begin{bmatrix} s & l & j \\ s & \lambda & j \\ 1 & 1 & 0 \end{bmatrix} = \begin{cases} -\dfrac{1}{\sqrt{6l}}\hat{j}\hat{1} & \text{if } j = l - \tfrac{1}{2}, \quad \lambda = l - 1 \\[2ex] \dfrac{1}{\sqrt{6(l+1)}}\hat{j}\hat{1} & \text{if } j = l + \tfrac{1}{2}, \quad \lambda = l + 1, \end{cases}$$

we finally have for the spin-$\frac{1}{2}$ free energy operator the very simple expression, analogous to the one obtained for the spin-0 field

$$\int \mathrm{d}^3 r : H_0 : = \sum_{\nu\nu' l j} \hat{j}\hat{t} : \left[\tilde{b}_{\nu l}^{[jt]} b_{\nu' l}^{[jt]}\right]^{[00]} :$$

$$\left(\int p^2 \mathrm{d}p \frac{M^2}{E} f_{\nu l}(p) f_{\nu' l}(p) + \int \mathrm{d}p \frac{p^4}{E} f_{\nu l}(p) f_{\nu' l}(p)\right)$$

$$= \sum_{\nu\nu' l j} \hat{j}\hat{t} : \left[\tilde{b}_{\nu l}^{[jt]} b_{\nu' l}^{[jt]}\right]^{[00]} : \int p^2 \mathrm{d}p\, f_{\nu l}(p) E f_{\nu' l}(p).$$

The multiplicity associated with the number of equivalent normal products is unity since according to (2.13) $:\psi^+\psi:$ is of the form $:(\tilde{b} + d)(b + \tilde{d}):$ which yields only one term $:\tilde{b}b:$.

6.2.3. *Spin-1 free energy*

The spin-1 free field hamiltonian is given in sect. 2.3.3,

$$\int \mathrm{d}^3 r : H_0 : = \tfrac{1}{2} \sum_{\kappa} \int \mathrm{d}^3 r : \left(\boldsymbol{E}_\kappa^2 + \boldsymbol{B}_\kappa^2 + M^2 \phi_\kappa^2 - M^2 \phi_4^2\right) : .$$

Using the expansions of sect. 5.2.3 for the electric field $\boldsymbol{E}_\kappa = -\pi_\kappa$, eqs. (5.19)–(5.21), and for the magnetic field $\boldsymbol{B}_\kappa$, eqs. (5.24)–(5.25), the hamiltonian can be evaluated by simple recouplings owing to the orthogonality of the multipolarities and the properties of the radial functions. This yields

$$\int \mathrm{d}^3r:H_0:=\sum_{\nu\nu'\kappa J}\hat{J}\hat{t}:\left[\tilde{A}^{[Jt]}_{\kappa\nu}A^{[Jt]}_{\kappa\nu'}\right]^{[00]}:\int p^2\mathrm{d}p\, f_{\nu J}(p)Ef_{\nu' J}(p).$$

The factor 2 arising from the two terms $:\tilde{A}A:$ and $:A\tilde{A}:$ inside the normal product cancels the factor $\frac{1}{2}$ in the original equation for H_0. The integrand is independent of κ. Thus it is seen that the spin-1 free field diagonal energies, in the discretized representation, can be classified according to the harmonic oscillator excitation quantum number $N=2\nu+J-2$.

6.2.4. *Summary of the free field energies*

We see that all cases lead to expressions containing the quantities

$$E^l_{\nu\nu'}=\int p^2\mathrm{d}p\, f_{\nu l}(p)Ef_{\nu' l}(p),$$

where E stands for the relativistic free particle energy $E=\sqrt{p^2+M^2}$. For $\nu'=\nu$, $E^l_{\nu\nu}$ is the mean value of the energy for a particle in state νl. These quantities differ from being invariant matrix elements by the factors like $\hat{l}\hat{t}$ in the expressions for H_0 below.

The invariant matrix elements of H_0 on the single-particle basis functions can now be written for spin-0, spin-$\frac{1}{2}$ and spin-1 particles respectively in the occupation number representation:

$$H_0=\sum_{\nu\nu' l}\hat{l}\hat{t}:\left[\tilde{a}^{[lt]}_{\nu}a^{[lt]}_{\nu'}\right]^{[00]}:E^l_{\nu\nu'},$$

$$H_0=\sum_{\nu\nu' lj}\hat{j}\hat{t}:\left[\tilde{b}^{[jt]}_{\nu l}b^{[jt]}_{\nu' l}+\tilde{d}^{[jt]}_{\nu l}d^{[jt]}_{\nu' l}\right]^{[00]}:E^l_{\nu\nu'},$$

$$H_0=\sum_{\nu\nu'\kappa J}\hat{J}\hat{t}:\left[\tilde{A}^{[Jt]}_{\kappa\nu}A^{[Jt]}_{\kappa\nu'}\right]^{[00]}:E^J_{\nu\nu'}.$$

To these we add the spin-$\frac{3}{2}$ free field energies, the computation of which is similar to that of the spin $\frac{1}{2}$,

$$H_0=\sum_{\nu\nu' lj}\hat{j}\hat{t}:\left[\tilde{B}^{[jt]}_{\nu l}B^{[jt]}_{\nu' l}+\tilde{D}^{[jt]}_{\nu l}D^{[jt]}_{\nu' l}\right]^{[00]}:E^l_{\nu\nu'}.$$

6.3. THE PION-NUCLEON INTERACTION

The pion-nucleon interaction is given in eqs. (2.22) and (2.23). The elementary vertex together with the notation used hereafter in the matrix elements is shown in fig. 6.1. We calculate only the pion creation vertex, according to the remark made in

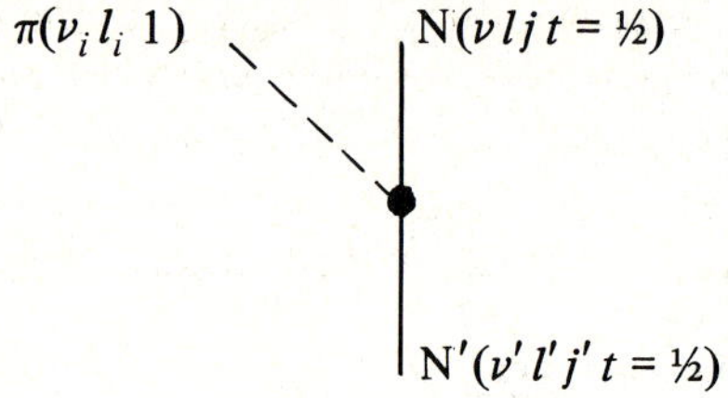

Fig. 6.1.

the introduction to this chapter. The full operator including pion absorption is discussed at the end of this section.

6.3.1. *The pseudo-scalar coupling*

The pseudo-scalar part of the interaction is of the form eq. (2.22),

$$H_{\pi N}^{PS} = ig_{\pi N}^{PS} \int d^3r : \bar{\psi}\gamma_5 \boldsymbol{\tau}\psi\boldsymbol{\varphi} : .$$

The matrix elements are calculated by evaluating separately the contributions from isospin space and from Minkowski space.

In isospin space, using the notation $t = \frac{1}{2}$ for the nucleon isospin and writing 1 for the pion isospin we have

$$H^{(\text{isospin})} = \hat{t}\big[\tilde{b}^{[t]}\tilde{\eta}^{[t]}\big]^{[0]}\hat{1}\big[\eta^{[1]}\tau^{[1]}\big]^{[0]}\hat{1}\big[\tilde{a}^{[1]}\eta^{[1]}\big]^{[0]}\hat{t}\big[b^{[t]}\eta^{[t]}\big]^{[0]},$$

where we have used the invariant form (4.34) for the isospin operator $\boldsymbol{\tau}$. In this form the isospin function $\eta^{[1]}$ plays the role of the unit vector in isospin space. Recall that the integer isospin function $\eta^{[1]}$ for the pion is self-conjugate and requires no tilde. As everywhere else, we now perform the scalar product $\boldsymbol{\tau}\cdot\boldsymbol{\varphi}$ in isospin space according to the recoupling diagram of fig. 6.2

$$\boldsymbol{\tau}\cdot\boldsymbol{\varphi} = \hat{1}^2\big[\eta^{[1]}\tau^{[1]}\big]^{[0]}\cdot\big[\tilde{a}^{[1]}\eta^{[1]}\big]^{[0]} = \hat{1}\big[\tilde{a}^{[1]}\tau^{[1]}\big]^{[0]}.$$

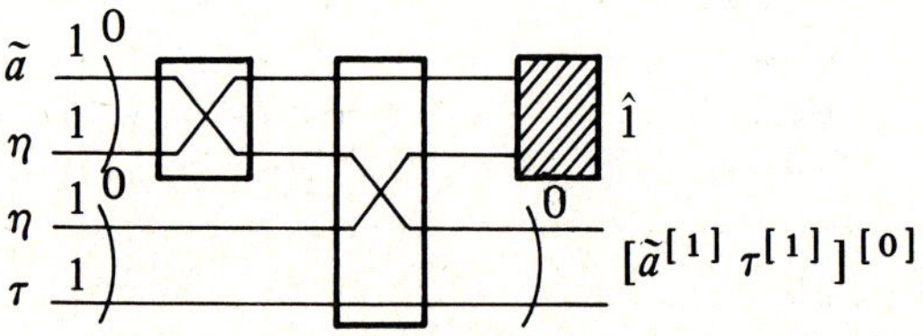

Fig. 6.2.

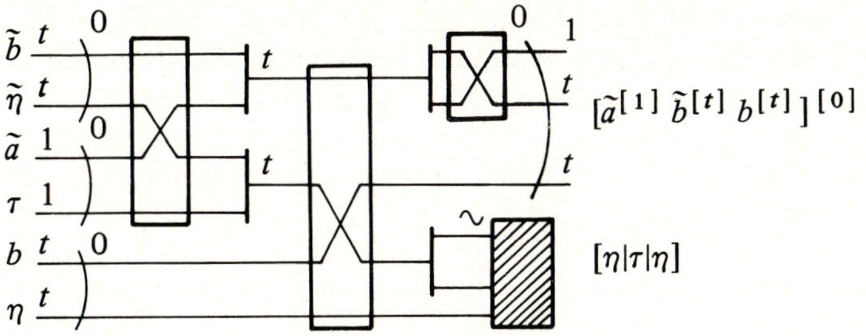

Fig. 6.3.

Thus evaluating the graph 6.3 with the end box value given by eq. (4.14)

$$H^{(\text{isospin})} = -i\sqrt{6}\,[\tilde{a}^{[1]}\tilde{b}^{[t]}b^{[t]}]^{[0]}.$$

The part of $H_{\pi N}^{PS}$ in Minkowski space is of the form

$$ig_{\pi N}^{PS} \sum_{\substack{\nu l j \\ \nu' l' j' \\ \nu_i l_i}} \int d^3 r\, \tfrac{1}{2}\sqrt{\tfrac{1}{2}}\, \hat{j}\hat{j}'\hat{l}_i \left[\tilde{b}_{\nu l}^{[j]} \begin{pmatrix} \tilde{Y}_l^{[j]}(\hat{\boldsymbol{r}})\, u_{\nu l}(r) \\ \tilde{Y}_\lambda^{[j]}(\hat{\boldsymbol{r}})\, v_{\nu l \lambda}(r) \end{pmatrix}^{T} \right]^{[0]}$$

$$\times \left[\tilde{a}_{\nu_i}^{[l_i]} \hat{r}^{[l_i]} g_{\nu_i l_i}(r)\right]^{[0]} \left[b_{\nu' l'}^{[j']} \begin{pmatrix} Y_{\lambda'}^{[j']}(\hat{\boldsymbol{r}})\, v_{\nu' l' \lambda'}(r) \\ -Y_{l'}^{[j']}(\hat{\boldsymbol{r}})\, u_{\nu' l'}(r) \end{pmatrix} \right]^{[0]}.$$

Note the transpose notation $(\ )^{T}$. Also, the action of the γ-matrices, i.e. the product $\gamma_4\gamma_5$, already has been carried out. This expression is calculated according to fig. 6.4.

The final result, including the contribution from the isospin, yields for the discretized expansion of the pion emission part of the interaction

$$H_{\pi N}^{PS} = \sum [\pi_i N | H_{\pi N}^{PS} | N'] : \left[\tilde{a}_{\nu_i}^{[l_i 1]} \tilde{b}_{\nu l}^{[jt]} b_{\nu' l'}^{[j' t]}\right]^{[00]} :,$$

where the elementary pseudo-scalar pion-nucleon invariant matrix element is

$$[\pi_i N | H_{\pi N}^{PS} | N'] = g_{\pi N}^{PS} \tfrac{1}{2}\sqrt{6}$$

$$\times \left(\begin{bmatrix} s & l & j \\ s & \lambda' & j' \\ 0 & l_i & l_i \end{bmatrix} [l_i | l | \lambda'] \int r^2 dr\, u_{\nu l}(r)\, g_{\nu_i l_i}(r)\, v_{\nu' l' \lambda'}(r) \right.$$

$$\left. - \begin{bmatrix} s & \lambda & j \\ s & l' & j' \\ 0 & l_i & l_i \end{bmatrix} [l_i | \lambda | l'] \int r^2 dr\, v_{\nu l \lambda}(r)\, g_{\nu_i l_i}(r)\, u_{\nu' l'}(r) \right). \quad (6.1)$$

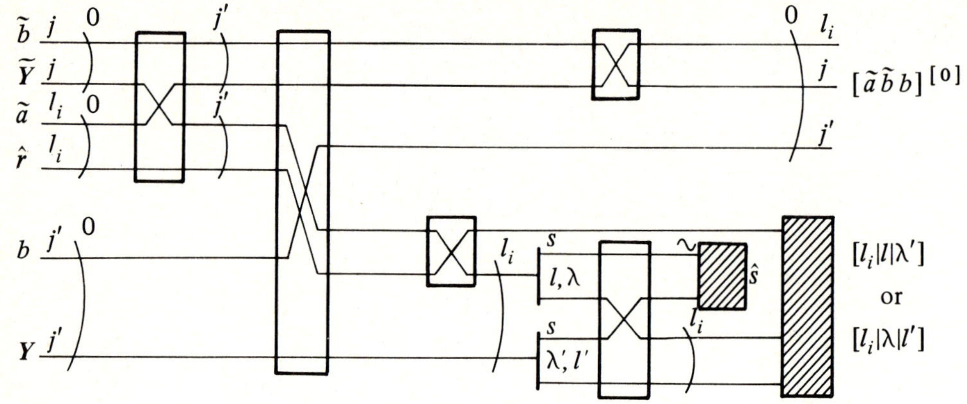

Fig. 6.4.

6.3.2. The pseudo-vector coupling

The pseudo-vector pion-nucleon interaction, eq. (2.23), is separated into the space-like and time-like parts of the operator

$$H^{\mathrm{PV}}_{\pi\mathrm{N}} = -ig^{\mathrm{PV}}_{\pi\mathrm{N}}\frac{1}{2M}\int \mathrm{d}^3r : (\bar{\psi}\gamma_i\gamma_5\boldsymbol{\tau}\psi\,\partial_i\boldsymbol{\varphi} + \bar{\psi}\gamma_4\gamma_5\boldsymbol{\tau}\psi\,\partial_4\boldsymbol{\varphi}):.$$

The isospin part is the same as calculated above. The contribution from Minkowski space is

$$-ig^{\mathrm{PV}}_{\pi\mathrm{N}}\frac{1}{2M}\sum_{\substack{\nu l j\\ \nu' l' j'\\ \nu_i l_i}}\int \mathrm{d}^3r\,\tfrac{1}{2}\sqrt{\tfrac{1}{2}}\,\hat{j}\hat{j}'\hat{l}_i\left(\left[\tilde{b}^{[j]}_{\nu l}\begin{pmatrix}\tilde{Y}^{[j]}_{l}(\hat{\boldsymbol{r}})u_{\nu l}(r)\\ \tilde{Y}^{[j]}_{\lambda}(\hat{\boldsymbol{r}})v_{\nu l\lambda}(r)\end{pmatrix}^{\mathrm{T}}\right]^{[0]}\right.$$

$$\times\begin{pmatrix} i\hat{1}[\boldsymbol{\sigma}^{[1]}\nabla^{[1]}]^{[0]} & 0\\ 0 & -i\hat{1}[\boldsymbol{\sigma}^{[1]}\nabla^{[1]}]^{[0]}\end{pmatrix}\left[\tilde{a}^{[l_i]}_{\nu_i}\hat{r}^{[l_i]}g_{\nu_i l_i}(r)\right]^{[0]}$$

$$\times\left[b^{[j']}_{\nu' l'}\begin{pmatrix}Y^{[j']}_{l'}(\hat{\boldsymbol{r}})u_{\nu' l'}(r)\\ -Y^{[j']}_{\lambda'}(\hat{\boldsymbol{r}})v_{\nu' l'\lambda'}(r)\end{pmatrix}\right]^{[0]}+\left[\tilde{b}^{[j]}_{\nu l}\begin{pmatrix}\tilde{Y}^{[j]}_{l}(\hat{\boldsymbol{r}})u_{\nu l}(r)\\ -\tilde{Y}^{[j]}_{\lambda}(\hat{\boldsymbol{r}})v_{\nu l\lambda}(r)\end{pmatrix}^{\mathrm{T}}\right]^{[0]}$$

$$\left.\times\left[\tilde{a}^{[l_i]}_{\nu_i}\hat{r}^{[l_i]}h_{\nu_i l_i}(r)\right]^{[0]}\left[b^{[j']}_{\nu' l'}\begin{pmatrix}Y^{[j']}_{\lambda'}(\hat{\boldsymbol{r}})v_{\nu' l'\lambda'}(r)\\ -Y^{[j']}_{l'}(\hat{\boldsymbol{r}})u_{\nu' l'}(r)\end{pmatrix}\right]^{[0]}\right).$$

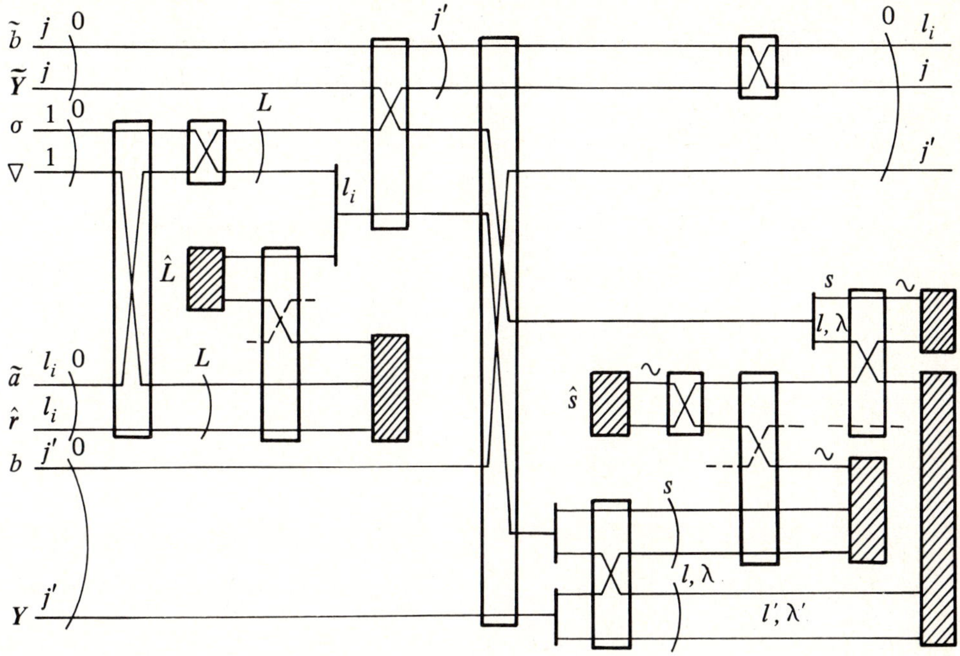

Fig. 6.5.

In the second term as usual the time derivation $\partial_4 = -i\partial_t$ has been performed in the Heisenberg picture before putting the time to zero. This changes the function $g_{\nu_i l_i}(r)$ into the function $\mp h_{\nu_i l_i}(r)$, see eqs. (3.4) and (3.6), where the sign comes from

$$\partial_4 e^{\mp iEt} = \mp E e^{\mp iEt},$$

with − for pion absorption and + for pion emission. Here we consider pion emission. The recoupling geometry is evaluated according to fig. 6.5 for the first term and fig. 6.4 for the second term.

The final result is

$$H_{\pi N}^{PV} = \sum \left[\pi_i N | H_{\pi N}^{PV} | N'\right] : \left[\tilde{a}_{\nu_i}^{[l_i 1]} \tilde{b}_{\nu l}^{[jt]} b_{\nu' l'}^{[j't]}\right]^{[00]} :,$$

where the invariant elementary pseudo-vector pion-nucleon matrix element is

$$[\pi_i \mathrm{N}|H_{\pi\mathrm{N}}^{\mathrm{PV}}|\mathrm{N}'] = g_{\pi\mathrm{N}}^{\mathrm{PV}} \frac{1}{2M} \Bigg\{ (-)^{j+l_i-j'} \frac{3}{2} \frac{\hat{j}}{\hat{l}_i} \sum_L \alpha_{Ll_i}$$

$$\times \left(\left(\begin{bmatrix} 1 & L & l_i \\ s & l' & j' \\ s & l & j \end{bmatrix} \frac{1}{\hat{l}} [l|L|l'] \int r^2 \mathrm{d}r\, u_{\nu l}(r) k_{\nu_i l_i L}(r) u_{\nu' l'}(r) \right.\right.$$

$$\left. + \begin{bmatrix} 1 & L & l_i \\ s & \lambda' & j' \\ s & \lambda & j \end{bmatrix} \frac{1}{\hat{\lambda}} [\lambda|L|\lambda'] \int r^2 \mathrm{d}r\, v_{\nu l\lambda}(r) k_{\nu_i l_i L}(r) v_{\nu' l' \lambda'}(r) \right)$$

$$- \sqrt{\tfrac{3}{2}} \left(\begin{bmatrix} s & l & j \\ s & \lambda' & j' \\ 0 & l_i & l_i \end{bmatrix} [l_i|l|\lambda'] \int r^2 \mathrm{d}r\, u_{\nu l}(r) h_{\nu_i l_i}(r) v_{\nu' l' \lambda'}(r) \right.$$

$$\left.\left. + \begin{bmatrix} s & \lambda & j \\ s & l' & j' \\ 0 & l_i & l_i \end{bmatrix} [l_i|\lambda|l'] \int r^2 \mathrm{d}r\, v_{\nu l\lambda}(r) h_{\nu_i l_i}(r) u_{\nu' l'}(r) \right) \right) \Bigg\}. \tag{6.2}$$

The derivative radial functions $k_{\nu l\lambda}(r)$ are given at the end of sect. 4.2.4 and the factors α_{Ll} by eq. (4.24).

The complete pion-nucleon vertex is the sum of the pseudo-scalar term (6.1) and the pseudo-vector term (6.2)

$$[\pi\mathrm{N}|H_{\pi\mathrm{N}}|\mathrm{N}'] = [\pi\mathrm{N}|H_{\pi\mathrm{N}}^{\mathrm{PS}}|\mathrm{N}'] + [\pi\mathrm{N}|H_{\pi\mathrm{N}}^{\mathrm{PV}}|\mathrm{N}']$$

and, adding pion absorption, the complete pion-nucleon interaction operator is

$$H_{\pi\mathrm{N}} = \sum \Big([\pi\mathrm{N}|H_{\pi\mathrm{N}}|\mathrm{N}'] : \big[\tilde{a}_{\nu_i}^{[l_i 1]} \tilde{b}_{\nu l}^{[jt]} b_{\nu' l'}^{[j't]} \big]^{[00]} :$$

$$+ [\mathrm{N}'|H_{\pi\mathrm{N}}|\pi\mathrm{N}] : \big[\tilde{b}_{\nu' l'}^{[j't]} a_{\nu_i}^{[l_i 1]} b_{\nu l}^{[jt]} \big]^{[00]} : \Big).$$

Note that since the last term in eq. (6.2) arises from the time derivative it changes sign for the pion absorption process. One may check from the signs arising from the change of indices under time-reversal that this way the real invariant matrix elements for the absorption and emission processes for both PS and PV couplings are related by

$$[\mathrm{N}'|H_{\pi\mathrm{N}}|\pi_i\mathrm{N}] = [\pi_i\mathrm{N}|H_{\pi\mathrm{N}}|\mathrm{N}'].$$

As pointed out in the introduction a choice of basis state vector phases will be given in Chapter 7 such that the hamiltonian matrix will be real, hence symmetric. Thus we may ignore for example the pion absorption part. From now on only one of each two time-reversal related processes will be given.

6.4. THE SIGMA-NUCLEON INTERACTION

The sigma-nucleon vertex is given in fig. 6.6. The interaction hamiltonian has the simple form of eq. (2.24)

$$H_{\sigma\mathrm{N}} = g_{\sigma\mathrm{N}} \int \mathrm{d}^3 r : \bar\psi \psi \varphi^{(\sigma)} : ,$$

where the σ-field $\varphi^{(\sigma)}(\boldsymbol{r})$ is identical to the pion field, except that it has zero isospin. This interaction is, for the σ creation part

$$H_{\sigma\mathrm{N}} = \sum [\sigma\mathrm{N}|H_{\sigma\mathrm{N}}|\mathrm{N}'] : \left[\tilde a^{[J0]}_{\nu_\sigma} \tilde b^{[jt]}_{\nu l} b^{[j't]}_{\nu' l'}\right]^{[00]} : .$$

The geometry of fig. 6.7 yields for the elementary invariant matrix element

$$[\sigma\mathrm{N}|H_{\sigma\mathrm{N}}|\mathrm{N}'] = g_{\sigma\mathrm{N}} \frac{1}{\sqrt{2}} \left(\begin{bmatrix} s & l & j \\ s & l' & j' \\ 0 & J & J \end{bmatrix} [J|l|l'] \int r^2 \mathrm{d}r\, u_{\nu l}(r)\, g_{\nu_\sigma J}(r)\, u_{\nu' l'}(r) \right.$$

$$\left. - \begin{bmatrix} s & \lambda & j \\ s & \lambda' & j' \\ 0 & J & J \end{bmatrix} [J|\lambda|\lambda'] \int r^2 \mathrm{d}r\, v_{\nu l \lambda}(r) g_{\nu_\sigma J}(r) v_{\nu' l' \lambda'}(r) \right). \tag{6.3}$$

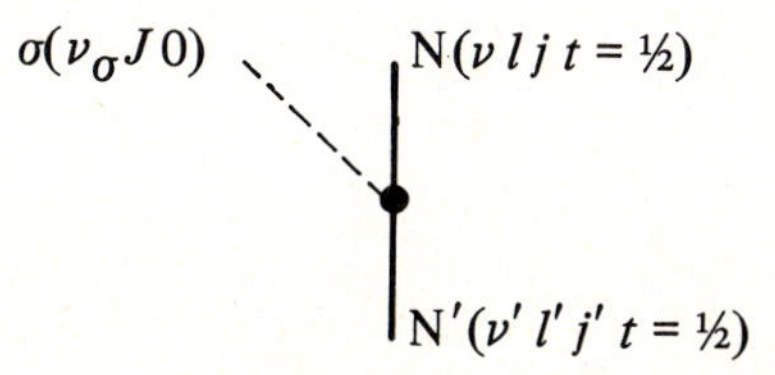

Fig. 6.6.

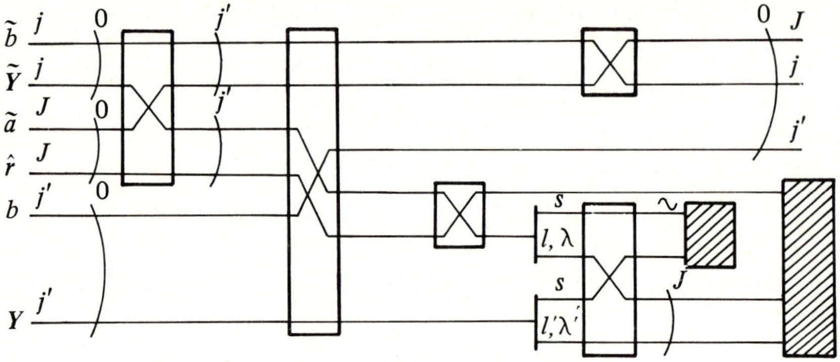

Fig. 6.7.

6.5. THE VECTOR MESON-NUCLEON INTERACTION

6.5.1 The vector coupling

We consider first the vector coupling part of the interaction between the nucleon and the omega and rho vector mesons, H_{VN}, given by eq. (2.25),

$$H^{\mathrm{V}}_{\omega\mathrm{N}} = -ig^{\mathrm{V}}_{\omega\mathrm{N}} \int \mathrm{d}^3 r : \bar{\psi}\gamma_\mu \psi \phi_\mu^{(\omega)} : ,$$

$$H^{\mathrm{V}}_{\rho\mathrm{N}} = -ig^{\mathrm{V}}_{\rho\mathrm{N}} \int \mathrm{d}^3 r : \bar{\psi}\gamma_\mu \tau \psi \phi_\mu^{(\rho)} : .$$

For compactness we combine these two cases introducing the vector meson isospin t_{h} with $t_{\mathrm{h}} = 0$ for the ω and $t_{\mathrm{h}} = 1$ for the ρ. The elementary vertex together with the notation is given in fig. 6.8. Again we only calculate the vector meson creation vertex.

For the ρ-meson the contribution of the isospin is the same as that of the pion given in sect. 6.3. Of course, for the ω-meson it is the overlap. For convenience this

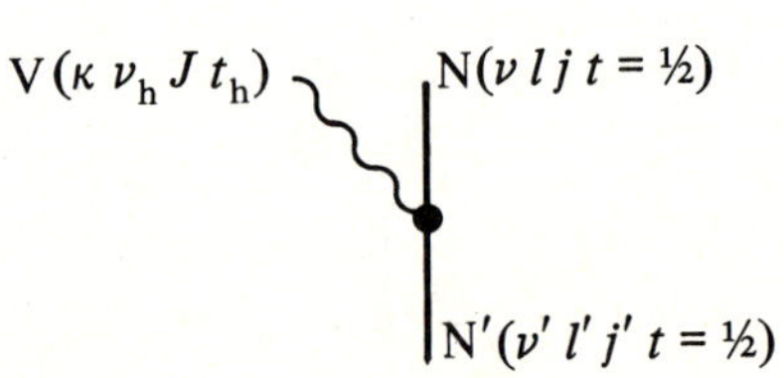

Fig. 6.8.

factor is written for both cases as

$$F(t_{\mathrm{h}}) = (-)^{t_{\mathrm{h}}}\left[t|\tau^{[t_{\mathrm{h}}]}|t\right] = \begin{cases} -i\sqrt{6} & \text{for } \rho, \\ \sqrt{2} & \text{for } \omega. \end{cases}$$

The remaining part separates into a space-like and a time-like term. The calculation of the space-like term makes use of the relation

$$\boldsymbol{\gamma}\cdot\boldsymbol{\phi} \propto \begin{pmatrix} 0 & -i\boldsymbol{\sigma} \\ i\boldsymbol{\sigma} & 0 \end{pmatrix}\cdot \boldsymbol{Y}^{[J]}_{LM} = i\gamma_4\gamma_5\left[\sigma^{[1]}\hat{r}^{[L]}\right]^{[J]}_M,$$

which is an application of eq. (4.35). The geometry of the space-like matrix element is given in fig. 6.9. Using again as in sect. 5.2.3 the notation $L = J + n$, $n = 0, \pm 1$, and computing separately the contributions for the different values n, we have

$$K(\mathrm{V},\mathrm{N},\mathrm{N}',n) = ig^{\mathrm{V}}_{\mathrm{VN}}(-)^{j+J-j'}F(t_{\mathrm{h}})\tfrac{1}{2}\sqrt{3}\,\hat{j}'$$
$$\times\left(\begin{bmatrix} s & l & j \\ 1 & L & J \\ s & \lambda' & j' \end{bmatrix}\frac{1}{\hat{\lambda}'}[l|L|\lambda']\int r^2\,\mathrm{d}r\, u_{\nu l}(r)\,\mathcal{W}_{\kappa\nu_{\mathrm{h}}Jn}(r)\,v_{\nu' l'\lambda'}(r)\right.$$
$$\left.+\begin{bmatrix} s & \lambda & j \\ 1 & L & J \\ s & l' & j' \end{bmatrix}\frac{1}{\hat{l}'}[\lambda|L|l']\int r^2\,\mathrm{d}r\, v_{\nu l\lambda}(r)\,\mathcal{W}_{\kappa\nu_{\mathrm{h}}Jn}(r)\,u_{\nu' l'}(r)\right),$$

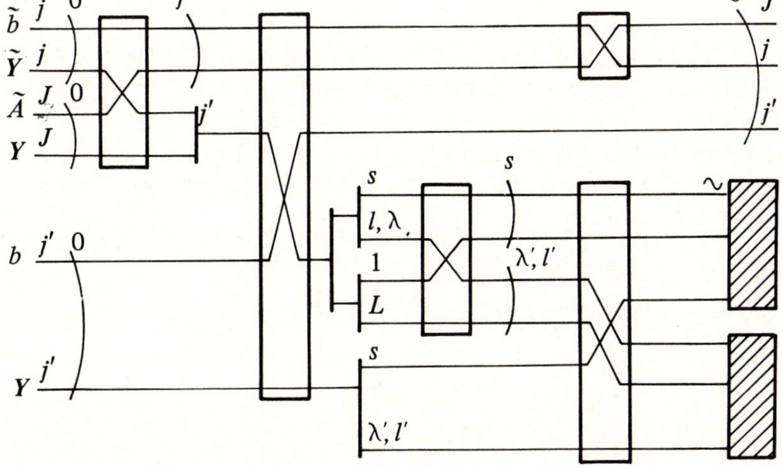

Fig. 6.9.

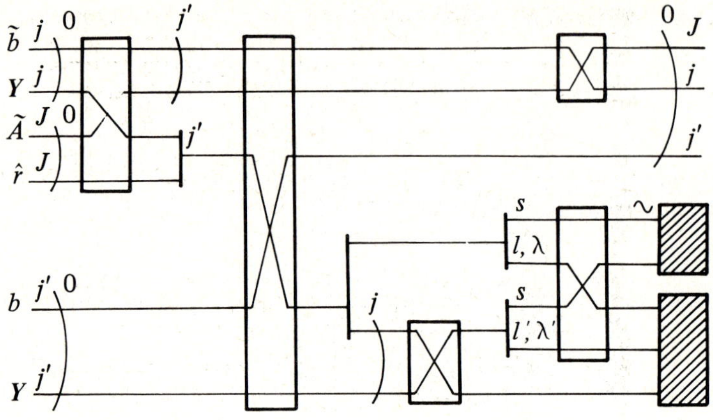

Fig. 6.10.

where V stands for the quantum numbers κ, ν_{h}, J, t_{h} of the vector meson. The radial wave packets $\mathcal{W}_{\kappa\nu_{\mathrm{h}}Jn}(r)$ for the vector meson are given in eq. (5.15).

The time-like part, i.e. the term with $\gamma_4\phi_4^{(+)}(\boldsymbol{r})$ is simpler and is shown in fig. 6.10. Only the longitudinal multipolarity $\kappa=\mathfrak{L}$ has a time-like part, which has $n=0$. The value of the graph is:

$$K_4(\mathrm{V},\mathrm{N},\mathrm{N}')=-i\sqrt{\tfrac{1}{2}}\,g_{\mathrm{VN}}^{\mathrm{V}}F(t_{\mathrm{h}})$$

$$\times\left(\begin{bmatrix} s & l & j\\ s & l' & j'\\ 0 & J & J\end{bmatrix}[l|l'|J]\int r^2\,\mathrm{d}r\,u_{\nu l}(r)\mathcal{X}_{\nu_{\mathrm{h}}J}(r)u_{\nu' l'}(r)\right.$$

$$\left.+\begin{bmatrix} s & \lambda & j\\ s & \lambda' & j'\\ 0 & J & J\end{bmatrix}[\lambda|\lambda'|J]\int r^2\,\mathrm{d}r\,v_{\nu l\lambda}(r)\mathcal{X}_{\nu_{\mathrm{h}}J}(r)v_{\nu' l'\lambda'}(r)\right).$$

The radial functions $\mathcal{X}_{\nu_{\mathrm{h}}J}(r)$ are defined in eq. (5.18).

The vector interaction is thus for the vector meson creation part

$$H_{\mathrm{VN}}^{\mathrm{V}}=\sum[\mathrm{VN}|H_{\mathrm{VN}}^{\mathrm{V}}|\mathrm{N}']:\left[\tilde{A}_{\kappa\nu_{\mathrm{h}}}^{[Jt_{\mathrm{h}}]}\tilde{b}_{\nu l}^{[jt]}b_{\nu' l'}^{[j't]}\right]^{[00]}:,$$

where the vector meson–nucleon matrix elements are given in terms of the multipolarities $\kappa=\mathfrak{M}$, $\mathfrak{E}$ and $\mathfrak{L}$ for total angular momentum J

$$[\mathrm{VN}|H_{\mathrm{VN}}^{\mathrm{V}}|\mathrm{N}']=\sum_n C_\kappa(n,J)K(\mathrm{V},\mathrm{N},\mathrm{N}',n)+K_4(\mathrm{V},\mathrm{N},\mathrm{N}')\delta_{\kappa\mathfrak{L}}. \qquad (6.4)$$

The coefficients C_κ are the unitary transformation (5.27).

6.5.2 *The tensor coupling*

We turn now to the tensor coupling (2.26), where M is the nucleon mass

$$H_{\mathrm{VN}}^{\mathrm{T}}=-g_{\mathrm{VN}}^{\mathrm{T}}\frac{1}{2M}\int\mathrm{d}^3r:\bar{\psi}\sigma_{\mu\nu}\psi(\partial_\mu\phi_\nu-\partial_\nu\phi_\mu):.$$

Here we recognize the strength fields, eq. (2.18),

$$\partial_i\phi_j - \partial_j\phi_i = \epsilon_{ijk}B_k,$$

$$\partial_k\phi_4 - \partial_4\phi_k = -iE_k = i\bar{\pi}_k.$$

Hence

$$H^{\mathrm{T}}_{\mathrm{VN}} = -g^{\mathrm{T}}_{\mathrm{VN}}\frac{1}{2M}\int \mathrm{d}^3r:2(\bar{\psi}\boldsymbol{\sigma}\cdot\boldsymbol{B}\psi - i\bar{\psi}\gamma_5\boldsymbol{\sigma}\cdot\bar{\boldsymbol{\pi}}\psi):.$$

The geometry of these two terms is again similar to that of fig. 6.9. The expansions of the field $\boldsymbol{B}$ are given by eqs. (5.24) and (5.25), and those of the field $\bar{\pi}$ by eqs. (5.19)–(5.21). Thus we obtain

$$K_{\boldsymbol{B}}(\mathrm{V},\mathrm{N},\mathrm{N}',n) = ig^{\mathrm{T}}_{\mathrm{VN}}\frac{1}{2M}(-)^{J+j-j'}F(t_{\mathrm{h}})\tfrac{1}{2}\sqrt{3}\,\hat{j}'$$

$$\times\left(\begin{bmatrix} s & l & j\\ 1 & L & J\\ s & l' & j'\end{bmatrix}\frac{1}{\hat{l}'}[l|L|l']\int r^2\mathrm{d}r\,u_{\nu l}(r)\mathscr{Z}_{\kappa\nu_{\mathrm{h}}Jn}(r)u_{\nu'l'}(r)\right.$$

$$\left.-\begin{bmatrix} s & \lambda & j\\ 1 & L & J\\ s & \lambda' & j'\end{bmatrix}\frac{1}{\hat{\lambda}'}[\lambda|L|\lambda']\int r^2\mathrm{d}r\,v_{\nu l\lambda}(r)\mathscr{Z}_{\kappa\nu_{\mathrm{h}}Jn}(r)v_{\nu'l'\lambda'}(r)\right),$$

$$K_{\boldsymbol{E}}(\mathrm{V},\mathrm{N},\mathrm{N}',n) = ig^{\mathrm{T}}_{\mathrm{VN}}\frac{1}{2M}(-)^{J+j-j'}F(t_{\mathrm{h}})\tfrac{1}{2}\sqrt{3}\,\hat{j}'$$

$$\times\left(\begin{bmatrix} s & \lambda & j\\ 1 & L & J\\ s & l' & j'\end{bmatrix}\frac{1}{\hat{l}'}[\lambda|L|l']\int r^2\mathrm{d}r\,v_{\nu l\lambda}(r)\mathscr{Y}_{\kappa\nu_{\mathrm{h}}Jn}(r)u_{\nu'l'}(r)\right.$$

$$\left.-\begin{bmatrix} s & l & j\\ 1 & L & J\\ s & \lambda' & j'\end{bmatrix}\frac{1}{\hat{\lambda}'}[l|L|\lambda']\int r^2\mathrm{d}r\,u_{\nu l}(r)\mathscr{Y}_{\kappa\nu_{\mathrm{h}}Jn}(r)v_{\nu'l'\lambda'}(r)\right).$$

The vector meson radial functions are defined in eqs. (5.22) and (5.26).

The operator of the tensor interaction for the meson creation part is thus

$$H^{\mathrm{T}}_{\mathrm{VN}} = \sum[\mathrm{VN}|H^{\mathrm{T}}_{\mathrm{VN}}|\mathrm{N}']:\left[\tilde{A}^{[Jt_{\mathrm{h}}]}_{\kappa\nu_{\mathrm{h}}}\tilde{b}^{[jt]}_{\nu l}b^{[j't]}_{\nu'l'}\right]^{[00]}:$$

where the matrix elements of the tensor interaction are given in terms of the multipolarities κ by

$$[\mathrm{VN}|H^{\mathrm{T}}_{\mathrm{VN}}|\mathrm{N}'] = \sum_n(D_\kappa(n,J)K_{\boldsymbol{B}}(\mathrm{V},\mathrm{N},\mathrm{N}',n) + C_\kappa(n,J)K_{\boldsymbol{E}}(\mathrm{V},\mathrm{N},\mathrm{N}',n)).$$

Here, the coefficients $C_\kappa(n, J)$ are defined in (5.27) and the $D_\kappa(n, J)$ are, eqs. (5.24) and (5.25),

$$D_{\mathfrak{M}}(-1, J) = \frac{\sqrt{J+1}}{\sqrt{2J+1}}, \qquad D_{\mathfrak{M}}(0, J) = 0, \qquad D_{\mathfrak{M}}(1, J) = \frac{\sqrt{J}}{\sqrt{2J+1}},$$

$$D_{\mathfrak{E}}(-1, J) = D_{\mathfrak{E}}(1, J) = 0, \qquad D_{\mathfrak{E}}(0, J) = 1,$$

$$D_{\mathfrak{L}}(n, J) = 0.$$

Note that with our choice of the phases for the field operators, for the ρ-meson both the space and time part of the invariant matrix elements are real and symmetric under time reversal, but for the ω-meson they are imaginary and antisymmetric. In that case reality of the secular matrix arises from the appropriate choice of the basis state phases discussed in sect. 4.4.2 and which will be taken into account in Chapter 7. When checking the properties of the invariant matrix elements under time reversal recall that $\phi_4^{(-)} = -\phi_4^{(+)+}$, while $\boldsymbol{\phi}^{(-)} = \boldsymbol{\phi}^{(+)+}$.

6.6. THE PION-PION INTERACTION

The pion-pion interaction hamiltonian, eq. (2.27) is of the form

$$H_\pi = g_\pi \int \mathrm{d}^3 r : \boldsymbol{\varphi}\boldsymbol{\varphi}\boldsymbol{\varphi}\boldsymbol{\varphi} : .$$

The various processes associated with the pion-pion vertex are shown in fig. 6.12. The indices π_i denote the quantum numbers of the pion, $(\nu_i l_i)$. Here intermediate quantum numbers LT are required to specify the configuration. In the evaluation of the invariant elementary matrix elements, we see that with our phase definition of the creation and annihilation part of the pion field, eqs. (5.6) and (5.7), and owing to the associativity of the triple product, there is no need to distinguish between the different processes of figs. 6.11 and 6.12. Hence, setting the short-hand notation

$$\theta_i = \theta_{\nu_i}^{[l_i 1]} = \tilde{a}_{\nu_i}^{[l_i 1]} + a_{\nu_i}^{[l_i 1]},$$

$$Z_i = \left[\theta_{\nu_i}^{[l_i 1]} \hat{r}^{[l_i]} \eta^{[1]}\right]^{[00]}, \tag{6.5}$$

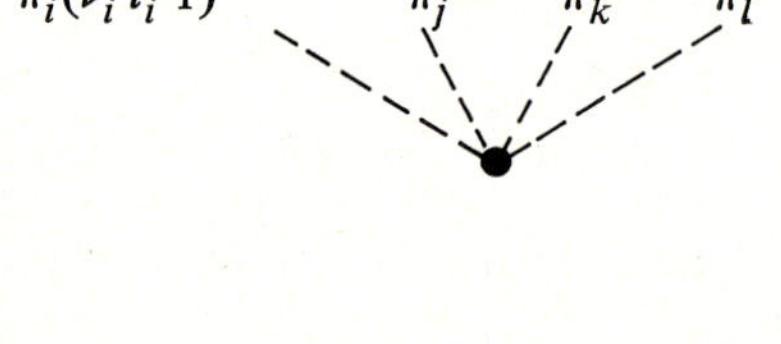

Fig. 6.11.

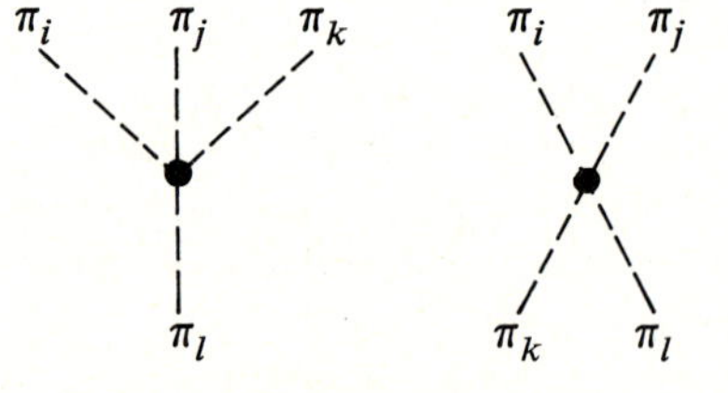

Fig. 6.12.

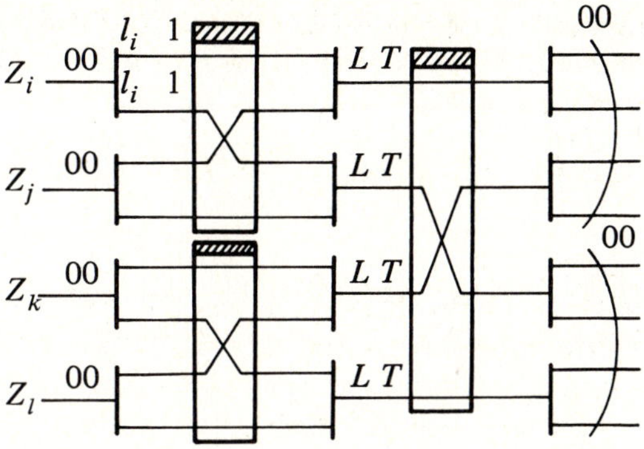

Fig. 6.13.

and substituting the expansions of the spin-0 fields, eq. (5.6), we have

$$H_\pi = \tfrac{1}{4} g_\pi \sum \hat{l}_i \hat{l}_j \hat{l}_k \hat{l}_l \hat{1}^4 \int \mathrm{d}^3 r\, g_{\nu_i l_i}(r) g_{\nu_j l_j}(r) g_{\nu_k l_k}(r) g_{\nu_l l_l}(r) P_0 \!:\! Z_i Z_j Z_k Z_l \!:\! .$$

In this expression the projection operator $\int \mathrm{d}^3 r$ makes the interaction a scalar in Minkowski space, while we must use the appropriate projection operator P_0 to make the interaction a scalar in isospin space.

The evaluation requires first the recoupling graph of fig. 6.13 which yields

$$:ZZZZ: \to \sum_{LT} \left[[\theta_i \theta_j]^{[LT]} [\theta_k \theta_l]^{[LT]} \right]^{[00]} \left[[(\hat{r}\,\eta)_i (\hat{r}\,\eta)_j]^{[LT]} [(\hat{r}\,\eta)_k (\hat{r}\,\eta)_l]^{[LT]} \right]^{[00]} .$$

The right-hand side is still fully symmetric in particle exchange, but not term by term.

For the isospin part, we recall that for example the field operator $P_0 \varphi^{(+)}_{\tau_i}(x) \varphi^{(+)}_{\tau_j}(x) \varphi^{(+)}_{\tau_k}(x) \varphi^{(+)}_{\tau_l}(x)$ creates four particles of charges τ_i, τ_j, τ_k, τ_l at point x with the appropriate coupling coefficients so that P_0 has the meaning $\sum_i \tau_i = 0$ (charge conservation) and total isospin equal zero (isospin conservation). Likewise the operator $P_0 \varphi^{(-)}_1(x) \varphi^{(+)}_1(x) \varphi^{(+)}_0(x) \varphi^{(+)}_0(x)$ annihilates a postively charged pion and creates three pions of charges $+1, 0, 0$ respectively, again with the condition of isospin conservation. It is represented in isospin space by the product $\eta^{(1)}_1 \eta^{[1]}_{-1} \eta^{[1]}_0 \eta^{[1]}_0$ with the appropriate coupling weight. More generally charge conservation is insured in each of the terms of the expression

$$F(T) = \left[[\eta^{[1]} \eta^{[1]}]^{[T]} [\eta^{[1]} \eta^{[1]}]^{[T]} \right]^{[0]}$$

which furthermore guarantees isospin conservation. Thus it defines the operator P_0. When evaluating this expression, we interpret the products of isospin wave functions in each term as unit weights and we obtain

$$F(T)=\hat{T},$$

where $T=0,1,2$.

This value for the isospin contribution ignores the symmetrization between the pions which involves the interplay of the orbital and isospin quantum numbers. Symmetrization will be achieved later on when evaluating the expectation values of the products of Fock space operators appearing in the full many-body matrix elements discussed in Chapter 7.

The space part is calculated according to fig. 6.14 and we finally get the complete interaction operator written as

$$H_{\pi}=\sum_{ijkl}\sum_{LT}\left[\pi_i\pi_j|H_{\pi}|\pi_k\pi_l,\ LT\right]:\left[\left[\theta_i\theta_j\right]^{[LT]}\theta_k\theta_l\right]^{[00]}:,$$

where θ_i are the linear combinations of Fock operators defined in eq. (6.5). We may omit writing the coupling of $\theta_k\theta_l$ to L, T since it is implied by the overall coupling to a scalar form. With the invariant elementary matrix elements given by

$$\left[\pi_i\pi_j|H_{\pi}|\pi_k\pi_l,\ LT\right]=g_{\pi}F(T)\hat{L}Q^{L}_{l_il_j}Q^{L}_{l_kl_l}\int r^2\,\mathrm{d}r\,g_{\nu_il_i}(r)g_{\nu_jl_j}(r)g_{\nu_kl_k}(r)g_{\nu_ll_l}(r).\tag{6.6}$$

The coefficients Q^L_{ab} for the reduction of a product of two spherical harmonics are defined in sect. 4.3.1, fig. 4.6. Note that here, owing to the associativity of the invariant triple product we have

$$\left[\pi_i\pi_j|H_{\pi}|\pi_k\pi_l\right]=\left[\pi_i\pi_j\pi_k|H_{\pi}|\pi_l\right]=\left[\pi_i\pi_j\pi_k\pi_l|H_{\pi}|\right],$$

where the indices are kept in the indicated order, which is not necessarily the

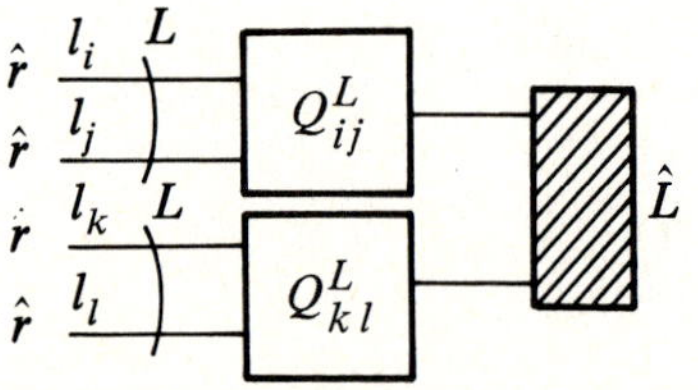

Fig. 6.14.

lexicographic order. Also note that in these expressions and in the invariant matrix elements (6.6) the pions are in general not properly symmetrized. It is only the complete vertex operator H_π which has the proper symmetries.

The sum in H_π is carried independently over all state indices i. Since the operators Z_i commute in the normal product and therefore can be rearranged to the desired order before doing the recoupling of fig. 6.13, the invariant matrix element is the same for all possible permutations of the state indices i, j, k, l, for a particular process. Hence we may account for this by replacing the general normal form: $\theta\theta\theta\theta$: by the product in which not only the annihilation operators are on the right but also individually the creation and annihilation operators each are written in the chosen lexicographic order. The product of operators thus written will be called "the well-ordered product". To account for the correct total number of terms contributing to a given well-ordered product we have to supply a multiplicity factor m, which we shall call "operator multiplicity". This factor is given by

$$\mathrm{m} = \frac{4!}{N_i!N_j!N_k!\ldots},$$

where the N_i are the numbers of fully equivalent operators, i.e. having the same character (creation or annihilation) and the same quantum numbers π_i. For example, for the term: $a_{\pi_1}a_{\pi_2}\tilde{a}^2_{\pi_1}$: we have $N_1 = 1$, $N_2 = 1$, $N_3 = 2$. Thus the terms of the interaction H_π will be written as well-ordered products multiplied by the appropriate operator multiplicities m. For example, a term with four particle creators, if they are in the same state π_1 is

$$[\pi_1\pi_1\pi_1\pi_1|H_\pi|][[\tilde{a}_1\tilde{a}_1]\tilde{a}_1\tilde{a}_1],$$

while if they are in three different states, $\pi_1\pi_2\pi_3\pi_3$, it is

$$12[\pi_1\pi_2\pi_3\pi_3|H_\pi|][[\tilde{a}_1\tilde{a}_2]\tilde{a}_3\tilde{a}_3],$$

or in four different states, $\pi_1\pi_2\pi_3\pi_4$, it is

$$24[\pi_1\pi_2\pi_3\pi_4|H_\pi|][[\tilde{a}_1\tilde{a}_2]\tilde{a}_3\tilde{a}_4].$$

Here, the indices are always taken in the lexicographic order. Likewise a process with two particles created in a same state π_1 and two annihilated in a same state, both if $\pi_2 = \pi_1$ or $\pi_2 \neq \pi_1$ is given by the contribution,

$$6[\pi_1\pi_1|H_\pi|\pi_2\pi_2][[\tilde{a}_1\tilde{a}_1]a_2a_2],$$

while for $\pi_1 \neq \pi_2$ we have

$$24[\pi_1\pi_2|H_\pi|\pi_1\pi_2][[\tilde{a}_1\tilde{a}_2]a_1a_2].$$

The question of multiplicities will be discussed again in full generality in sect. 7.2.7 in the context of the complete many-body matrix elements. The overall multiplicity M given in that section contains the operator multiplicity m defined here.

6.7. THE RHO-PION INTERACTION

The hamiltonian of the rho-pion interaction is of the form, eq. (2.28),

$$H_{\rho\pi} = -ig_{\rho\pi}\int \mathrm{d}^3r :\phi_\mu\left(\varphi\overleftrightarrow{\partial}_\mu \tau\varphi\right):,$$

with

$$f\overleftrightarrow{\partial}_\mu g = f(\partial_\mu g) - (\partial_\mu f)g.$$

The various processes associated with the creation of the ρ are presented in fig. 6.15. Note that the geometry is the same in all these processes owing to the associativity of the triple product. The only difference arises in the time derivative operation which yields different signs for the creation part or the annihilation part of the pion field.

The geometry of the isospin part is similar to the one shown in fig. 6.3, with $t = 1$ instead of the $t = \frac{1}{2}$ of the nucleon. Since it so happens that, from (4.15) and (4.14) respectively,

$$\left[1|\tau^{[1]}|1\right] = \left[\tfrac{1}{2}|\tau^{[1]}|\tfrac{1}{2}\right],$$

the numerical value of the isospin factor is again $-i\sqrt{6}$.

The remaining part separates into a space-like term ($\mu = 1, 2, 3$) and a time-like term ($\mu = 4$). The calculation of the space-like term makes use of the usual scalar product relation, eq. (4.35),

$$\hat{1}\left[e^{[1]}\,\nabla^{[1]}\right]^{[0]}\cdot\left[A^{[J]}Y_L^{[J]}\right]^{[0]} = \left[A^{[J]}\,\nabla^{[1]}\hat{r}^{[L]}\right]^{[0]}.$$

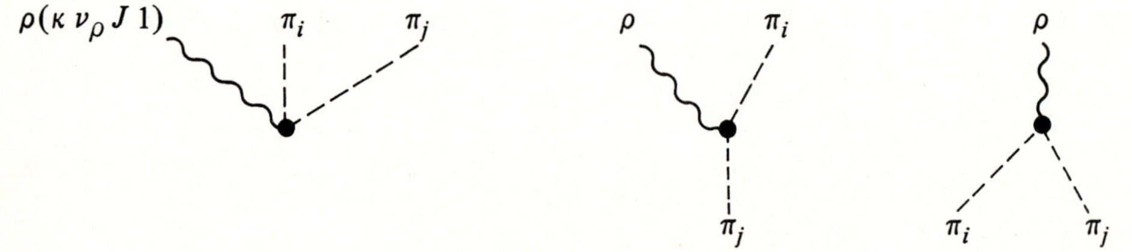

Fig. 6.15.

The geometry is shown in figure 6.16 and it yields for the two terms of the interaction, including the isospin factor

$$K(\rho, \pi_i, \pi_j, n) = -g_{\rho\pi}\frac{\sqrt{6}}{2}$$

$$\times \sum_{\Lambda} \frac{1}{\hat{\Lambda}} \left(\begin{bmatrix} 1 & L & J \\ l_j & l_i & J \\ \Lambda & \Lambda & 0 \end{bmatrix} \alpha_{\Lambda l_j} [\Lambda|L|l_i] \int r^2 \mathrm{d}r\, \mathcal{W}_{\kappa\nu_\rho Jn}(r) g_{\nu_i l_i}(r) k_{\nu_j l_j \Lambda}(r) \right.$$

$$\left. - \begin{bmatrix} 1 & L & J \\ l_i & l_j & J \\ \Lambda & \Lambda & 0 \end{bmatrix} \alpha_{\Lambda l_i} [\Lambda|L|l_j] \int r^2 \mathrm{d}r\, \mathcal{W}_{\kappa\nu_\rho Jn}(r) k_{\nu_i l_i \Lambda}(r) g_{\nu_j l_j}(r) \right), \tag{6.7}$$

where the index ρ denotes all the quantum numbers for the ρ-meson, $\kappa\nu_\rho J$ and $t = 1$. The functions $\mathcal{W}$ are defined in eq. (5.15). As usual $L = J + n$. The functions $k_{\nu l\lambda}(r)$ have been given at the end of sect. 4.2.4.

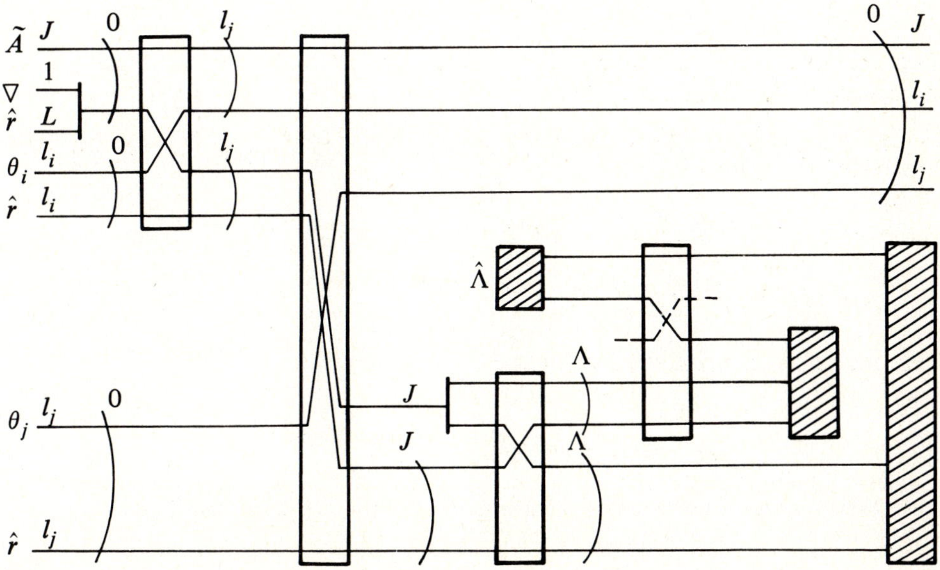

Fig. 6.16.

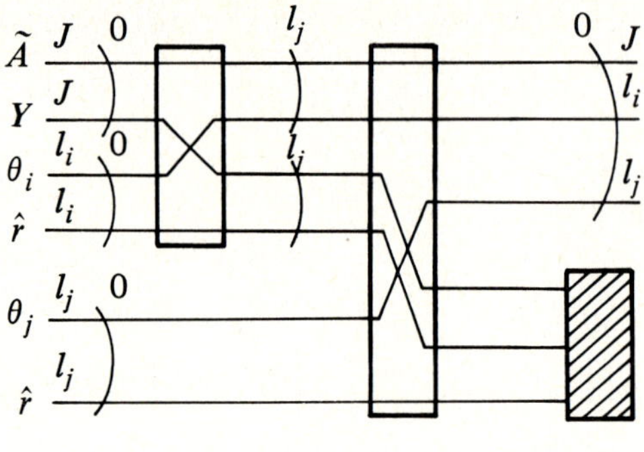

Fig. 6.17.

The geometry of the time-like part is shown in fig. 6.17. In this matrix element the symmetrized time derivative acting on the pion fields yields factors of the type

$$\mathrm{e}^{\pm iE_i t}\overleftrightarrow{\partial}_4 \mathrm{e}^{\pm iE_j t} = (\pm E_j \mp E_i)\mathrm{e}^{it(\pm E_i \pm E_j)},$$

which distinguish between the various processes associated with the vertex. The pion energy factors E_i or E_j transform the radial function $g_{\nu_i l_i}(r)$, eq. (3.4), into the function $h_{\nu_i l_i}(r)$, eq. (3.6). Thus the result of fig. 6.17 yields, including the isospin factor, for the time-like part, which has $\kappa = \mathcal{L}$, $n = 0$,

$$K_4^P(\rho, \pi_i, \pi_j) = -g_{\rho\pi}\frac{\sqrt{6}}{2}[J|l_i|l_j]\int r^2 \mathrm{d}r\, \mathcal{X}_{\nu_\rho J}(r)\Gamma^P_{\nu_i l_i \nu_j l_j}, \tag{6.8}$$

where the functions $\mathcal{X}$ are defined in (5.18) and where the superscript P denotes the process and Γ^P is defined as

$$\Gamma^1_{\nu_i l_i \nu_j l_j}(r) = -g_{\nu_i l_i}(r)h_{\nu_j l_j}(r) + h_{\nu_i l_i}(r)g_{\nu_j l_j}(r) \qquad \text{for pion pair annihilation},$$

$$\Gamma^2_{\nu_i l_i \nu_j l_j}(r) = -g_{\nu_i l_i}(r)h_{\nu_j l_j}(r) - h_{\nu_i l_i}(r)g_{\nu_j l_j}(r) \qquad \text{for pion scattering},$$

$$\Gamma^3_{\nu_i l_i \nu_j l_j}(r) = g_{\nu_i l_i}(r)h_{\nu_j l_j}(r) - h_{\nu_i l_i}(r)g_{\nu_j l_j}(r) \qquad \text{for pion pair production}.$$

Then, the rho-pion interaction is for the rho creation part

$$H_{\rho\pi} = \sum [\rho|H_{\rho\pi}|\pi_i\pi_j]^P : \left[\tilde{A}^{[J1]}_{\kappa\nu_\rho}\theta^{[l_i 1]}_{\nu_i}\theta^{[l_j 1]}_{\nu_j}\right]^{[00]} :,$$

where the invariant matrix elements are given for the different processes by

$$[\rho|H_{\rho\pi}|\pi_i\pi_j]^P = \sum_n C_\kappa(n, J)K(\rho, \pi_i, \pi_j, n) + K_4^P(\rho, \pi_i, \pi_j)\delta_{\kappa\mathcal{L}}. \tag{6.9}$$

In the expression (6.9) the sum over n is done as for eq. (6.4) with the coefficients C_κ, eq. (5.27). Again the operator multiplicity factor m can be introduced in order that the sum over the two pion indices be limited to well-ordered products as discussed in sect. 6.6. In that case $\mathsf{m} = 2$ for all cases except if $\pi_i = \pi_j$ in a pair creation or pair annihilation process, where $\mathsf{m} = 1$.

In Chapter 7 we also will use the following notation: for pion pair annihilation

$$[\rho|H_{\rho\pi}|\pi_i\pi_j] = [\rho|H_{\rho\pi}|\pi_i\pi_j]^{P=1},$$

for pion scattering

$$[\rho\pi_i|H_{\rho\pi}|\pi_j] = [\rho|H_{\rho\pi}|\pi_i\pi_j]^{P=2},$$

and for pion pair creation

$$[\rho\pi_i\pi_j|H_{\rho\pi}|] = [\rho|H_{\rho\pi}|\pi_i\pi_j]^{P=3}.$$

Again note that the invariant matrix elements (6.7) and (6.8) which involve a ρ vector meson are real and symmetric (space and time parts) under time reversal, as in sect. 6.5 for the ρNN′ case. In the case of the time part, K_4^P, this property results from the sign changes arising from both the pion field time derivative and the behavior of the fourth component of the vector field under time reversal.

6.8. THE OMEGA-PION INTERACTION

The omega-pion interaction hamiltonian is of the form, eq. (2.29),

$$H_{\omega\pi} = ig_{\omega\pi}\left(\frac{1}{2M_\omega}\right)^2 \int \mathrm{d}^3r :(\varepsilon_{\alpha\beta\gamma}\phi_\alpha\partial_\beta\partial_\gamma)(\varphi\varphi\varphi):,$$

where the indices α, β, γ are the coordinates x, y, z. The various processes associated with the creation of the ω are represented in fig. 6.18. With the self-conjugation property of the pion wave functions and the definition of our coupling scheme they are all described by the same matrix element.

We recall, from sect. 2.4, that the antisymmetric product in the Minkowski space is restricted to $\alpha, \beta, \gamma \neq 4$. Since this way the operator is antisymmetric in space, it also must be antisymmetric in isospin space, with charge conservation. The three states of the pions must be distinct both in space (owing to the $A \times B$ structure of the derivatives) and in isospin space (recalling that the 3-j coefficient $\begin{pmatrix} 1 & 1 & 1 \\ 0 & 0 & 0 \end{pmatrix}$ vanishes).

Thus the interaction $H_{\omega\pi}$ can be written

$$H_{\omega\pi} = ig_{\omega\pi}\left(\frac{1}{2M_\omega}\right)^2 \int \mathrm{d}^3r : \boldsymbol{\phi}\cdot(\nabla \times \nabla)P_0(\varphi\varphi\varphi):,$$

where P_0 denotes that the pion fields are coupled into a scalar quantity in isospin

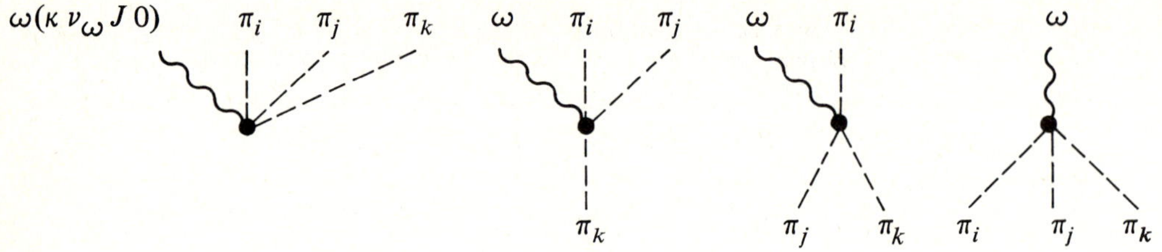

Fig. 6.18.

space for charge conservation. The differential operators are distributed amongst the fields according to the totally antisymmetric form (2.30) in the pion state indices 1, 2, 3

$$\nabla \times \nabla\varphi_1\varphi_2\varphi_3 = (\nabla\varphi_1)\times(\nabla\varphi_2)\varphi_3 - (\nabla\varphi_1)\varphi_2\times(\nabla\varphi_3) + \varphi_1(\nabla\varphi_2)\times(\nabla\varphi_3)\,. \tag{6.10}$$

For the treatment of the isospin, the operator P_0 projects on the $T=0$ state and can be written as the totally antisymmetric invariant, eqs. (4.9)–(4.10),

$$P_0 = \sqrt{6}\left[\eta_1^{[1]}\eta_2^{[1]}\eta_3^{[1]}\right]^{[0]}.$$

The evaluation of the isospin part yields

$$P_0\hat{1}^3\left[\theta_1^{[1]}\eta_1^{[1]}\right]^{[0]}\left[\theta_2^{[1]}\eta_2^{[1]}\right]^{[0]}\left[\theta_3^{[1]}\eta_3^{[1]}\right]^{[0]} = \sqrt{6}\left[\theta_1^{[1]}\theta_2^{[1]}\theta_3^{[1]}\right]^{[0]},$$

where we have introduced the Fock operators θ_i defined in eq. (6.5).

For the space part, substituting the expansions of the fields, before distribution of the gradients, the interaction is of the form

$$\sum \tfrac{1}{2}\sqrt{\tfrac{1}{2}}\,\hat{J}\hat{l}_i\hat{l}_j\hat{l}_k \int \mathrm{d}^3r\, \mathcal{W}_{\kappa\nu Jn}(r)$$

$$\times\left[\tilde{A}_{\kappa\nu}^{[J]}\left[\nabla^{[1]}\nabla^{[1]}\right]^{[1]}\hat{r}^{[J+n]}\right]^{[0]}\left[\theta_{\nu_i}^{[l_i]}g_{\nu_i l_i}(r)\hat{r}^{[l_i]}\right]^{[0]}$$

$$\times\left[\theta_{\nu_j}^{[l_j]}g_{\nu_j l_j}(r)\hat{r}^{[l_j]}\right]^{[0]}\left[\theta_{\nu_k}^{[l_k]}g_{\nu_k l_k}(r)\hat{r}^{[l_k]}\right]^{[0]}.$$

Here the two gradient operators replace the unit vector in the vector spherical harmonics according to eq. (4.35). These two differential operators of course act only on the pion fields. In order to perform the calculation the pion fields are directly inserted next to the gradient operators acting on them, as shown in fig. 6.19. We

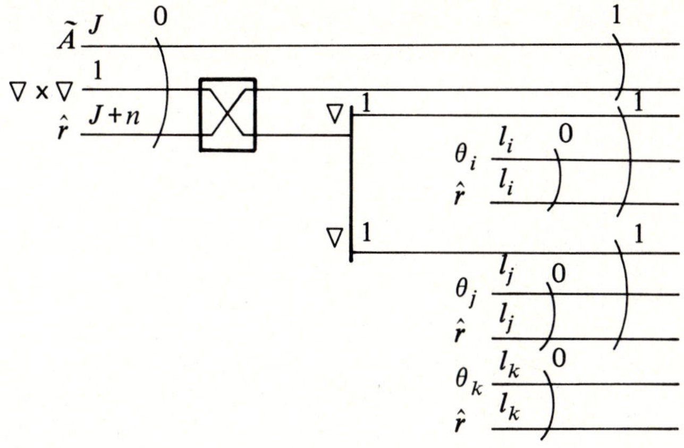

Fig. 6.19.

introduce the sub-graphs $B_{\nu l \lambda}$ shown in fig. 6.20,

$$B_{\nu_i l_i \lambda_i} = \frac{\alpha_{\lambda_i l_i}}{\hat{l}_i \hat{1}} k_{\nu_i l_i \lambda_i}(r).$$

The continuation of the calculation is given in fig. 6.21 and including the contribution from the pion isospins we have

$$K(\omega, \pi_i, \pi_j, \pi_k, R, n) = -i \sum_{\Lambda \lambda_i \lambda_j} (-)^n \tfrac{1}{2} \alpha_{\lambda_i l_i} \alpha_{\lambda_j l_j} \hat{l}_k$$

$$\times \begin{bmatrix} l_i & \lambda_i & 1 \\ l_j & \lambda_j & 1 \\ R & \Lambda & 1 \end{bmatrix} \begin{bmatrix} R & \Lambda & 1 \\ l_k & l_k & 0 \\ J & J+n & 1 \end{bmatrix} Q^{\Lambda}_{\lambda_i \lambda_j} Q^{J+n}_{\Lambda l_k}$$

$$\times \int r^2 \mathrm{d}r\, \mathcal{W}_{\kappa \nu J n}(r) k_{\nu_i l_i \lambda_i}(r) k_{\nu_j l_j \lambda_j}(r) g_{\nu_k l_k}(r).$$

We now write the complete matrix element distributing the action of the two

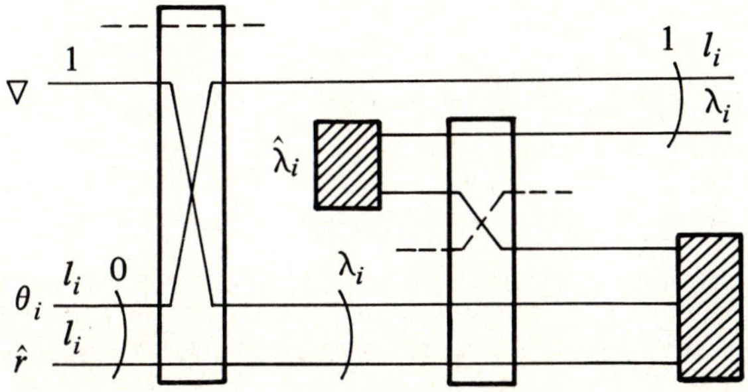

Fig. 6.20.

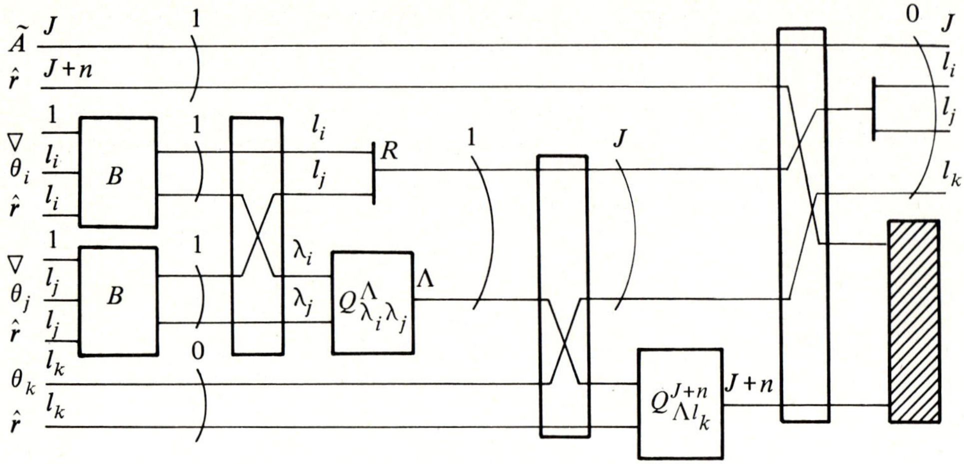

Fig. 6.21.

gradients according to eq. (6.10) over the three pion fields, two of the pions being coupled to L,

$$[\omega|H_{\omega\pi}|\pi_i\pi_j\pi_k] = g_{\omega\pi}\left(\frac{1}{2M_\omega}\right)^2 \sum_n C_\kappa(n, J)$$

$$\times \sum_R \left(K(\omega, \pi_i, \pi_j, \pi_k, R, n)\delta_{RL} \right.$$

$$- \begin{bmatrix} l_i & l_k & R \\ l_j & 0 & l_j \\ L & l_k & J \end{bmatrix} K(\omega, \pi_i, \pi_k, \pi_j, R, n)$$

$$\left. + (-)^{l_i+l_j-L} \begin{bmatrix} l_j & l_k & R \\ l_i & 0 & l_i \\ L & l_k & J \end{bmatrix} K(\omega, \pi_j, \pi_k, \pi_i, R, n) \right). \tag{6.11}$$

In the sum, for the longitudinal multipolarity $\kappa = \mathfrak{L}$ the term associated with $n = 0$ is absent since there is no time-like contribution. Furthermore we have as in sect. 6.6:

$$[\omega|H_{\omega\pi}|\pi_i\pi_j\pi_k] = [\omega\pi_i|H_{\omega\pi}|\pi_j\pi_k] = [\omega\pi_i\pi_j|H_{\omega\pi}|\pi_k] = [\omega\pi_i\pi_j\pi_k|H_{\omega\pi}|].$$

The ω-pion interaction hamiltonian is thus for the vector meson emission part

$$H_{\omega\pi} = \sum [\omega|H_{\omega\pi}|\pi_i\pi_j\pi_k] : \left[\tilde{A}^{[J]}_{\kappa\nu_\omega} \left[\theta^{[l_i 1]}_{\nu_i} \theta^{[l_j 1]}_{\nu_j} \right]^{[L1]} \theta^{[l_k 1]}_{\nu_k} \right]^{[00]} : .$$

When writing this expression in terms of well-ordered operator products the operator multiplicity factor m, sect. 6.6, is always present since the three pion states i, j, k, are always distinct. The complete operator $H_{\omega\pi}$ above is of course symmetric in all pion indices.

Note that the invariant matrix element (6.11) which involves an omega vector meson field is imaginary and antisymmetric under time reversal, as in sect. 6.5 for the $\omega NN'$ vertex. Reality of the final matrix can be obtained by a proper phase choice of the basis as given in Chapter 7.

6.9. THE SIGMA-PION INTERACTION

The sigma meson with $s=0$, $t=0$, couples to two pions. The interaction hamiltonian is simply

$$H_{\sigma\pi}=g_{\sigma\pi}(2M_\sigma)\int \mathrm{d}^3r : \varphi^{(\sigma)}\varphi\varphi : .$$

All possible processes have the same matrix element since the fields are real and self-conjugate, and they form an invariant triple product. Writing for the pions as in eq. (6.5)

$$\theta_{\nu_i}^{[l_i1]}=\tilde{a}_{\nu_i}^{[l_i1]}+a_{\nu_i}^{[l_i1]}$$

and for the sigma

$$\theta_{\nu}^{[l0]}=\tilde{a}_{\nu}^{[l0]}+a_{\nu}^{[l0]}$$

the complete interaction is

$$H_{\sigma\pi}=\sum\left[\sigma|H_{\sigma\pi}|\pi_i\pi_j\right] : \left[\theta_{\nu}^{[l0]}\theta_{\nu_i}^{[l_i1]}\theta_{\nu_j}^{[l_j1]}\right] : .$$

Here the invariant matrix element is for all processes

$$\left[\sigma|H_{\sigma\pi}|\pi_i\pi_j\right]=g_{\sigma\pi}(2M_\sigma)\left(\sqrt{\tfrac{1}{2}}\right)^3 \hat{1}\left[l|l_i|l_j\right]\int r^2\,\mathrm{d}r\, g_{\nu l}(r)\, g_{\nu_i l_i}(r)\, g_{\nu_j l_j}(r).$$

In this expression the function $g_{\nu l}(r)$ is given in eq. (3.4), where of course the mass is the sigma mass.

6.10. THE PION-NUCLEON-DELTA INTERACTION

The pion-nucleon-delta interaction is of the form, eq. (2.69), for the delta annihilation part,

$$H_{\pi N\Delta}=-g_{\pi N\Delta}\sqrt{\frac{1}{2M_\Delta}}\int \mathrm{d}^3r : (\partial_\mu\varphi)\bar{\psi}\gamma_5 \boldsymbol{C}^+ D_\mu^+ \Phi : .$$

Here the operator $\boldsymbol{C}$ changes the isospin from $\frac{3}{2}$ to $\frac{1}{2}$ and conserves the charge. The processes associated with this interaction are represented in fig. 6.22 which also specifies the notation. The spin-$\frac{3}{2}$, isospin-$\frac{3}{2}$ field is given in eq. (5.34). The structure

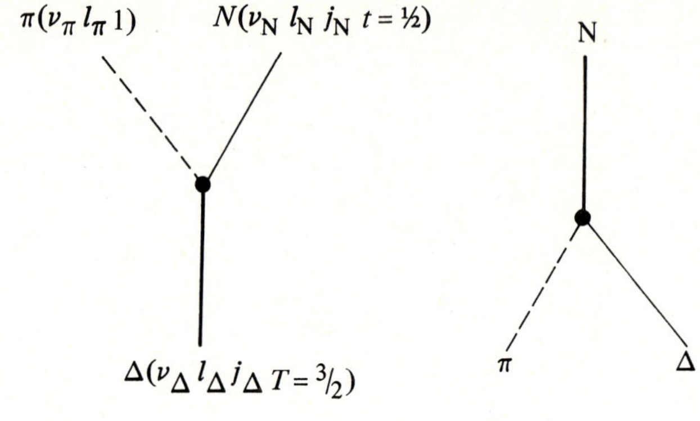

Fig. 6.22.

of the matrix element is the following. Writing

$$\psi = \begin{pmatrix} L \\ -S \end{pmatrix}, \qquad \Phi = \begin{pmatrix} a \\ 0 \\ b \\ 0 \end{pmatrix},$$

and from (2.70)

$$D_k^+ = \begin{pmatrix} 0 & G_k^+ & -F_k^+ & 0 \\ F_k^+ & 0 & 0 & -G_k^+ \end{pmatrix},$$

we have

$$\bar{\psi}\gamma_5 = (-S^+ - L^+),$$

$$-\bar{\psi}\gamma_5 D_k^+ \Phi = L^+ F_k^+ a - S^+ F_k^+ b.$$

There is no contribution from D_4, eq. (2.70), in our transverse gauge. The recoupling graph is given in fig. 6.23. Writing $t = \frac{1}{2}$, $T = \frac{3}{2}$ and $s = \frac{1}{2}$, $S = \frac{3}{2}$, the invariant matrix element is, with the spin changing matrices of section 5.2.4,

$$[\pi \mathrm{N}|H_{\pi\mathrm{N}\Delta}|\Delta] = g_{\pi\mathrm{N}\Delta}\frac{1}{\sqrt{8}}\frac{1}{2M_\Delta}[t|\tilde{C}^{[1]}|T][s|\tilde{F}^{[1]}|S]\sum_{\Lambda}\frac{1}{\hat{1}}\frac{1}{\hat{\Lambda}}\alpha_{\Lambda l_\pi}$$

$$\times\left(\begin{bmatrix} s & l_{\mathrm{N}} & j_{\mathrm{N}} \\ S & l_\Delta & j_\Delta \\ 1 & \Lambda & l_\pi \end{bmatrix}[\Lambda|l_{\mathrm{N}}|l_\Delta]\int r^2\,\mathrm{d}r\, k_{\nu_\pi l_\pi \Lambda}(r)\, u_{\nu_{\mathrm{N}} l_{\mathrm{N}}}(r)\, u_{\nu_\Delta l_\Delta}(r)\right.$$

$$-\sum_{\lambda_\Delta}\beta_{l_\Delta \lambda_\Delta j_\Delta}\begin{bmatrix} s & \lambda_{\mathrm{N}} & j_{\mathrm{N}} \\ S & \lambda_\Delta & j_\Delta \\ 1 & \Lambda & l_\pi \end{bmatrix}[\Lambda|\lambda_{\mathrm{N}}|\lambda_\Delta]$$

$$\left.\times\int r^2\,\mathrm{d}r\, k_{\nu_\pi l_\pi \Lambda}(r)\, v_{\nu_{\mathrm{N}} l_{\mathrm{N}} \lambda_{\mathrm{N}}}(r)\, v_{\nu_\Delta l_\Delta \lambda_\Delta}(r)\right).$$

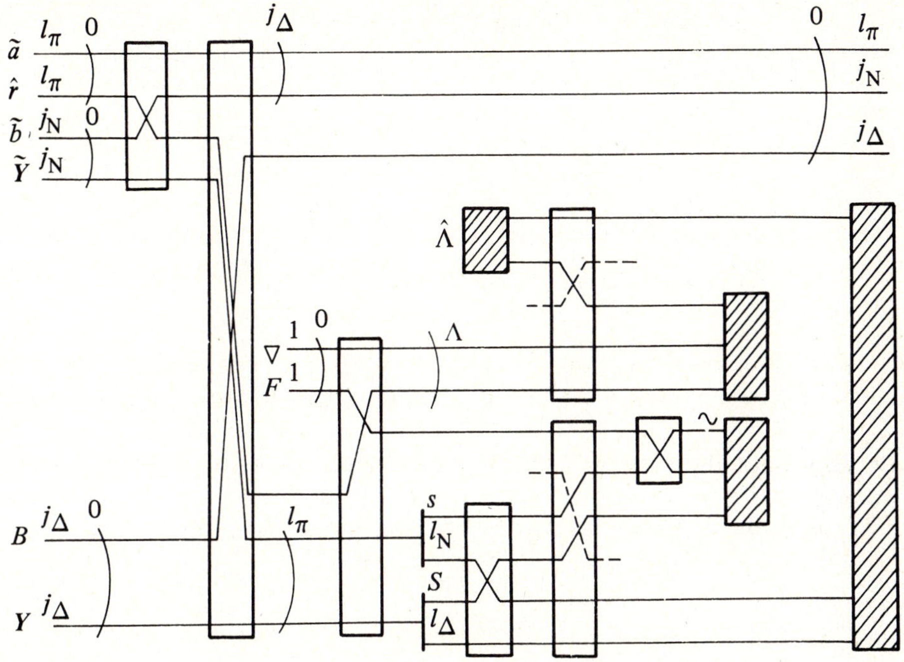

Fig. 6.23.

The lower component amplitudes $\beta_{l_\Delta \lambda_\Delta j_\Delta}$ are defined in sect. 5.2.4. Furthermore in this expression the invariant matrix element of the isospin changing operator $\boldsymbol{C}$ is defined and evaluated according to eq. (4.11)

$$\langle \tfrac{1}{2}\tfrac{1}{2}|\tilde{C}^{[1]}_{-1}|\tfrac{3}{2}\tfrac{3}{2}\rangle = i = (-)^{\frac{1}{2}+\frac{1}{2}}(-)^{\frac{1}{2}-1+\frac{3}{2}}\begin{pmatrix} \frac{1}{2} & 1 & \frac{3}{2} \\ -\frac{1}{2} & -1 & \frac{3}{2} \end{pmatrix}\left[\tfrac{1}{2}|\tilde{C}^{[1]}|\tfrac{3}{2}\right].$$

Hence

$$\left[\tfrac{1}{2}|\tilde{C}^{[1]}|\tfrac{3}{2}\right] = 2i.$$

Likewise from the expression (2.71) and eq. (4.11), we have for the spin changing operator $\boldsymbol{F}$

$$\left[\tfrac{1}{2}|\tilde{F}^{[1]}|\tfrac{3}{2}\right] = 2i.$$

The $\pi N\Delta$ interaction thus is, for the $\Delta \to N$ and $N \to \Delta$ parts, for both pion absorption and emission with θ_i defined by (6.5)

$$H_{\pi N\Delta} = \sum [\pi N|H_{\pi N\Delta}|\Delta]\Big(:\Big[\theta^{[l_\pi 1]}_{\nu_\pi}\tilde{b}^{[j_N t]}_{\nu_N l_N}B^{[j_\Delta T]}_{\nu_\Delta l_\Delta}\Big]^{[00]}:+:\Big[\tilde{B}^{[j_\Delta T]}_{\nu_\Delta l_\Delta}\theta^{[l_\pi 1]}_{\nu_\pi}b^{[j_N t]}_{\nu_N l_N}\Big]^{[00]}:\Big)$$

owing to the reality of the invariant matrix elements.

6.11. THE PION-DELTA INTERACTION

The pion-delta interaction given in sect. 2.6.2 has a pseudo-scalar term

$$H_{\pi\Delta}^{\mathrm{PS}} = ig_{\pi\Delta}^{\mathrm{PS}}(2M_\Delta)\int \mathrm{d}^3r : \bar{\Phi}\gamma_5 \boldsymbol{C}\Phi\varphi : ,$$

and a pseudo-vector term

$$H_{\pi\Delta}^{\mathrm{PV}} = -ig_{\pi\Delta}^{\mathrm{PV}}\frac{1}{2M_\Delta}\int \mathrm{d}^3r : \left(\bar{\Phi}\gamma_\mu\gamma_5\boldsymbol{C}\Theta + \bar{\Theta}\gamma_\mu\gamma_5\boldsymbol{C}\Phi\right)\partial_\mu\varphi : .$$

Here, as in the previous section, the isospin operator is denoted $\boldsymbol{C}$ so as not to confuse it with the notation τ_k used for the blocks of the γ-matrices introduced in eq. (2.53).

The elementary vertex is given in fig. 6.24. We calculate only the pion creation interaction. In order to evaluate the action of the γ-matrices we write for the "final" spin-$\frac{3}{2}$ field, in the transverse gauge of sect. 2.6.2, the form (2.65)

$$\Phi^+ = \begin{pmatrix} a \\ 0 \\ b \\ 0 \end{pmatrix}^+ ,$$

and for the "initial" field

$$\Phi = \begin{pmatrix} f \\ 0 \\ g \\ 0 \end{pmatrix} .$$

6.11.1. *The pseudo-scalar coupling*

We begin with the evaluation of the pseudo-scalar term. Here the integrand is of the form, eq. (2.52)

$$i\bar{\Phi}\gamma_5\boldsymbol{C}\Phi\varphi = i\begin{pmatrix} a \\ 0 \\ b \\ 0 \end{pmatrix}^+ \gamma_4\gamma_5\boldsymbol{C}\begin{pmatrix} f \\ 0 \\ g \\ 0 \end{pmatrix}\varphi = i\begin{pmatrix} a \\ 0 \\ b \\ 0 \end{pmatrix}^+ \begin{pmatrix} 0 & -K \\ K & 0 \end{pmatrix}\boldsymbol{C}\begin{pmatrix} f \\ 0 \\ g \\ 0 \end{pmatrix}\varphi$$

$$= -i\left(\begin{pmatrix} a \\ 0 \end{pmatrix}^+ \boldsymbol{C}\begin{pmatrix} g \\ 0 \end{pmatrix} - \begin{pmatrix} b \\ 0 \end{pmatrix}^+ \boldsymbol{C}\begin{pmatrix} f \\ 0 \end{pmatrix}\right)\varphi ,$$

with the definitions of sect. 2.6.1.

The geometry for the matrix element is that of the pseudo-scalar pion-nucleon interaction, sect. 6.3.1. The evaluation of the isospin part is given by fig. 6.3 and yields together with the value (4.15) for the invariant $[\frac{3}{2}|C^{[1]}|\frac{3}{2}]$ an overall factor

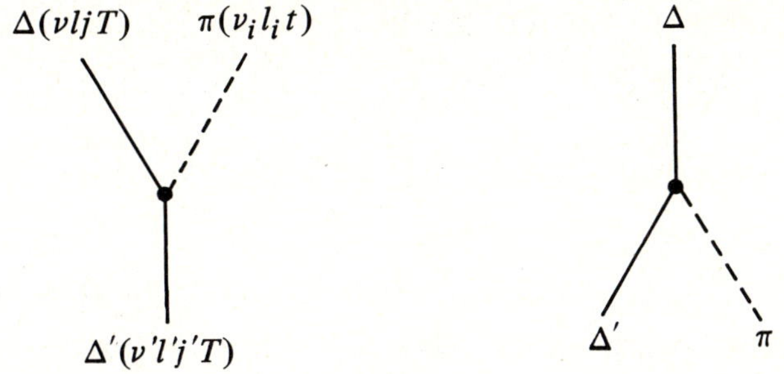

Fig. 6.24.

$-i\sqrt{15}$. In orbital space the evaluation of fig. 6.4 with the definitions of the delta field (5.34) yields including the isospin part, using again the notation $S = \frac{3}{2}$,

$$\left[\pi_i \Delta | H^{\mathrm{PS}}_{\pi\Delta} | \Delta'\right] = -g^{\mathrm{PS}}_{\pi\Delta}\sqrt{\frac{15}{2}}$$

$$\times \sum_{\lambda\lambda'} \left(\begin{bmatrix} S & l & j \\ S & \lambda' & j' \\ 0 & l_i & l_i \end{bmatrix} [l_i \| l \| \lambda'] \beta_{l'\lambda' j'} \int r^2 \mathrm{d}r\, u_{\nu l}(r) g_{\nu_i l_i}(r) v_{\nu' l' \lambda'}(r) \right.$$

$$\left. - \begin{bmatrix} S & \lambda & j \\ S & l' & j' \\ 0 & l_i & l_i \end{bmatrix} [l_i \| \lambda \| l'] \beta_{l\lambda j} \int r^2 \mathrm{d}r\, v_{\nu l \lambda}(r) g_{\nu_i l_i}(r) u_{\nu' l'}(r) \right). \tag{6.12}$$

6.11.2. *The pseudo-vector coupling*

We now turn to the pseudo-vector interaction. Its integrand is of the form

$$-i\left(\bar{\Phi}\gamma_\mu\gamma_5 C\Theta + \bar{\Theta}\gamma_\mu\gamma_5 C\Phi\right)\partial_\mu\varphi = -i\left(\Phi^+\gamma_4\gamma_\mu\gamma_5\gamma_\nu\vec{\partial}_\nu C\Phi - \Phi^+\gamma_4\overleftarrow{\partial}_\nu\gamma_\nu\gamma_\mu\gamma_5 C\Phi\right)\partial_\mu\varphi,$$

where use is made of

$$\Theta = \gamma_\nu \vec{\partial}_\nu \Phi,$$

$$\bar{\Theta} = -\bar{\Phi}\overleftarrow{\partial}_\nu \gamma_\nu,$$

from sect. 2.6.1. Substituting

$$\gamma_4\gamma_5 = -\gamma_5\gamma_4 = \begin{pmatrix} 0 & -K \\ K & 0 \end{pmatrix},$$

$$\gamma_4\gamma_5\gamma_4 = \begin{pmatrix} 0 & I \\ I & 0 \end{pmatrix},$$

the integrand writes

$$-i\begin{pmatrix} b \\ 0 \\ -a \\ 0 \end{pmatrix}^{+} \left(\boldsymbol{\gamma}\cdot(\nabla\varphi)\boldsymbol{\gamma}\cdot\vec{\nabla} + \overleftarrow{\nabla}\cdot\boldsymbol{\gamma}\boldsymbol{\gamma}\cdot(\nabla\varphi) + \boldsymbol{\gamma}\cdot(\nabla\varphi)\gamma_4\vec{\partial}_4 + \overleftarrow{\partial}_4\gamma_4\boldsymbol{\gamma}\cdot(\nabla\varphi)\right.$$

$$\left. + \gamma_4(\partial_4\varphi)\boldsymbol{\gamma}\cdot\vec{\nabla} + \overleftarrow{\nabla}\cdot\boldsymbol{\gamma}\gamma_4(\partial_4\varphi) + \gamma_4(\partial_4\varphi)\gamma_4\vec{\partial}_4 + \overleftarrow{\partial}_4\gamma_4\gamma_4(\partial_4\varphi)\right)\boldsymbol{C}\begin{pmatrix} f \\ 0 \\ g \\ 0 \end{pmatrix}. \tag{6.13}$$

As can be seen from this expression derivative radial functions will be required. To the gradient of the radial function of the pion given at the end of sect. 4.2.4

$$k_{\nu lL}(r) = \sqrt{\frac{2}{\pi}} \int p^2 \mathrm{d}p \frac{1}{\sqrt{E}} p f_{\nu l}(p) j_L(pr),$$

we add the definitions for the gradient of the radial functions of the large and small components of the delta spinors (5.34):

$$U_{\nu l\Lambda}(r) = \sqrt{\frac{2}{\pi}} \int p^2 \mathrm{d}p \sqrt{\frac{E+M}{E}}\, p f_{\nu l}(p) j_\Lambda(pr),$$

$$V_{\nu l\Lambda}(r) = \sqrt{\frac{2}{\pi}} \int p^2 \mathrm{d}p \sqrt{\frac{E-M}{E}}\, p f_{\nu l}(p) j_\Lambda(pr).$$

Likewise to the radial functions of the time-derived pion field, eq. (3.6)

$$h_{\nu l}(r) = \sqrt{\frac{2}{\pi}} \int p^2 \mathrm{d}p \sqrt{E}\, f_{\nu l}(p) j_l(pr),$$

we add for the delta spinors

$$y_{\nu l}(r) = \sqrt{\frac{2}{\pi}} \int p^2 \mathrm{d}p \sqrt{E(E+M)}\, f_{\nu l}(p) j_l(pr),$$

$$z_{\nu l\lambda}(r) = \sqrt{\frac{2}{\pi}} \int p^2 \mathrm{d}p \sqrt{E(E-M)}\, f_{\nu l}(p) j_\lambda(pr).$$

In the evaluation of (6.13), the most complicated are the first two terms. We write out the first term in full detail, with eq. (2.53) and omitting here the isospin operator $\boldsymbol{C}$:

$$M_1 = -i\begin{pmatrix} b\\ 0\\ -a\\ 0\end{pmatrix}^{+} \boldsymbol{\gamma}\cdot(\nabla\varphi)\boldsymbol{\gamma}\cdot\nabla\begin{pmatrix} f\\ 0\\ g\\ 0\end{pmatrix}$$

$$= i\begin{pmatrix} b\\ 0\\ -a\\ 0\end{pmatrix}^{+}\begin{pmatrix} -\hat{\boldsymbol{\tau}} & -\boldsymbol{\sigma}\\ \boldsymbol{\sigma} & \hat{\boldsymbol{\tau}}\end{pmatrix}\cdot(\nabla\varphi)\begin{pmatrix} -\hat{\boldsymbol{\tau}} & -\boldsymbol{\sigma}\\ \boldsymbol{\sigma} & \hat{\boldsymbol{\tau}}\end{pmatrix}\cdot\nabla\begin{pmatrix} f\\ 0\\ g\\ 0\end{pmatrix}$$

$$= i\begin{pmatrix} b\\ 0\\ -a\\ 0\end{pmatrix}^{+}\begin{pmatrix} (\hat{\boldsymbol{\tau}}\cdot(\nabla\varphi)\hat{\boldsymbol{\tau}}\cdot\nabla - \boldsymbol{\sigma}\cdot(\nabla\varphi)\boldsymbol{\sigma}\cdot\nabla) & (\hat{\boldsymbol{\tau}}\cdot(\nabla\varphi)\boldsymbol{\sigma}\cdot\nabla - \boldsymbol{\sigma}\cdot(\nabla\varphi)\hat{\boldsymbol{\tau}}\cdot\nabla)\\ (-\boldsymbol{\sigma}\cdot(\nabla\varphi)\hat{\boldsymbol{\tau}}\cdot\nabla + \hat{\boldsymbol{\tau}}\cdot(\nabla\varphi)\boldsymbol{\sigma}\cdot\nabla) & (-\boldsymbol{\sigma}\cdot(\nabla\varphi)\boldsymbol{\sigma}\cdot\nabla + \hat{\boldsymbol{\tau}}\cdot(\nabla\varphi)\hat{\boldsymbol{\tau}}\cdot\nabla)\end{pmatrix}\begin{pmatrix} f\\ 0\\ g\\ 0\end{pmatrix}$$

$$= i\begin{pmatrix} a\\ 0\end{pmatrix}^{+}[\boldsymbol{\tau}\cdot(\nabla\varphi)\boldsymbol{\tau}\cdot\nabla + \boldsymbol{\sigma}\cdot(\nabla\varphi)\boldsymbol{\sigma}\cdot\nabla]\begin{pmatrix} g\\ 0\end{pmatrix} - i\begin{pmatrix} b\\ 0\end{pmatrix}^{+}[\boldsymbol{\tau}\cdot(\nabla\varphi)\boldsymbol{\tau}\cdot\nabla + \boldsymbol{\sigma}\cdot(\nabla\varphi)\boldsymbol{\sigma}\cdot\nabla]\begin{pmatrix} f\\ 0\end{pmatrix},$$

where we have used from sect. 2.6.1:

$$\hat{\boldsymbol{\tau}}\cdot \boldsymbol{A}\hat{\boldsymbol{\tau}}\cdot \boldsymbol{B} = -\boldsymbol{\tau}\cdot\boldsymbol{A}\boldsymbol{\tau}\cdot\boldsymbol{B}.$$

Owing to the transverse gauge, terms linear in $\boldsymbol{\tau}$ are absent in the final result.

The geometry of the invariant matrix element M_1 is shown in fig. 6.25 for the orbital part. Writing $s=\frac{1}{2}$, $S=\frac{3}{2}$, for the spins of the lower and upper components of the spin-$\frac{3}{2}$ field we obtain, including the isospin contribution $-i\sqrt{15}$

$$M_1 = \sqrt{\frac{15}{8}}\left(\frac{1}{2M_\Delta}\right)^2\frac{\hat{j}'}{\hat{l}_i\hat{1}}(-)^{j+l_i+j'}\sum_{\lambda\lambda'}\sum_{\Lambda L}\alpha_{Ll_i}\frac{1}{\hat{\Lambda}}$$

$$\times\left(\alpha_{\Lambda l'}Q^{\Lambda}_{L\lambda}\left(8\begin{bmatrix} S & \lambda & j\\ 1 & L & l_i\\ s & \Lambda & j'\end{bmatrix}\begin{bmatrix} 1 & 1 & 0\\ S & l' & j'\\ s & \Lambda & j'\end{bmatrix} - 5\begin{bmatrix} S & \lambda & j\\ 1 & L & l_i\\ S & \Lambda & j'\end{bmatrix}\begin{bmatrix} 1 & 1 & 0\\ S & l' & j'\\ S & \Lambda & j'\end{bmatrix}\right)\beta_{l\lambda j}\right.$$

$$\times\int r^2\,\mathrm{d}r\, v_{\nu l\lambda}(r)k_{\nu_i l_i L}(r)U_{\nu' l'\Lambda}(r)$$

$$-\alpha_{\Lambda\lambda'}Q^{\Lambda}_{Ll}\left(8\begin{bmatrix} S & l & j\\ 1 & L & l_i\\ s & \Lambda & j'\end{bmatrix}\begin{bmatrix} 1 & 1 & 0\\ S & \lambda' & j'\\ s & \Lambda & j'\end{bmatrix} - 5\begin{bmatrix} S & l & j\\ 1 & L & l_i\\ S & \Lambda & j'\end{bmatrix}\begin{bmatrix} 1 & 1 & 0\\ S & \lambda' & j'\\ S & \Lambda & j'\end{bmatrix}\right)\beta_{l'\lambda' j'}$$

$$\left.\times\int r^2\,\mathrm{d}r\, u_{\nu l}(r)k_{\nu_i l_i L}(r)V_{\nu' l'\Lambda}(r)\right).$$

Here we have used the quantity $Q^{\Lambda}_{L\lambda}$ defined in sect. 4.3.1 and the relations from

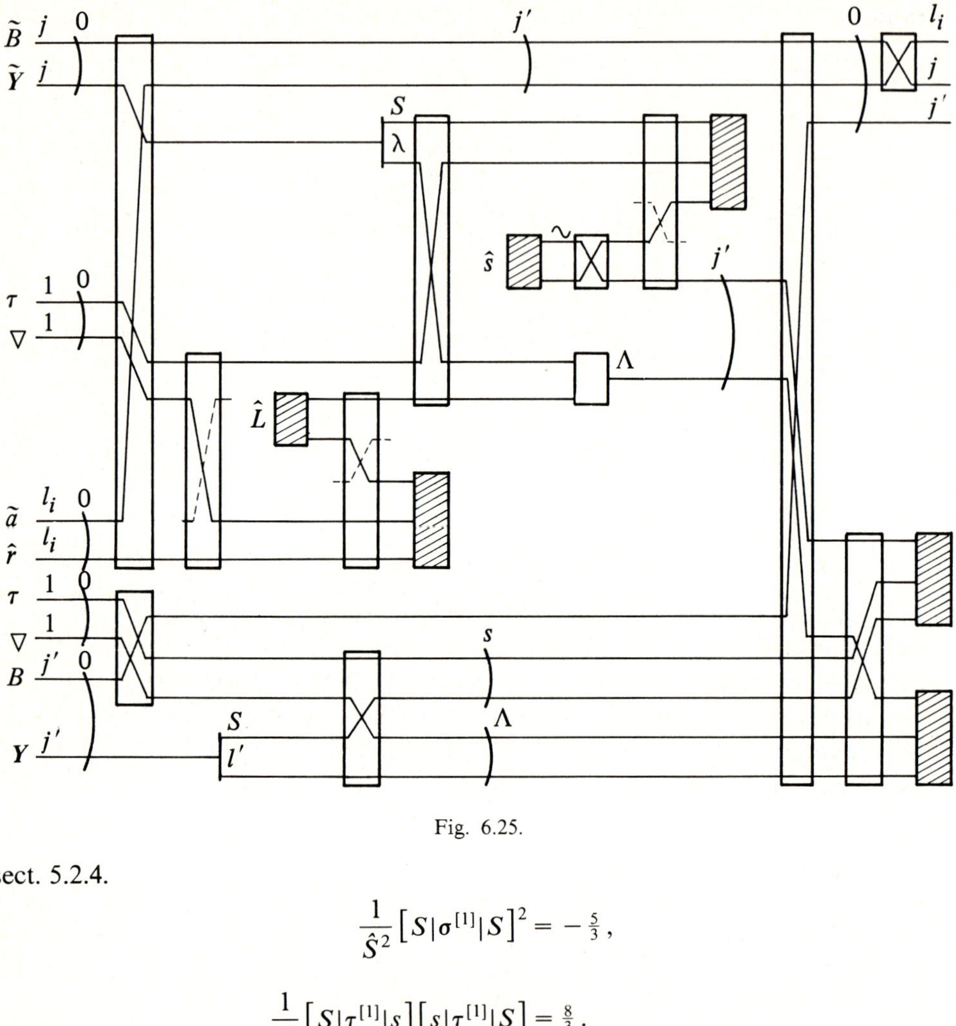

Fig. 6.25.

sect. 5.2.4.

$$\frac{1}{\hat{S}^2}\left[S|\sigma^{[1]}|S\right]^2 = -\tfrac{5}{3},$$

$$\frac{1}{\hat{s}^2}\left[S|\tau^{[1]}|s\right]\left[s|\tau^{[1]}|S\right] = \tfrac{8}{3}.$$

We now simply list the other terms:

$$M_2 = -i\begin{pmatrix} b \\ 0 \\ -a \\ 0 \end{pmatrix}^{+} \overleftarrow{\nabla}\cdot\boldsymbol{\gamma\gamma}\cdot(\nabla\varphi)\boldsymbol{C}\begin{pmatrix} f \\ 0 \\ g \\ 0 \end{pmatrix}$$

$$= -i\begin{pmatrix} b \\ 0 \end{pmatrix}^{+}\left(\overleftarrow{\nabla}\cdot\boldsymbol{\sigma\sigma}\cdot(\nabla\varphi) + \overleftarrow{\nabla}\cdot\boldsymbol{\tau\tau}\cdot(\nabla\varphi)\right)\boldsymbol{C}\begin{pmatrix} f \\ 0 \end{pmatrix}$$

$$+i\begin{pmatrix} a \\ 0 \end{pmatrix}^{+}\left(\overleftarrow{\nabla}\cdot\boldsymbol{\sigma\sigma}\cdot(\nabla\varphi) + \overleftarrow{\nabla}\cdot\boldsymbol{\tau\tau}\cdot(\nabla\varphi)\right)\boldsymbol{C}\begin{pmatrix} g \\ 0 \end{pmatrix},$$

Hence,

$$M_2 = \sqrt{\frac{15}{8}} \left(\frac{1}{2M_\Delta}\right)^2 \frac{\hat{j}}{\hat{l}_i \hat{1}} (-)^{j+l_i+j'} \sum \alpha_{Ll_i} \frac{1}{\hat{\Lambda}}$$

$$\times \left(\alpha_{\Lambda\lambda} Q^{\Lambda}_{Ll'} \left(8 \begin{bmatrix} S & \lambda & j \\ 1 & 1 & 0 \\ s & \Lambda & j \end{bmatrix} \begin{bmatrix} 1 & L & l_i \\ S & l' & j' \\ s & \Lambda & j \end{bmatrix} - 5 \begin{bmatrix} S & \lambda & j \\ 1 & 1 & 0 \\ S & \Lambda & j \end{bmatrix} \begin{bmatrix} 1 & L & l_i \\ S & l' & j' \\ S & \Lambda & j \end{bmatrix} \right) \beta_{l\lambda j} \right.$$

$$\times \int r^2 \mathrm{d}r \, V_{\nu l \Lambda}(r) k_{\nu_i l_i L}(r) u_{\nu' l'}(r)$$

$$- \alpha_{\Lambda l} Q^{\Lambda}_{L\lambda'} \left(8 \begin{bmatrix} S & l & j \\ 1 & 1 & 0 \\ s & \Lambda & j \end{bmatrix} \begin{bmatrix} 1 & L & l_i \\ S & \lambda' & j' \\ s & \Lambda & j \end{bmatrix} - 5 \begin{bmatrix} S & l & j \\ 1 & 1 & 0 \\ S & \Lambda & j \end{bmatrix} \begin{bmatrix} 1 & L & l_i \\ S & \lambda' & j' \\ S & \Lambda & j \end{bmatrix} \right) \beta_{l'\lambda' j'}$$

$$\left. \times \int r^2 \mathrm{d}r \, U_{\nu l \Lambda}(r) k_{\nu_i l_i L}(r) v_{\nu' l' \lambda'}(r) \right),$$

where the sum is over the indices λ, λ', L and Λ. Likewise,

$$M_3 = -i \begin{pmatrix} b \\ 0 \\ -a \\ 0 \end{pmatrix}^{+} (\boldsymbol{\gamma} \cdot (\nabla\varphi) \gamma_4 \vec{\partial}_4 + \overleftarrow{\partial}_4 \gamma_4 \boldsymbol{\gamma} \cdot (\nabla\varphi)) \mathbf{C} \begin{pmatrix} f \\ 0 \\ g \\ 0 \end{pmatrix}$$

$$= -\begin{pmatrix} b \\ 0 \end{pmatrix}^{+} \boldsymbol{\sigma} \cdot (\nabla\varphi)(\vec{\partial}_4 - \overleftarrow{\partial}_4) \mathbf{C} \begin{pmatrix} g \\ 0 \end{pmatrix} + \begin{pmatrix} a \\ 0 \end{pmatrix}^{+} \boldsymbol{\sigma} \cdot (\nabla\varphi)(\vec{\partial}_4 - \overleftarrow{\partial}_4) \begin{pmatrix} f \\ 0 \end{pmatrix}$$

$$= 5\sqrt{\frac{1}{2}} \left(\frac{1}{2M_\Delta}\right)^2 \frac{1}{\hat{l}_i} (-)^{j+l_i+j'} \sum \alpha_{Ll_i}$$

$$\times \left(\begin{bmatrix} S & l & j \\ 1 & L & l_i \\ S & l' & j' \end{bmatrix} [l|L|l'] \int r^2 \mathrm{d}r (u_{\nu l}(r) y_{\nu' l'}(r) + y_{\nu l}(r) u_{\nu' l'}(r)) h_{\nu_i l_i}(r) \right.$$

$$\left. - \begin{bmatrix} S & \lambda & j \\ 1 & L & l_i \\ S & \lambda' & j' \end{bmatrix} [\lambda|L|\lambda'] \beta_{l\lambda j} \beta_{l'\lambda' j'} \int r^2 \mathrm{d}r (v_{\nu l \lambda}(r) z_{\nu' l' \lambda'}(r) + z_{\nu l \lambda}(r) v_{\nu' l' \lambda'}(r)) h_{\nu_i l_i}(r) \right).$$

Here we have used, from sect. 5.2.4,

$$[S|\sigma^{[1]}|S] = 2i\sqrt{\tfrac{5}{3}}\ .$$

The next term is

$$M_4 = -i\begin{pmatrix} b \\ 0 \\ -a \\ 0 \end{pmatrix}^{+} (\gamma_4(\partial_4\varphi)\boldsymbol{\gamma}\cdot\vec{\nabla} + \overleftarrow{\nabla}\cdot\boldsymbol{\gamma}\gamma_4(\partial_4\varphi))\boldsymbol{C}\begin{pmatrix} f \\ 0 \\ g \\ 0 \end{pmatrix}$$

$$= \begin{pmatrix} a \\ 0 \end{pmatrix}^{+} \boldsymbol{\sigma}\cdot(\vec{\nabla} - \overleftarrow{\nabla})\boldsymbol{C}\begin{pmatrix} f \\ 0 \end{pmatrix}\partial_4\varphi - \begin{pmatrix} b \\ 0 \end{pmatrix}^{+} \boldsymbol{\sigma}\cdot(\vec{\nabla} - \overleftarrow{\nabla})\boldsymbol{C}\begin{pmatrix} g \\ 0 \end{pmatrix}\partial_4\varphi$$

$$-5\sqrt{\frac{3}{2}}\left(\frac{1}{2M_\Delta}\right)^2 \sum \left(\begin{bmatrix} 1 & 1 & 0 \\ S & l' & j' \\ S & \Lambda & j' \end{bmatrix} \begin{bmatrix} S & l & j \\ S & \Lambda & j' \\ 0 & l_i & l_i \end{bmatrix} \alpha_{\Lambda l'} Q^{\Lambda}_{ll_i} \right.$$

$$\times \int r^2 \mathrm{d}r\, u_{\nu l}(r) h_{\nu_i l_i}(r) U_{\nu' l' \Lambda}(r)$$

$$- \begin{bmatrix} 1 & 1 & 0 \\ S & \lambda' & j' \\ S & \Lambda & j' \end{bmatrix} \begin{bmatrix} S & \lambda & j \\ S & \Lambda & j' \\ 0 & l_i & l_i \end{bmatrix} \alpha_{\Lambda\lambda'} Q^{\Lambda}_{\lambda l_i} \beta_{l\lambda j} \beta_{l'\lambda' j'} \int r^2 \mathrm{d}r\, v_{\nu l\lambda}(r) h_{\nu_i l_i}(r) V_{\nu' l' \Lambda}(r)$$

$$- \begin{bmatrix} 1 & 1 & 0 \\ S & l & j \\ S & \Lambda & j \end{bmatrix} \begin{bmatrix} S & \Lambda & j \\ S & l' & j' \\ 0 & l_i & l_i \end{bmatrix} \alpha_{\Lambda l} Q^{\Lambda}_{l' l_i} \int r^2 \mathrm{d}r\, U_{\nu l \Lambda}(r) h_{\nu_i l_i}(r) u_{\nu' l'}(r)$$

$$\left. + \begin{bmatrix} 1 & 1 & 0 \\ S & \lambda & j \\ S & \Lambda & j \end{bmatrix} \begin{bmatrix} S & \Lambda & j \\ S & \lambda' & j' \\ 0 & l_i & l_i \end{bmatrix} \alpha_{\Lambda\lambda} Q^{\Lambda}_{\lambda' l_i} \beta_{l\lambda j} \beta_{l'\lambda' j'} \int r^2 \mathrm{d}r\, V_{\nu l \Lambda}(r) h_{\nu_i l_i}(r) v_{\nu' l' \lambda'}(r) \right).$$

And, finally, for the last term

$$M_5 = -i\begin{pmatrix} b \\ 0 \\ -a \\ 0 \end{pmatrix}^{+} (\gamma_4(\partial_4\varphi)\gamma_4(\vec{\partial}_4 + \overleftarrow{\partial}_4))\boldsymbol{C}\begin{pmatrix} f \\ 0 \\ g \\ 0 \end{pmatrix}$$

$$= i\begin{pmatrix} a \\ 0 \end{pmatrix}^{+} (\vec{\partial}_4 + \overleftarrow{\partial}_4)\boldsymbol{C}\begin{pmatrix} g \\ 0 \end{pmatrix}\partial_4\varphi - i\begin{pmatrix} b \\ 0 \end{pmatrix}^{+} (\vec{\partial}_4 + \overleftarrow{\partial}_4)\boldsymbol{C}\begin{pmatrix} f \\ 0 \end{pmatrix}\partial_4\varphi$$

$$= \sqrt{\frac{15}{2}} \left(\frac{1}{2M_\Delta}\right)^2 \sum \left(\begin{bmatrix} S & l & j \\ S & \lambda' & j' \\ 0 & l_i & l_i \end{bmatrix} [l|\lambda'|l_i] \beta_{l'\lambda' j'} \right.$$

$$\times \int r^2 \mathrm{d}r \left(y_{\nu l}(r) v_{\nu' l' \lambda'}(r) - u_{\nu l}(r) z_{\nu' l' \lambda'}(r) \right) h_{\nu_i l_i}(r)$$

$$- \begin{bmatrix} S & \lambda & j \\ S & l' & j' \\ 0 & l_i & l_i \end{bmatrix} [\lambda|l'|l_i] \beta_{l\lambda j}$$

$$\left. \times \int r^2 \mathrm{d}r \left(z_{\nu l \lambda}(r) u_{\nu' l'}(r) - v_{\nu l \lambda}(r) y_{\nu' l'}(r) \right) h_{\nu_i l_i}(r) \right).$$

Thus adding all the terms, the invariant pseudo-vector interaction matrix element is

$$[\pi_i \Delta | H_{\pi\Delta}^{\mathrm{PV}} | \Delta'] = g_{\pi\Delta}^{\mathrm{PV}} (M_1 + M_2 + M_3 + M_4 + M_5).$$

Herewith the hamiltonian is, for pion emission,

$$H_{\pi\Delta} = \sum [\pi_i \Delta | H_{\pi\Delta} | \Delta'] : \left[\tilde{a}_{\nu_i}^{[l_i t]} \tilde{B}_{\nu l}^{[jT]} B_{\nu' l'}^{[j'T]} \right]^{[00]} : ,$$

with

$$[\pi_i \Delta | H_{\pi\Delta} | \Delta'] = [\pi_i \Delta | H_{\pi\Delta}^{\mathrm{PS}} | \Delta'] + [\pi_i \Delta | H_{\pi\Delta}^{\mathrm{PV}} | \Delta'].$$

This invariant matrix element is real and symmetric under hermitian conjugation

$$[\Delta' | H_{\pi\Delta} | \pi_i \Delta] = [\pi_i \Delta | H_{\pi\Delta} | \Delta'].$$

6.12. RADIAL INTEGRALS AND CONSERVATION OF MOMENTUM

In all the invariant matrix elements of the vertex interaction operators of this chapter, the radial integrals can be evaluated directly in the position space representation in which they have been given. These integrals assure of course momentum conservation at the vertex. This now will be demonstrated.

Substituting for the radial functions $g_{\nu l}(r)$, $h_{\nu l}(r)$, $u_{\nu l}(r)$, etc., their Bessel transforms such as in eqs. (3.4), (3.6), (5.9) etc., we see that the three-field radial integrals are of the form

$$\int r^2 \mathrm{d}r \int p_1^2 \mathrm{d}p_1 F_1(p_1) j_{l_1}(p_1 r) \int p_2^2 \mathrm{d}p_2 F_2(p_2) j_{l_2}(p_2 r) \int p_3^2 \mathrm{d}p_3 F_3(p_3) j_{l_3}(p_3 r),$$

and a similar expression for the four-field vertices.

The momentum conservation requirement is visible in the integral, given by A. D. Jackson and L. C. Maximon, SIAM J. Math. Anal. *3* (1972) p. 446,

$$\Delta^{l_1 l_2 l_3}_{p_1 p_2 p_3} = \int r^2 \mathrm{d}r\, j_{l_1}(p_1 r) j_{l_2}(p_2 r) j_{l_3}(p_3 r)$$

$$= \pi^2 \frac{\delta(p_1 p_2 p_3)}{p_1 p_2 p_3} \frac{1}{[l_1 | l_2 | l_3]} \left[\hat{p}_1^{[l_1]} \hat{p}_2^{[l_2]} \hat{p}_3^{[l_3]}\right]^{[0]}.$$

Here the discontinuous function

$$\delta(p_1 p_2 p_3) = \begin{cases} 1 & \text{if } p_1 p_2 p_3 \text{ form a non-degenerate triangle} \\ \tfrac{1}{2} & \text{if } p_1 p_2 p_3 \text{ form a degenerate triangle} \\ 0 & \text{if } p_1 p_2 p_3 \text{ do not form a triangle.} \end{cases}$$

Furthermore $\hat{\boldsymbol{p}}_1, \hat{\boldsymbol{p}}_2, \hat{\boldsymbol{p}}_3$ are the directions of the triangle defined by $\boldsymbol{p}_1 + \boldsymbol{p}_2 + \boldsymbol{p}_3 = 0$. Thus momentum conservation is insured.

If one wants to use the above closed form for actual calculations one can evaluate the integral as follows. Orienting the triangle such that $\boldsymbol{p}_3$ lies along the axis Oz and the plane of the triangle is the Ox-Oz plane we obtain the value of the invariant

$$\left[\hat{p}_1^{[l_1]} \hat{p}_2^{[l_2]} \hat{p}_3^{[l_3]}\right]^{[0]} = \sum_m \begin{pmatrix} l_1 & l_2 & l_3 \\ m & -m & 0 \end{pmatrix} Y_m^{[l_1]}(\theta_1, \varphi_1) Y_{-m}^{[l_2]}(\theta_2, \varphi_2) Y_0^{[l_3]}(0,0)$$

with

$$\varphi_1 = \varphi_2 = 0,$$

$$\cos\theta_1 = \frac{1}{2} \frac{p_2^2 - p_1^2 - p_3^2}{p_1 p_3}, \qquad \cos\theta_2 = \frac{1}{2} \frac{p_1^2 - p_2^2 - p_3^2}{p_2 p_3}.$$

For the four-field vertices the corresponding discontinuous function

$$\square^{l_1 l_2 l_3 l_4}_{p_1 p_2 p_3 p_4} = \int r^2 \mathrm{d}r\, j_{l_1}(p_1 r) j_{l_2}(p_2 r) j_{l_3}(p_3 r) j_{l_4}(p_4 r)$$

is evaluated in terms of the above three-field integral by introducing the delta function

$$\frac{\delta(r - r')}{r^2} = \frac{2}{\pi} \int p^2 \mathrm{d}p\, j_\lambda(pr) j_\lambda(pr')$$

and we obtain

$$\Box^{l_1 l_2 l_3 l_4}_{p_1 p_2 p_3 p_4} = \frac{2}{\pi} \int p^2 \, \mathrm{d}p \, \Delta^{\lambda l_1 l_2}_{p p_1 p_2} \Delta^{\lambda l_3 l_4}_{p p_3 p_4} .$$

Here λ must be chosen such that both the triangular rules $(\lambda l_1 l_2)$ and $(\lambda l_3 l_4)$ are fulfilled, in order not to obtain indeterminate values for the functions Δ. We see that the discontinuous character of $\Box$ is again such that $\boldsymbol{p}_1 + \boldsymbol{p}_2 + \boldsymbol{p}_3 + \boldsymbol{p}_4 = 0$ must be fulfilled.

For practical applications, in particular in the case of the four-field vertices, it may be more convenient to evaluate the radial integrals directly in position space, as they are given in the expressions of the earlier sections of this chapter.

CHAPTER 7

Many-body hamiltonian matrix

7.1. INTRODUCTION

The hamiltonian field operators have been expressed in Chapter 6 in the Fock space representation as expansions in the discretized quantum numbers of space localized wave-packets. They are now utilized to calculate the matrix elements of the secular matrix of the system.

The calculation of a many-body matrix element in general involves separating off the non-interacting particles and performing their overlaps. This way, one is left with the interacting particles and their elementary vertices. The relativistic case, in the present treatment, is not different in that respect from the usual non-relativistic treatment. The new aspects arising with relativistic systems are only that we have to deal both with bosons and fermions at the same time, and with the n-body operator of the center-of-mass position. This last problem is treated in Chapter 8.

In this chapter, in sect. 7.2, we discuss the general structure of the many-body matrix elements, the application of the Wick theorem for the calculation of expectation values of coupled operators, and the multiplicities associated with the use of the well-ordered products in the operators and of normalized basis vectors.

The phase freedom of the basis state vectors discussed in sect. 4.4.2, is now used to obtain a real symmetric hamiltonian matrix. Two choices for the phases associated with each vertex of the different processes considered in Chapter 6 are listed. Owing to the reality of the hamiltonian matrix only one of the two elements representing time-reversed processes need to be treated.

The remaining part of the chapter is devoted to the treatment of a number of examples, both to demonstrate explicitly how calculations are carried through and to give a collection of formulae for commonly encountered cases. In particular, the pion cloud is treated in great detail. The examples include all the interactions in the meson-baryon Hilbert spaces given in Chapter 5.

Owing to its simplicity we do not give explicit formulae for the many-body matrix elements of H_0 since they are diagonal in all quantum numbers except for the radial quantum numbers. They are easily obtained following the steps of sect. 7.2.

7.2. STRUCTURE OF THE MANY-BODY MATRIX ELEMENTS

7.2.1. Basis vectors

In general a configuration contains several kinds of particles, for example in the meson-baryon models: nucleons, pions, rho-mesons etc. Furthermore, the particles

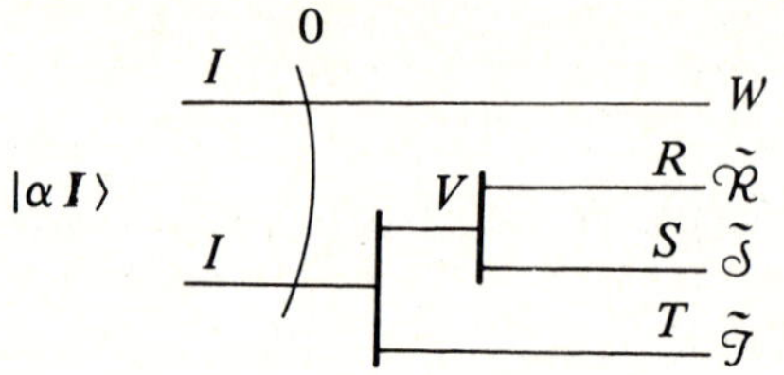

Fig. 7.1.

of one kind may be distributed among several single-particle states. We shall use the following nomenclature: the ensemble of particles of a given kind is referred to as a "type", while in a given type the ensemble of particles in the same orbital is referred to as a "group".

In this section we write generally for a basis vector of such a configuration, assuming for example that it contains three types:

$$|\alpha I\rangle = (-)^{2I}\hat{I}\left[W_\alpha^{[I]}\tilde{\mathfrak{F}}_\alpha^{\ [I]}\right]^{[0]}|0\rangle$$

$$= (-)^{2I}\hat{I}\left[W_\alpha^{[I]}\left[[\tilde{\mathfrak{R}}^{[R]}\tilde{\mathfrak{S}}^{[S]}]^{[V]}\tilde{\mathfrak{T}}^{\ [T]}\right]_\alpha^{[I]}\right]^{[0]}|0\rangle. \tag{7.1}$$

The configuration $|\alpha I\rangle$ is given in its invariant form as defined in sect. 4.6.2. The diagram representing its coupling scheme is shown in fig. 7.1. Here the $W_\alpha^{[I]}$ are the basis state amplitudes describing the orientation of the system normalized as in eq. (4.33)

$$\left[W_\alpha^{[I]}|W_\alpha^{[I]}\right] = \frac{1}{\hat{I}}.$$

The operators $\tilde{\mathfrak{R}}, \tilde{\mathfrak{S}}, \tilde{\mathfrak{T}}$ are products of creation operators normalized according to eq. (4.48)

$$\langle 0|[\mathfrak{R}^{[R]}\tilde{\mathfrak{R}}^{[R]}]^{[0]}|0\rangle = (-)^{2R}\hat{R}.$$

With these definitions the basis vectors (7.1) are properly orthonormalized, as shown in sect. 4.6.2,

$$\langle\alpha I|\beta I\rangle = \delta_{\alpha\beta}.$$

This normalization can be verified on the diagram of fig. 7.2. We recognize in this figure the circled recoupling boxes of fig. 4.32 associated with the commutation sign for crossing operator lines, and the vacuum expectation boxes of fig. 4.30.

7.2.2. *Many-body matrix elements, overview*

We now turn to the general form of the secular problem the matrix of which is

$$M_{\alpha\beta} = \langle\alpha I|H|\beta I\rangle = \langle\beta I|H|\alpha I\rangle^*.$$

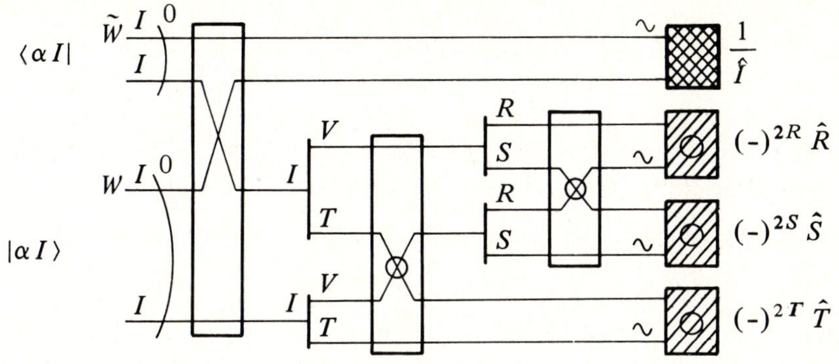

Fig. 7.2.

In detail, the general solution of the secular problem (3.10) on the basis vectors $|\alpha I\rangle$ given in invariant form by eq. (7.1) is of the form

$$\Psi_n = \sum_\alpha X_\alpha^n |\alpha I\rangle = (-)^{2I} \sum_\alpha X_\alpha^n \hat{I} \left[W_\alpha^{[I]} \tilde{\mathfrak{F}}_\alpha^{[I]} \right]^{[0]} |0\rangle ,$$

with the amplitudes X_α^n solution of the secular problem

$$(-)^{2I} \sum_\beta \left[W_\alpha^{[I]} | W_\beta^{[I]} \right] \langle 0 | \left[\mathfrak{F}_\alpha^{[I]} | H | \mathfrak{F}_\beta^{[I]} \right] | 0 \rangle X_\beta^n = E_n X_\alpha^n .$$

According to sect. 4.4.2 the amplitudes $W_{\alpha M}^{[I]}$ carry the information on the polarization of the state Ψ_n, and their dependence on a basis configuration α is simply a phase factor. We recall that

$$W_{\alpha M}^{[I]} = \xi_\alpha W_M^{[I]} = (-i)^{\delta_\alpha} W_M^{[I]} .$$

Thus

$$\left[W_\alpha^{[I]} | W_\beta^{[I]} \right] = \xi_\alpha^* \xi_\beta \frac{1}{\hat{I}} = i^{\delta_\alpha - \delta_\beta} \frac{1}{\hat{I}} ,$$

and

$$\Psi_n = (-)^{2I} \sum_\alpha X_\alpha^n (-i)^{\delta_\alpha} \hat{I} \left[W^{[I]} \tilde{\mathfrak{F}}_\alpha^{[I]} \right]^{[0]} |0\rangle .$$

The secular problem for the amplitudes X^n with this notation now takes the form

$$\sum_\beta M_{\alpha\beta} X_\beta^n = E_n X_\alpha^n ,$$

with

$$M_{\alpha\beta} = (-)^{2I}\frac{1}{\hat{I}}\eta_{\alpha\beta}\langle 0|\left[\mathfrak{F}_{\alpha}^{[I]}|H|\mathfrak{F}_{\beta}^{[I]}\right]|0\rangle .$$

We see that in our notation the phase freedom associated with the different possible choices for the phases of the basis state vectors appears in the factor

$$\eta_{\alpha\beta} = i^{\delta_\alpha - \delta_\beta} .$$

The interaction operators for the energy in the Fock representation have been given in Chapter 6 in the form

$$H = \sum_{\zeta} H_{\zeta} : \Theta_{\zeta} : ,$$

where the sum ζ is over the different possible processes such as NN$'\pi$, $\omega 3\pi$, etc.... The operators H_ζ are given in that chapter as

$$H_{\zeta} = \sum_{\mathsf{ab}} \left[\mathsf{a}|H_\zeta|\mathsf{b}\right] . \Theta_{\zeta(\mathsf{ab})} :$$

where a and b denote the discretized particle states entering the elementary vertex ζ. Here

$$\Theta_\zeta = [\Theta_{\mathsf{R}}\Theta_{\mathsf{S}}\Theta_{\mathsf{T}}]_{\zeta}^{[0]}$$

is a product of the particle operators entering the vertex. The indices $\mathsf{R}, \mathsf{S}, \mathsf{T}$ refer to the three types of our example.

The complete many-body matrix of H_ζ between the invariant basis configurations (7.1) is according to the recoupling of fig. 7.3 in full detail

$$M_{\alpha\beta} = \langle \alpha I|H_\zeta|\beta I\rangle = (-)^{2I}\frac{1}{\hat{I}}\sum_{\mathsf{ab}}\eta_{\mathsf{ab}}\left[\mathsf{a}|H_\zeta|\mathsf{b}\right]$$

$$\times \langle 0|\left[[\mathfrak{R}\mathfrak{S}\mathfrak{T}]_{\alpha}^{[I]} : \Theta_{\zeta(\mathsf{ab})} : [\tilde{\mathfrak{R}}\tilde{\mathfrak{S}}\tilde{\mathfrak{T}}]_{\beta}^{[I]}\right]^{[0]}|0\rangle . \qquad (7.2)$$

In this expression the overlaps of the spectator particles are real and the phase η_{ab} arises only from the interacting particles. The last factor is an expectation value on the vacuum, which contains the many-body aspects of the matrix. It is real and symmetric under hermitian conjugation with the rules of sect. 4.6 and as shown in the next sections. Thus the property under hermitian conjugation of $M_{\alpha\beta}$ is entirely determined by the behaviour of the vertex quantities $\eta_{\mathsf{ab}}[\mathsf{a}|H_\zeta|\mathsf{b}]$, which we now discuss.

With our definitions for the field operators, all the invariant elementary matrix elements $[\mathsf{a}|H_\zeta|\mathsf{b}]$ calculated in Chapter 6 are real and symmetric

$$\left[\mathsf{b}|H_\zeta|\mathsf{a}\right] = \left[\mathsf{a}|H_\zeta|\mathsf{b}\right] ,$$

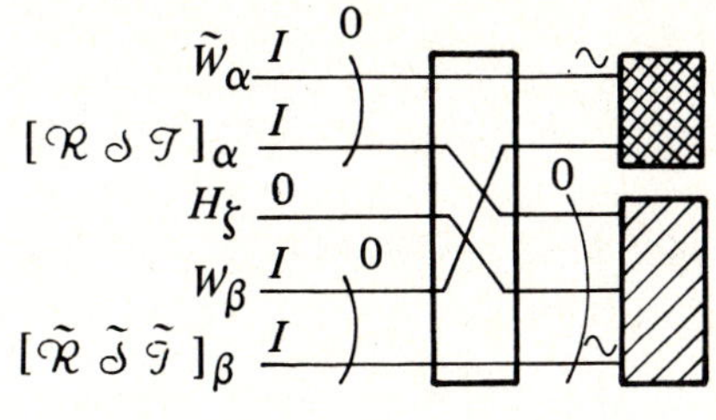

Fig. 7.3.

except those with a vector meson participant for which we have

$$\left[\mathsf{b}|H_\zeta|\mathsf{a}\right] = (-)^{1+t_h}\left[\mathsf{b}|H_\zeta|\mathsf{a}\right].$$

In words, with a ρ-meson they are real and symmetric under hermitian conjugation, while with an ω-meson they are imaginary thus antisymmetric, as seen in Chapter 6. This is due to the intrinsic property of the vector meson field, whose phase behaves as i^{1+t_h} under hermitian conjugation with our conventions for the fields, sect. 5.2.3.

In order to achieve for computational convenience a real symmetric secular matrix $M_{\alpha\beta}$, we use now the freedom allowed by the definition of the phases of the representation vectors, i.e. the choice of the phases ξ_α of the amplitudes $W_\alpha^{[I]}$.

The simplest phase choice for a real $M_{\alpha\beta}$ with the fields defined in Chapter 5 consists in associating to each basis configuration a phase

$$\xi_\alpha = 1,$$

except those which have a vector meson, where according to the above remarks we adopt

$$\xi_\alpha = (-i)^{1+t_h}.$$

With this choice, for all elementary processes considered in Chapter 6,

$$\eta_{\mathsf{ab}} = 1,$$

except for those where an omega meson is emitted $(+i)$ or annihilated $(-i)$

$$\eta_{\mathsf{ab}} = \pm i.$$

Another choice, useful particularly for computing transistion matrix elements, consists in defining the basis vector phases such that they achieve not only a real hamiltonian matrix but also invariant forms of given reality (real or imaginary) for "real" polarized particle states along the lines of sect. 4.4.2. Namely, to each particle entering the vertex we associate the following phases: for each spin-0 particle

$$\xi = (-i)^{l+t},$$

for each spin-1 particle

$$\xi = (-i)^{1+l_h+S+t_h} = (-i)^{J_h+\varepsilon_\kappa+t_h},$$

and for each half-integer spin ($\frac{1}{2}$ or $\frac{3}{2}$) particle

$$\xi = (-i)^l.$$

For the spin-0 particles the basis state invariant forms are chosen real with l the orbital angular momentum and t the isospin. For the spin-1 particles the invariant forms are chosen pure imaginary with J_h the total angular momentum and l_h an orbital angular momentum defined as $l_h = J_h$ for the magnetic multipolarity and the time part of the longitudinal mode and as $l_h = J_h - 1$ for the electric and the space part of the longitudinal multipolarities. We also have for the spin index S of the vector fields the value $S = 1$ for the space parts and $S = 0$ for the time part. Thus $\varepsilon_\kappa = 0$ for the magnetic multipolarity and $\varepsilon_\kappa = 1$ for the electric and longitudinal multipolarities. For half-integer spins the invariant forms of the upper components are chosen real and the lower component invariant forms are then imaginary, with l the orbital angular momentum of the upper components. This is due to the fact that spinors have mixed reality characters as already discussed in sect. 4.4.2.

In table 7.1 we give the values of the phase η_{ab} attached to the vertices with this second phase choice for all the elementary matrix elements of Chapter 6.

To indicate this phase freedom, we retain in the following sections explicitly in all the formal expressions of the many-body matrix elements the phase symbol η at each vertex.

7.2.3. Separation of types

In the Fock representation, the many-body aspects of eq. (7.2) are contained in the evaluation of the vacuum expectation values of a product of particle operators. This is done by applying the Wick theorem. We now shall develop the form the Wick theorem takes when working with coupled operators involving several particle types.

The evaluation of the vacuum expectation value is performed in the following steps:

(i) separation of particle types;

(ii) separation of groups within a type;

(iii) evaluation of the vacuum expectation values within each group.

We consider each of these steps in turn and develop the general rules to be followed when performing recoupling operations with non-commuting operators, i.e. drawing and evaluating graphs having lines which represent non-commuting variables.

The first step of the evaluation of the vacuum expectation values, eq. (7.2), consists in bringing together all particle operators of a given type. They may be originating in the bra, the ket, and the interaction operator. In our restricted case the

TABLE 7.1
Phases η for "real" polarized states

$[\mathrm{a} \mid H_0 \mid \mathrm{b}]$	η_{ab}
$[\pi \mathrm{N} \mid H_{\pi\mathrm{N}} \mid \mathrm{M}]$	$i^{l_\pi + l_\mathrm{N} - l_\mathrm{M} + 1}$
$[\sigma \mathrm{N} \mid H_{\sigma\mathrm{N}} \mid \mathrm{M}]$	$i^{J_\mathrm{h} + l_\mathrm{N} - l_\mathrm{M}}$
$[\mathrm{VN} \mid H_{V\mathrm{N}} \mid \mathrm{M}]$	$i^{J_\mathrm{h} + \varepsilon_\kappa + l_\mathrm{N} - l_\mathrm{M} + t_\mathrm{h}}$
$[\pi_i \pi_j \mid H_\pi \mid \pi_k \pi_l]$	$i^{l_i + l_j - l_k - l_l}$
$[\pi_i \pi_j \pi_k \mid H_\pi \mid \pi_l]$	$i^{l_i + l_j + l_k - l_l + 2}$
$[\pi_i \pi_j \pi_k \pi_l \mid H_\pi \mid]$	$i^{l_i + l_j + l_k + l_l}$
$[\rho \mid H_{\rho\pi} \mid \pi_i \pi_j]$	$i^{J_\mathrm{h} + \varepsilon_\kappa - l_i - l_j - 1}$
$[\rho \pi_i \mid H_{\rho\pi} \mid \pi_j]$	$i^{J_\mathrm{h} + \varepsilon_\kappa + l_i - l_j + 1}$
$[\rho \pi_i \pi_j \mid H_{\rho\pi} \mid]$	$i^{J_\mathrm{h} + \varepsilon_\kappa + l_i + l_j - 1}$
$[\omega \mid H_{\omega\pi} \mid \pi_i \pi_j \pi_k]$	$i^{J_\mathrm{h} + \varepsilon_\kappa - l_i - l_j - l_k + 1}$
$[\omega \pi_i \mid H_{\omega\pi} \mid \pi_j \pi_k]$	$i^{J_\mathrm{h} + \varepsilon_\kappa + l_i - l_j - l_k - 1}$
$[\omega \pi_i \pi_j \mid H_{\omega\pi} \mid \pi_k]$	$i^{J_\mathrm{h} + \varepsilon_\kappa + l_i + l_j - l_k + 1}$
$[\omega \pi_i \pi_j \pi_k \mid H_{\omega\pi} \mid]$	$i^{J_\mathrm{h} + \varepsilon_\kappa + l_i + l_j + l_k - 1}$
$[\sigma \mid H_{\sigma\pi} \mid \pi_i \pi_j]$	$i^{l_\sigma - l_i - l_j - 2}$
$[\sigma \pi_i \mid H_{\sigma\pi} \mid \pi_j]$	$i^{l_\sigma + l_i - l_j}$
$[\sigma \pi_i \pi_j \mid H_{\sigma\pi} \mid]$	$i^{l_\sigma + l_i + l_j + 2}$
$[\pi \mathrm{N} \mid H_{\pi\mathrm{N}\Delta} \mid \Delta]$	$i^{l_\pi + l_\mathrm{N} - l_\Delta + 1}$
$[\mathrm{N} \mid H_{\pi\mathrm{N}\Delta} \mid \pi\Delta]$	$i^{l_\mathrm{N} - l_\pi - l_\Delta - 1}$
$[\pi\Delta \mid H_{\pi\Delta} \mid \Delta]$	$i^{l_\pi + l_\Delta - l'_\Delta + 1}$

$\varepsilon_\kappa = 0$ for $\kappa = \mathfrak{M}$.
$\varepsilon_\kappa = 1$ for $\kappa = \mathfrak{E}, \mathfrak{L}$.

needed geometry is shown in fig. 7.4. In this figure it is seen that it is convenient to first bring the interaction operator lines to the bottom of the diagram. Of course, before performing the vacuum expectation values later on, the operators $\Theta_\mathrm{R}^{[r]}$ will have to be recoupled into their proper position, i.e. between the basis state vectors. This of course introduces not only a reordering phase arising from the coupling

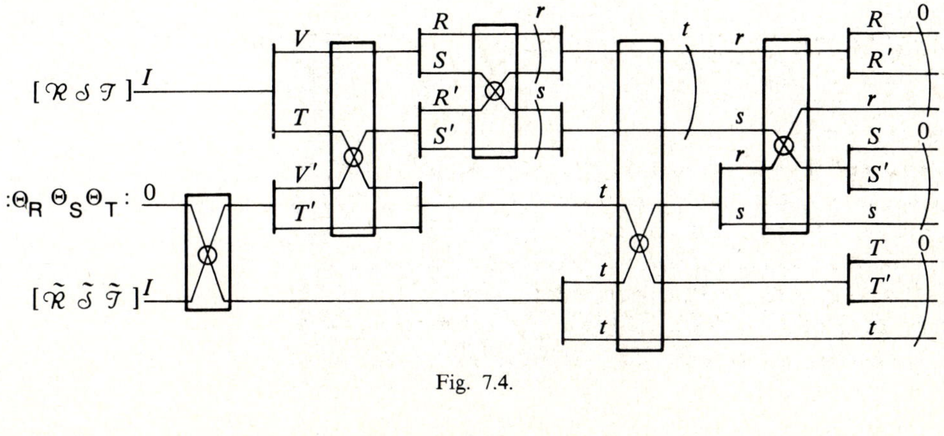

Fig. 7.4.

(which here is $+1$) but it may also introduce phases arising from the anticommutation properties of the fermion operators. Hence the circled crossing boxes contribute a phase ε, as defined in the context of fig. 4.32.

In this first step we only reorder the operators. No contractions are as yet to be performed. All contractions will be carried out later on after recoupling the interaction operators into their proper positions within each group. These recouplings again will bring anticommutation phases which will exactly cancel any superfluous phases introduced in previous steps. This procedure insures automatic book-keeping of the Fock space phases.

The evaluation of the diagram 7.4 yields

$$\big[[\mathfrak{R}^{[R]}\mathfrak{S}^{[S]}]^{[V]}\mathfrak{T}^{[T]}\big]^{[I]} : \big[[\Theta_{\mathsf{R}}^{[r]}\Theta_{\mathsf{S}}^{[s]}]^{[t]}\Theta_{\mathsf{T}}^{[t]}\big]^{[0]} : \big[[\tilde{\mathfrak{R}}^{[R']}\tilde{\mathfrak{S}}^{[S']}]^{[V']}\tilde{\mathfrak{T}}^{[T']}\big]^{[I]}$$

$$= \varepsilon \begin{bmatrix} V & T & I \\ V' & T' & I \\ t & t & 0 \end{bmatrix} \begin{bmatrix} R & S & V \\ R' & S' & V' \\ r & s & t \end{bmatrix} \frac{1}{\hat{r}} \big[\mathfrak{R}^{[R]}\tilde{\mathfrak{R}}^{[R']} : \Theta_{\mathsf{R}}^{[r]} :\big]^{[0]} \frac{1}{\hat{s}} \big[\mathfrak{S}^{[S]}\tilde{\mathfrak{S}}^{[S']} : \Theta_{\mathsf{S}}^{[s]} :\big]^{[0]}$$

$$\times \frac{1}{\hat{t}} \big[\mathfrak{T}^{[T]}\tilde{\mathfrak{T}}^{[T']} : \Theta_{\mathsf{T}}^{[t]} :\big]^{[0]} .$$

7.2.4. Separation of groups

We are now faced with the task of evaluating the vacuum expectation values for a product of particle operators for a given type R. We thus consider the product

$$\big[\mathfrak{R}_{\alpha}^{[R]}\tilde{\mathfrak{R}}_{\alpha'}^{[R']} : \Theta_{\mathsf{R}}^{[r]} :\big] .$$

Here the double dots represent the well-ordered normal product as defined in sect. 6.6 for the operators constituting $\Theta_{\mathsf{R}}^{[r]}$ acting respectively on the particles in the orbitals $j, k, l \ldots$

$$:\Theta_{\mathsf{R}}^{[r]}: \, = \, :\Big[\big[\Theta_{\{j\}}^{[A]}\Theta_{\{k\}}^{[B]}\big]^{[C]}\Theta_{\{l\}}^{[D]}\Big]^{[r]}: .$$

They are ordered with the annihilators on the right, group by group. Furthermore, the state indices are arranged in the standard lexicographic order. If needed the resulting multiplicity is accounted for by a factor m, as explained in sect. 6.6, 6.7 and 6.8, and as will be fully discussed in sect. 7.2.6.

As defined above, for particles of a given type, $\mathfrak{R}_{\alpha}^{[R]}$ is made up of several groups, i.e. n_j particles in orbital state j, n_k particles in k, etc. Here $j, k, \ldots$ stand for all quantum numbers which specify the orbitals. For our example we again limit ourselves to the configuration spaces of sect. 5.4, where at most three groups participate

$$\tilde{\mathfrak{R}}_{\alpha}^{[R]} = \Big[\big[\tilde{\mathfrak{A}}_{n_j\alpha_j}^{[J]}\tilde{\mathfrak{A}}_{n_k\alpha_k}^{[K]}\big]^{[M]}\tilde{\mathfrak{A}}_{n_l\alpha_l}^{[L]}\Big]^{[R]} .$$

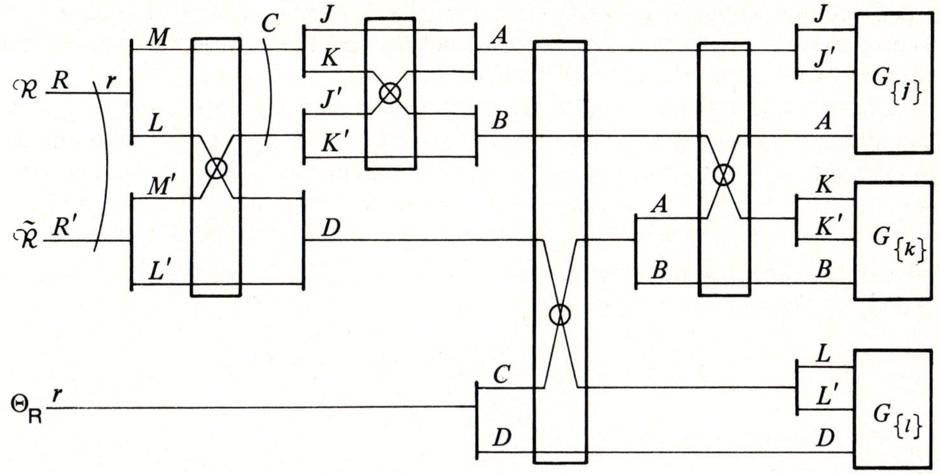

Fig. 7.5.

Here the indices α_j denote the orthonormal basis states within a group. The operators $\tilde{\mathscr{Q}}^{[J]}_{n_j\alpha_j}$ and their hermitian conjugates have been introduced in sect. 4.6.2.

The separation of the groups is given by fig. 7.5 and yields

$$\left[\mathfrak{R}^{[R]}_{\alpha}\tilde{\mathfrak{R}}^{[R']}_{\alpha'}:\Theta^{[r]}_{\mathrm{R}}:\right]^{[0]}=\varepsilon\sum\begin{bmatrix} M & L & R\\ M' & L' & R'\\ C & D & r\end{bmatrix}\begin{bmatrix} J & K & M\\ J' & K' & M'\\ A & B & C\end{bmatrix}\hat{r}G_{\{j\}}G_{\{k\}}G_{\{l\}}$$

with

$$G_{\{j\}}=\frac{1}{\hat{A}}\left[\mathscr{Q}^{[J]}_{n_j\alpha_j}\tilde{\mathscr{Q}}^{[J']}_{n'_j\alpha'_j}:\Theta^{[A]}_{\{j\}}:\right]^{[0]}.$$

The final boxes $G_{\{j\}}$ have been defined in the fig. 7.5 so as to have the normal product $:\Theta^{[A]}_{\{j\}}:$ still placed at the bottom. Before evaluating the vacuum expectation values the operators $\Theta^{[A]}_{\{j\}}$ will have to be recoupled into the middle position, between $\mathscr{Q}$ and $\tilde{\mathscr{Q}}$.

7.2.5. Evaluation of expectation values

At this point the detailed form of the operator $:\Theta_{\{j\}}:$ operating within one group $\{j\}$ must be introduced. The complexity of the many-body aspect of the problem emerges at this point. We therefore classify the operators $:\Theta_{\{j\}}:$ according to the number of particle operators they contain:
the unit operator

$$:\Theta^{[0]}_{\{j\}}:=\Theta^{[0]}=1\,,$$

the one-particle operator

$$:\Theta^{[j]}_{\{j\}}: = (\tilde{a}^{[j]} + a^{[j]}),$$

the many-particle operator, $x > 1$

$$:\Theta^{[\,]}_{\{j\}}: = :(\tilde{a}^{[j]} + a^{[j]})^x:.$$

Of this last operator we consider here the two important terms: the one-body "scattering" operator

$$:\Theta^{[A]}_{\{j\}}: = :[\tilde{a}^{[j]} a^{[j]}]^{[A]}:$$

and the two-body operator

$$:\Theta^{[A]}_{\{j\}}: = :[[\tilde{a}^{[j]}\tilde{a}^{[j]}][a^{[j]}a^{[j]}]]^{[A]}:.$$

In the interactions of Chapter 6 at most $x = 4$ appears in the π^4 interaction.

We now perform the vacuum expectation values on the vacuum for the operators $G_{\{j\}}$ of sect. 7.2.4. According to our procedure, before performing any contractions the operator $\Theta^{[A]}_{\{j\}}$ is first recoupled into its proper position, i.e. between the state operators $\mathcal{A}$ and $\tilde{\mathcal{A}}$. Thus this step yields a recoupling coefficient and a crossing phase for anticommuting operators. Finally the expectation value is calculated according to the results of sects. 4.6.2 and 4.6.3, in particular eqs. (4.53) and (4.54).

For the unit operator, the expectation value on the vacuum reduces to the overlap (4.48)

$$\langle G_{\{j\}} \rangle = \langle 0 | [\mathcal{A}^{[J]}_{n\alpha} \tilde{\mathcal{A}}^{[J']}_{n'\alpha'}]^{[0]} | 0 \rangle = (-)^{2J} \hat{J} \delta_{JJ'} \delta_{\alpha\alpha'} \delta_{nn'}.$$

For the one-particle operator we consider first the creation part

$$G_{\{j\}} = \frac{1}{\hat{j}} [\mathcal{A}^{[J]}_{n\alpha} \tilde{\mathcal{A}}^{[J']}_{n-1\alpha'} \tilde{a}^{[j]}]^{[0]}.$$

According to the general rules $\tilde{a}$ must be recoupled into the middle position. However in order to use the formula (4.53), $\tilde{a}$ must then be returned to the position on the right. Thus the expectation value can be calculated directly from the graph of fig. 7.6 which yields

$$\langle G_{\{j\}} \rangle = (-)^{2J} \frac{\mathsf{M}}{\hat{j}} \hat{J} (j^{n-1}(\alpha' J') j | \} j^n (\alpha J)).$$

We shall denote by M the multiplicity which will be given in full generality in sect. 7.2.6. In this case $\mathsf{M} = \sqrt{n}$.

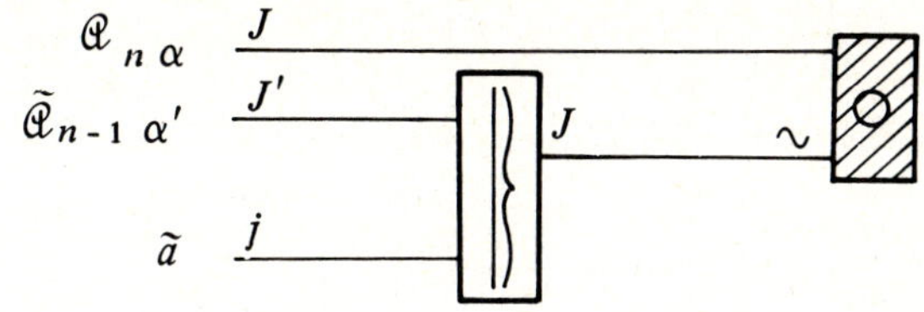

Fig. 7.6.

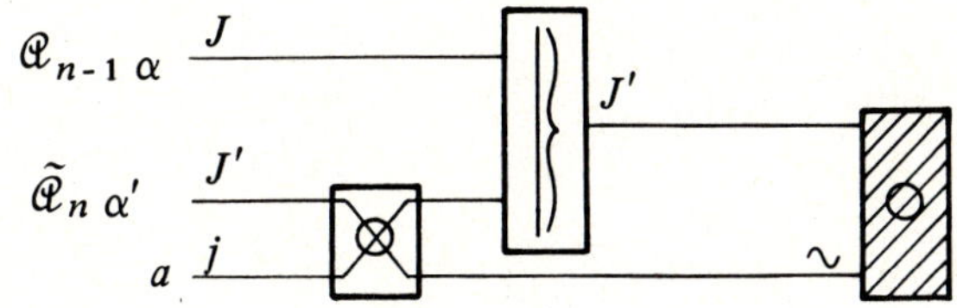

Fig. 7.7.

For the annihilation part we have the graph of fig. 7.7

$$G_{\{j\}} = \varepsilon(-)^{J'+j-J}\frac{1}{\hat{j}}\left[\mathcal{A}^{[J]}_{n-1\alpha}a^{[j]}\tilde{\mathcal{A}}^{[J']}_{n\alpha'}\right]^{[0]},$$

$$\langle G_{\{j\}}\rangle = (-)^{J'-j+J}\frac{\mathsf{M}}{\hat{j}}\hat{J}'\left(j^{n-1}(\alpha J)\,j|\}j^{n}(\alpha' J')\right),$$

where again first the proper order of the operators has been restored and where use has been made of the relation (4.53) for combining the operators by means of the CFP's. The multiplicity is again denoted by M, and $\mathsf{M} = \sqrt{n}$.

The creation and annihilation of two particles is treated along the same lines. We have for the creation part

$$G_{\{j\}} = \frac{1}{\hat{A}}\left[\mathcal{A}^{[J]}_{n\alpha}\tilde{\mathcal{A}}^{[J']}_{n-2\alpha'}\left[\tilde{a}^{[j]}\tilde{a}^{[j]}\right]^{[A]}\right]^{[0]}.$$

As above the product $[\tilde{a}\tilde{a}]$ can be retained at the right in view of eq. (4.54) and fig. 7.8 yields

$$\langle G_{\{j\}}\rangle = (-)^{2J}\frac{\mathsf{M}}{\hat{A}}\hat{J}\left(j^{n-2}(\alpha' J')\,j^{2}A\|\}j^{n}(\alpha J)\right)$$

with $\mathsf{M} = \sqrt{n(n-1)}$. Similarly, we have from fig. 7.9, for the annihilation part

$$G_{\{j\}} = (-)^{J'+A-J}\frac{1}{\hat{A}}\left[\mathcal{A}^{[J]}_{n-2\alpha}\left[a^{[j]}a^{[j]}\right]^{[A]}\tilde{\mathcal{A}}^{[J']}_{n\alpha'}\right]^{[0]},$$

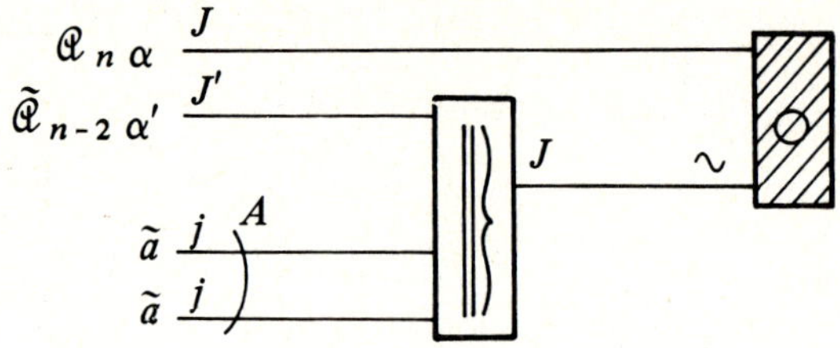

Fig. 7.8.

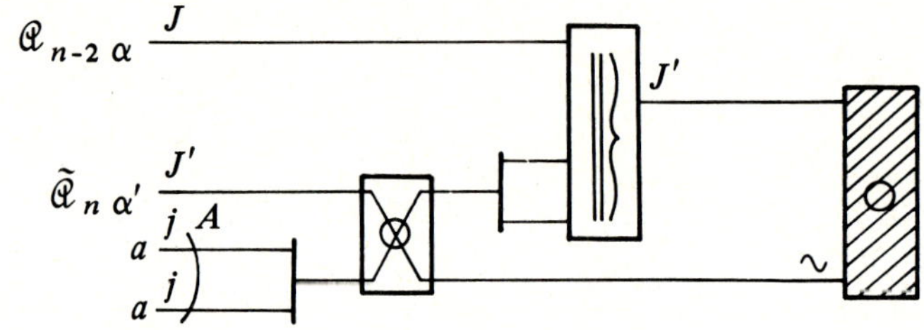

Fig. 7.9.

and with (4.54)

$$\langle G_{\{j\}}\rangle = (-)^{J'-A+J}\frac{\mathsf{M}}{\hat{A}}\hat{J}'(j^{n-2}(\alpha J)\,j^2A\|\}j^n(\alpha' J')),$$

with $\mathsf{M} = \sqrt{n(n-1)}$.

The one-body scattering operator is of the form

$$G_{\langle j\rangle} = \frac{1}{\hat{A}}\left[\mathcal{Q}_{n\alpha}^{[J]}\tilde{\mathcal{Q}}_{n\alpha'}^{[J']}\left[\tilde{a}^{[j]}a^{[j]}\right]^{[A]}\right]^{[0]}$$

and fig. 7.10 yields

$$\langle G_{\langle j\rangle}\rangle = \varepsilon(-)^{2j-A}\frac{\mathsf{M}}{\hat{A}}\sum_{\beta K}(-)^{2K}\hat{K}(j^n(\alpha J)\,j|\}j^{n+1}(\beta K))$$

$$\times(j^n(\alpha' J')\,j|\}j^{n+1}(\beta K))\begin{bmatrix} J & J' & A \\ j & j & A \\ K & K & 0 \end{bmatrix},$$

where the sign ε is $\varepsilon = (-)^{2j}$ and $\mathsf{M} = 2(n+1)$.

In the case of the two-body operator

$$G_{\langle j\rangle} = \frac{1}{\hat{A}}\left[\mathcal{Q}_{n\alpha}^{[J]}\tilde{\mathcal{Q}}_{n\alpha'}^{[J']}\left[\left[\tilde{a}^{[j]}\tilde{a}^{[j]}\right]^{[p]}\left[a^{[j]}a^{[j]}\right]^{[p']}\right]^{[A]}\right]^{[0]}$$

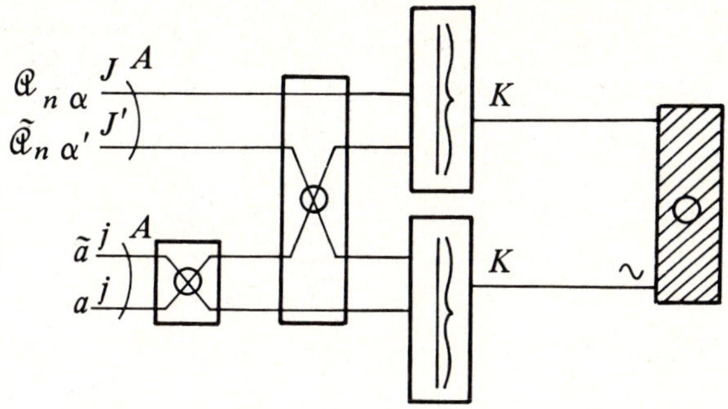

Fig. 7.10.

we have from fig. 7.11, with $\mathsf{M} = 6(n+2)(n+1)$,

$$\langle G_{\langle j \rangle} \rangle = (-)^{p+p'-A} \frac{\mathsf{M}}{\hat{A}} \sum_{\beta K} (-)^{2K} \hat{K} \left(j^n(\alpha J) j^2(p) | \} j^{n+2}(\beta K) \right)$$

$$\times \left(j^n(\alpha' J') j^2(p') | \} j^{n+2}(\beta K) \right) \begin{bmatrix} J & J' & A \\ p & p' & A \\ K & K & 0 \end{bmatrix} .$$

These few examples should suffice to demonstrate the general method of evaluating the vacuum expectation values. The treatment of other cases of $:\Theta_i:$ presents no further difficulty. Some of such additional examples will arise in the cases treated in full below. The multiplicities M are given in full generality in sect. 7.2.6.

We conclude this section with a word of caution. When evaluating a vacuum expectation value, for example in the case of fig. 7.7, one should not proceed as shown in fig. 7.12. This would be erroneous: when working with coupled operators, one must remember to combine all creation operators and annihilation operators

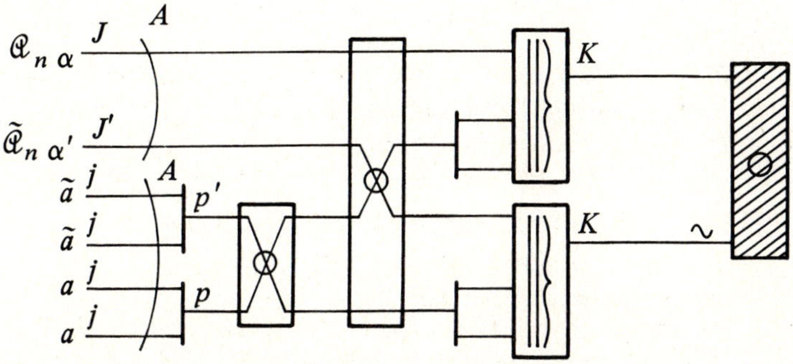

Fig. 7.11.

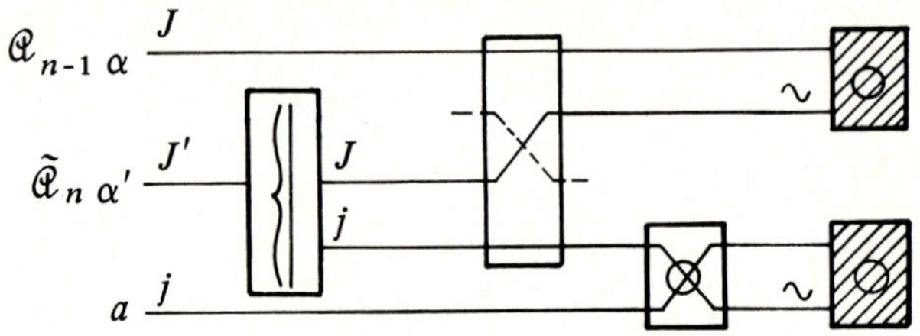

Fig. 7.12.

into products such as in eqs. (4.51) or (4.52) before performing the vacuum expectation values.

7.2.6. *Multiplicities in the Fock representation*

The problem of the multiplicities of the interaction operator associated with the use of well-ordered products in Fock space has already been touched in sects. 6.6, 6.7, and 6.8. Likewise, numerical factors are associated with the normalization of the states in Fock space as in particular in eqs. (4.53) and (4.54). Multiplicity factors have also appeared in the expressions of sect. 7.2.5 for the expectation values $G_{\{j\}}$. We give here the general form of the multiplicity factors. The discussion is done for bosons. For fermions it would follow very similar lines, except for anticommutation.

In general a normal product involving one particle kind is given by a complete sum over all state indices α, β, $\gamma\ldots$ of the participating particles

$$:\Theta_{\mathrm{R}}: = \sum_{\alpha\beta\gamma\ldots} :\theta_\alpha\theta_\beta\theta_\gamma\ldots: = \sum_{\alpha\beta\gamma\ldots} :(a_\alpha+\tilde{a}_\alpha)(a_\beta+\tilde{a}_\beta)(a_\gamma+\tilde{a}_\gamma)\ldots:.$$

A particular process specifies the number of creation and annihilation operators, for example

$$:\Theta_{\mathrm{R}}: = \sum_{\alpha\beta\gamma} :a_\alpha\tilde{a}_\beta\tilde{a}_\gamma + \tilde{a}_\alpha a_\beta\tilde{a}_\gamma + \tilde{a}_\alpha\tilde{a}_\beta a_\gamma:.$$

Performing the normal ordering and introducing the lexicographic order $j, k, l\ldots$ for the creation operators on one hand and for the annihilation operators on the other hand we obtain a sum of well-ordered products each having a multiplicity m:

$$:\Theta_{\mathrm{R}}: = \sum_{jkl} \mathrm{m}:\tilde{a}_j^{N_j}\tilde{a}_k^{N_k}a_l^{N_l}\ldots:.$$

For a term containing N operators with the partition $N_j, N_k, N_l\ldots$ such that $N = N_j + N_k + N_l + \cdots$ and where a given index j denotes fully equivalent particle operators, i.e. having the same state index and the same character, creation or annihilation, the operator multiplicity is:

$$\mathsf{m} = \frac{N!}{N_j! N_k! N_l! \cdots}.$$

For example, this operator multiplicity has been discussed in detail in sect. 6.6 for the pion-pion interaction.

Turning now to the expectation values of N fully equivalent particle operators between normalized state vectors, we find, according to eqs. (4.51) and (4.52), omitting CFP's and recoupling coefficients,

$$\langle 0|\mathcal{Q}_n : \tilde{a}^N : \tilde{\mathcal{Q}}_{n-N}|0\rangle = \langle 0|\mathcal{Q}_{n-N} : a^N : \tilde{\mathcal{Q}}_n|0\rangle$$

$$= \sqrt{n(n-1)\ldots(n-N+1)} = \sqrt{\frac{n!}{(n-N)!}}.$$

Thus the complete factor including both the operator multiplicity m and the state normalization factors, denoted the multiplicity M, is for the general expectation value for a well-ordered product, omitting as above CFP's and recouplings,

$$\mathsf{M} = \langle 0|\mathcal{Q}_{n_j}\mathcal{Q}_{n_k}\ldots : \tilde{a}_j^{N_j}\tilde{a}_k^{N_k}\ldots a_j^{M_j}a_k^{M_k}\ldots : \tilde{\mathcal{Q}}_{m_j}\tilde{\mathcal{Q}}_{m_k}\ldots|0\rangle$$

$$= \sqrt{\frac{n_j! n_k! \ldots}{(n_j - N_j)!(n_k - N_k)!\ldots}} \frac{(N_j + N_k \ldots + M_j + M_k \ldots)!}{N_j! N_k! \ldots M_j! M_k! \ldots}$$

$$\times \sqrt{\frac{m_j! m_k! \ldots}{(m_j - M_j)!(m_k - M_k)!\ldots}}. \tag{7.3}$$

Many examples of multiplicity factors M will be found in the matrix elements given below.

For configurations containing two particles only, sometimes symmetrization or antisymmetrization is performed below directly by calculating all permutations of the particles and introducing the state normalization

$$\mathfrak{N}(\alpha\beta) = \frac{1}{\sqrt{1 + \delta_{\alpha\beta}}},$$

where the symbols $\delta_{\alpha\beta}$ are Kronecker delta functions in all quantum numbers of the particle states α and β.

7.2.7. Space representation

The occupation number representation of the previous sections will be generally used to construct the matrix elements of the many-body systems. It is particularly well adapted to cases where the particle number is not conserved. However, in a few instances, in particular for more-than-two-body operators and for number-conserving operators, the wave function representation is sometimes more convenient for showing the structure of the matrix elements and for deriving the multiplicities. This will be the case for the full n-body center-of-mass position operator treated in the next chapter. Hence for completeness we give in this section a number of relations in the space representation. In this representation the basis vectors are fully symmetrized or antisymmetrized product wave functions of numbered particle coordinates, in position or momentum space. The interaction operators are symmetrized functions of these coordinates.

Since the details of the coupling to good angular momentum are not needed in the determination of the multiplicities and structures of the matrix elements, we here will omit all angular momentum quantum numbers and recoupling coefficients. Otherwise the expressions would be much too cumbersome to show in a transparent manner the underlying symmetry structure of the many-body states.

The space representation is not well adapted to operators which change particle number. Therefore, we shall not give their expression in this representation, and we shall restrict the discussion to one- and two-body number-conserving operators. The extension to full n-body operators will be given in Chapter 8. We first discuss the basis states. Then we use these expressions for obtaining the multiplicities and structure of the various matrix elements.

For a given type of particles, let us consider a general basis vector containing three groups. We use the notation of double underlining to indicate a normalized fully symmetrized state (i.e. symmetrized for bosons and antisymmetrized for fermions), as in sect. 4.6.3:

$$\psi = | \underline{\underline{j^n k^m l^p}} \rangle .$$

When the state is only normalized and symmetrized relatively to the particles in the group j^n it is denoted by

$$\psi = | \underline{\underline{j^n}}\, k^m l^p \rangle .$$

A single underlining is used to denote a non-normalized but fully symmetrized state

$$\psi = | \underline{j^n k^m l^p} \rangle .$$

Similarly when the state is symmetrized but non-normalized in the group j^n only we write

$$\psi = | \underline{j^n}\, k^m l^p \rangle .$$

When the state is fully symmetrized in all particles but normalized only with respect to the groups j^n and k^m separately, we write

$$\psi = |\, \underline{\underline{\underline{j}}^{\,n} \underline{\underline{k}}^{m} l^p} \rangle .$$

Finally, the non-normalized non-symmetrized wave function is

$$\psi = |\, j^n k^m l^p \rangle .$$

Omitting recouplings and summations, we have with this notation the following decompositions of a wave function:

$$\begin{aligned}
|\, \underline{\underline{j}}^{\,n} \alpha \rangle &= (\beta | \} j^n \alpha) \left[|\, \underline{\underline{j}}^{\,n-1} \beta \rangle |\, j \rangle \right] \\
&= (\beta | \} j^n \alpha) |\, \underline{\underline{j}}^{\,n-1} j \beta \rangle \\
&= \frac{1}{\sqrt{n}} (\beta | \} j^n \alpha) |\, \underline{\underline{\underline{j}}^{\,n-1} j}\, \beta \rangle \\
&= (\beta \gamma \| \} j^n \alpha) |\, \underline{\underline{j}}^{\,n-2} \underline{\underline{j}}^{2} \beta \gamma \rangle \\
&= \sqrt{\frac{2}{n(n-1)}}\, (\beta \gamma \| \} j^n \alpha) |\, \underline{\underline{\underline{j}}^{\,n-2} \underline{\underline{j}}^{2}}\, \beta \gamma \rangle \\
&= \sqrt{\frac{1}{n(n-1)}}\, (\beta \gamma \| \} j^n \alpha) |\, \underline{\underline{\underline{j}}^{\,n-2} \underline{j}^{2}}\, \beta \gamma \rangle ,
\end{aligned}$$

where in the last expression the single underline of the sub-group j^2 means that it is symmetrized but non-normalized. In all of these expressions the indices α, β, γ denote all the quantum numbers needed to specify the respective states, and of course summations are implied over β and γ. The examples of table 7.2 will be useful.

The general formula for the symmetrization between the different groups is

$$|\, \underline{\underline{j^n k^m l^p \ldots}} \rangle = \frac{1}{\sqrt{\begin{pmatrix} n+m+p \ldots \\ n \quad m \quad p \ldots \end{pmatrix}}} |\, \underline{\underline{\underline{j}}^{\,n} \underline{\underline{k}}^{m} \underline{\underline{l}}^{p} \ldots} \rangle ,$$

where the multinomial coefficient is given by

$$\begin{pmatrix} n+m+p+\cdots \\ n \quad m \quad p \quad \ldots \end{pmatrix} = \frac{(n+m+p+\cdots)!}{n!m!p!\ldots} .$$

TABLE 7.2
Decompositions of wave functions

$$|\underline{\underline{\underline{j^n}}\,j}\rangle = \sqrt{n+1}\,|\underline{\underline{j^n}}\,j\rangle$$

$$|\underline{\underline{j^n k}}\rangle = \frac{1}{\sqrt{n+1}}|\underline{\underline{\underline{j^n}}\,k}\rangle$$
$$= \frac{1}{\sqrt{n+1}}(\,|\}j^n)|\underline{\underline{\underline{j^{n-1}}}\,jk}\rangle$$
$$= \frac{1}{\sqrt{n+1}}(\,||\}j^n)|\underline{\underline{\underline{j^{n-2}}}\,\underline{\underline{j^2}}\,k}\rangle$$

$$|\underline{\underline{j^n k^2}}\rangle = \sqrt{\frac{2}{(n+2)(n+1)}}\,|\underline{\underline{\underline{j^n}}\,\underline{\underline{k^2}}}\rangle$$
$$= \frac{1}{\sqrt{2(n+2)(n+1)}}|\underline{\underline{\underline{j^n}}\,k^2}\rangle$$
$$= \frac{1}{\sqrt{2(n+2)(n+1)}}(\,|\}j^n\,)|\underline{\underline{\underline{j^{n-1}}}\,j\underline{k^2}}\rangle$$
$$= \frac{1}{\sqrt{2(n+2)(n+1)}}(\,||\}j^n\,)|\underline{\underline{\underline{j^{n-2}}}\,j^2\underline{k^2}}\rangle$$

$$|\underline{\underline{jkl}}\rangle = \frac{1}{\sqrt{3!}}|\underline{jkl}\rangle$$

$$|\underline{\underline{j^n kl}}\rangle = \frac{1}{\sqrt{(n+2)(n+1)}}|\underline{\underline{j^n\,kl}}\rangle$$
$$= \frac{1}{\sqrt{(n+2)(n+1)}}(\,|\}j^n\,)|\underline{\underline{\underline{j^{n-1}}}\,jkl}\rangle$$
$$= \frac{1}{2\sqrt{(n+2)(n+1)}}(\,||\}j^n\,)|\underline{\underline{\underline{j^{n-2}}}\,\underline{j^2}kl}\rangle$$

After the wave functions we now consider the matrix elements. The one-body operator in the space representation is symmetrized in the particle coordinates

$$T = \sum_i T(i)\,.$$

Its matrix element between symmetrized normalized wave functions is generally made up of several terms; some of them are identical and can be written as a

TABLE 7.3
One-body matrix elements

$$\langle \underline{\underline{j^n k}}|T|\underline{\underline{j^n k}}\rangle = n(j^n\{|\ )(\ |\}j^n)\langle j|T|j\rangle + \langle k|T|k\rangle$$

$$\langle \underline{\underline{j^n k^2}}|T|\underline{\underline{j^n k^2}}\rangle = n(j^n\{|\ )(\ |\}j^n)\langle j|T|j\rangle + 2\langle k|T|k\rangle$$

$$\langle \underline{\underline{j^n kl}}|T|\underline{\underline{j^n kl}}\rangle = n(j^n\{|\ )(\ |\}j^n)\langle j|T|j\rangle + \langle k|T|k\rangle + \langle l|T|l\rangle$$

$$\langle \underline{\underline{j}}^n|T|\underline{\underline{j^{n-1}k}}\rangle = \sqrt{n}\,(j^n\{|\ )\langle j|T|k\rangle$$

$$\langle \underline{\underline{j^n k}}|T|\underline{\underline{j^{n-1}k^2}}\rangle = \sqrt{2n}\,(j^n\{|\ )\langle j|T|k\rangle$$

$$\langle \underline{\underline{j^n k}}|T|\underline{\underline{j^{n-1}kl}}\rangle = \sqrt{n}\,(j^n\{|\ )\langle j|T|l\rangle$$

$$\langle \underline{\underline{j^n k^2}}|T|\underline{\underline{j^n kl}}\rangle = \sqrt{2}\,\langle k|T|l\rangle$$

multiplicity factor M times a one-body matrix element. For example

$$\langle \underline{\underline{j}}^n\alpha|T|\,\underline{\underline{j}}^n\beta\rangle = \mathsf{M}\sum_{\gamma}(j^n\alpha\{|\gamma)(\gamma|\}j^n\beta)\langle \underline{\underline{j}}^{n-1}\gamma|\,\underline{\underline{j}}^{n-1}\gamma\rangle\langle j|T|\,j\rangle\,.$$

Here $\mathsf{M} = n$. Of course the overlap $\langle \underline{\underline{j}}^{n-1}\gamma|\,\underline{\underline{j}}^{n-1}\gamma\rangle = 1$. If $\alpha = \beta$ the normalization of the CFP's yields

$$\langle \underline{\underline{j}}^n\alpha|T|\,\underline{\underline{j}}^n\alpha\rangle = n\langle j|T|\,j\rangle\,.$$

Table 7.3 contains as examples a list of matrix elements of a one-body operator between states of different partition.

Similarly a list of matrix elements for a symmetrized two-body operator

$$T = \sum_{i<j} T(i, j)$$

is given in table 7.4. Note the distribution of the single underline in the elementary two-body matrix elements. Also recall that the state $|j^2\rangle$ when coupled to good angular momentum is always normalized and symmetrized. All these relations can be used to obtain the multiplicity factors in the many-body matrix elements which will be calculated in the following sections. The general form for these multiplicities has also been given in sect. 7.2.6.

7.2.8. General notation for many-body configurations

In the next sections we give as examples tables of the matrix elements of the hamiltonian for the configurations of the model spaces of sect. 5.4. They correspond to the description of various light nuclear systems ($B = 0, 1, 2$) in the picture of the

TABLE 7.4
Two-body matrix elements

$$\langle \underline{\underline{j^n}}|T|\underline{\underline{j^n}}\rangle = \tfrac{1}{2}n(n-1)(j^n\{\|\ )(\ \|\}j^n)\langle j^2|T|j^2\rangle$$

$$\langle \underline{\underline{j^nk}}|T|\underline{\underline{j^nk}}\rangle = \tfrac{1}{2}n(n-1)(j^n\{\|\ )(\ \|\}j^n)\langle j^2|T|j^2\rangle$$
$$+n(j^n\{|\ )(\ |\}j^n)\langle jk|T|\underline{jk}\rangle$$

$$\langle \underline{\underline{j^nk^2}}|T|\underline{\underline{j^nk^2}}\rangle = \langle k^2|T|k^2\rangle + \tfrac{1}{2}n(n-1)(j^n\{\|\ )(\ \|\}j^n)\langle j^2|T|j^2\rangle$$
$$+n(j^n\{|\ )(\ |\}j^n)\langle jk|T|\underline{jk}\rangle$$

$$\langle \underline{\underline{j^nkl}}|T|\underline{\underline{j^nkl}}\rangle = \langle kl|T|\underline{kl}\rangle + \tfrac{1}{2}n(n-1)(j^n\{\|\ )(\ \|\}j^n)\langle j^2|T|j^2\rangle$$
$$+n(j^n\{|\ )(\ |\}j^n)\langle jk|T|\underline{jk}\rangle$$
$$+n(j^n\{|\ )(\ |\}j^n)\langle jl|T|\underline{jl}\rangle$$

$$\langle \underline{\underline{j^n}}|T|\underline{\underline{j^{n-1}k}}\rangle = \sqrt{n}(n-1)(j^n\{\|\ )(\ |\}j^{n-1})\langle j^2|T|jk\rangle$$

$$\langle \underline{\underline{j^n}}|T|\underline{\underline{j^{n-2}k^2}}\rangle = \sqrt{\tfrac{1}{2}n(n-1)}\,(j^n\{\|\ )\langle j^2|T|k^2\rangle$$

$$\langle \underline{\underline{j^n}}|T|\underline{\underline{j^{n-2}kl}}\rangle = \sqrt{n(n-1)}\,(j^n\{\|\ )\langle j^2|T|kl\rangle$$

$$\langle \underline{\underline{j^nk}}|T|\underline{\underline{j^{n-1}k^2}}\rangle = \sqrt{2n}\,(j^n\{|\ )\langle jk|T|k^2\rangle$$
$$+2\sqrt{2n}(n-1)(j^n\{\|\ )(\ |\}j^{n-1})\langle j^2|T|jk\rangle$$

$$\langle \underline{\underline{j^nk}}|T|\underline{\underline{j^{n-1}kl}}\rangle = \tfrac{1}{2}\sqrt{n}\,(j^n\{|\ )\langle \underline{jk}|T|\underline{kl}\rangle$$
$$+\tfrac{1}{2}\sqrt{n}(n-1)(j^n\{\|\ )(\ |\}j^{n-1})\langle j^2|T|\underline{jl}\rangle$$

$$\langle \underline{\underline{j^nk^2}}|T|\underline{\underline{j^nkl}}\rangle = \sqrt{2}\,\langle k^2|T|kl\rangle + n(j^n\{|\ )(\ |\}j^n)\langle \underline{jk}|T|\underline{jl}\rangle$$

mesonic degrees of freedom. These examples, with up to six 1s pions in the pion cloud and one vector meson, cover most of the situations one may encounter when treating the relativistic nuclear problem. In these tables a compact and systematic notation will be used. We give here the general rules for that notation.

As far as the basis states of Chapter 5 are concerned CFP's arise only for the isospin of the $1s^n$ pion cloud. They will be denoted by $\mathrm{CFP}_1^n(T, R)$ and $\mathrm{CFP}_2^n(T, R, t)$ as defined in eqs. (4.55) and (4.56).

As in Chapter 5, the indices $\pi, \rho, \omega, \sigma, \mathrm{N}$ denote all the quantum numbers of the particles, for example $\pi_1(\nu_1 l_1)$, $\rho(\kappa\nu_\rho J)\ldots\mathrm{N}(\nu lj)$. In our model spaces the basis states contain at most one heavy boson. They contain at most two nucleons denoted $\mathrm{N}(\nu_N l_N j_N)$ and $\mathrm{M}(\nu_M l_M j_M)$ respectively. They contain several pions states denoted $\pi_1(\nu_1 l_1)$, $\pi_2(\nu_2 l_2)\ldots\pi_i(\nu_i l_i)$. The same notation is used for the bra and the ket, except that the ket symbols carry primes: π_i', L_{12}', N' etc...

The coupling scheme is the one adopted in sect. 5.4., eq. (5.39). Namely, the types are coupled in the order:

$$[\alpha]^{[IT]} = \Big[\big[[\text{heavy boson}]^{[Jt_h]}[\text{pion cloud}]^{[LP\,TP]}\big]^{[JB\,TB]}[\text{nucleons}]^{[JF\,TF]}\Big]^{[IT]}. \tag{7.4}$$

The pion cloud has the standard coupling (5.38), recalling that we have at most three groups:

$$[\text{pion cloud}]^{[LP\,TP]} = \Big[\big[[\text{group 1}]^{[LP_1\,TP_1]}[\text{group 2}]^{[LP_2\,TP_2]}\big]^{[LP_{12}\,TP_{12}]} \times [\text{group 3}]^{[LP_3\,TP_3]}\Big]^{[LP\,TP]}. \tag{7.5}$$

For the spin and isospin, we use throughout in the expressions the symbols s, t if they have half-integer values, and directly their values if they are integers.

The elementary invariant matrix elements are always multiplied by the phases η which are discussed in sect. 7.2.2. Recall that these phases are determined only by the interacting particles at the vertex. They are independent of the spectator particles. As a rule, for simplification, the initial and final particle state indices for these phases are omitted ($\eta_{\mathrm{ab}} \to \eta$). The phase choices given in sect. 7.2.2 achieve a real symmetric hamiltonian matrix.

The invariant matrix elements at the vertices are given directly by the results of Chapter 6.

In the listing of the many-body matrix elements we will omit the recoupling graphs most of the time, except for a few complicated cases. However, every case will be accompanied by a reaction graph. It defines the initial and final configurations and specifies which single-particle states are involved in the different elementary processes possible between the initial and final configurations.

The selection rules imposed by the overlaps of the spectator particles will always be visible on the reaction graph. They are taken into account in the arguments of the formulae. Sometimes they also will be indicated in form of Kronecker deltas. However, for brevity, these delta functions generally will be omitted as the reaction graphs define completely the configurations.

No examples are explicitly given which include spin-$\frac{3}{2}$ fields. If need be they are simply obtained starting from the different formal expressions with spin-$\frac{1}{2}$ fields and replacing the elementary invariant matrix elements by those of the $\pi N\Delta$ and $\pi\Delta\Delta$ processes.

7.3. THE MANY-BODY MATRIX ELEMENTS OF THE PION-NUCLEON INTERACTION

7.3.1. The nucleon system

We give the matrix elements of the interaction of a nucleon with a pion cloud for all the configurations of Chapter 5. The notation for the elementary vertex is given in fig. 6.1 and the expression for the elementary invariant matrix element of $H_{\pi N}$ is given in sect. 6.2. The complete matrix element will be denoted M thereafter.

As a first example, the matrix element M for pion emission by a nucleon with the total quantum numbers $j't$ and final and initial basis state vectors

$$\langle \pi N| = \hat{j}'\hat{t}\langle 0|\left[\tilde{W}^{[j't]}\left[a_{\nu_i}^{[l_i 1]} b_{\nu l}^{[jt]}\right]^{[j't]}\right]^{[00]},$$

$$|N'\rangle = (-)^{2j'+2t}\,\hat{j}'\hat{t}\left[W^{[j't]}\tilde{b}_{\nu' l'}^{[j't]}\right]^{[00]}|0\rangle,$$

is given by eq. (7.2)

$$M = \langle \pi N|H_{\pi N}|N'\rangle = (-)^{2j'+2t}\frac{1}{\hat{j}'\hat{t}}\eta[\pi N|H_{\pi N}|N']$$

$$\times \langle 0|\left[a_{\nu_i}^{[l_i 1]} b_{\nu l}^{[jt]}\right]^{[j't]} : \left[\tilde{a}_{\nu_i}^{[l_i 1]}\tilde{b}_{\nu l}^{[jt]} b_{\nu' l'}^{[j't]}\right]^{[00]} : \tilde{b}_{\nu' l'}^{[j't]}|0\rangle$$

$$= \frac{1}{\hat{j}'\hat{t}}\eta[\pi N|H_{\pi N}|N'].$$

With the phase choices of sect. 7.2.2 M is real and the element of a symmetric matrix as can be checked by the evaluation of $M = \langle N'|H_{\pi N}|\pi N\rangle^*$. Likewise all many-body matrix elements with a pion-nucleon vertex will be real and symmetric. Henceforth we list only the upper half of the hamiltonian matrix.

We now consider in turn the case of one, two and three groups in the pion cloud which surrounds the nucleon. For the group $1s^n$ the formulae hold for an arbitrary number n of pions.

(i) One pion group in the final state.

$1s^n N/1s^{n-1}N'$, fig. 7.13 for the $1s^n$ pion cloud ($\pi_1 = 1s$):

$$M = \sqrt{n}\,\frac{1}{\hat{j}'\hat{t}}\mathrm{CFP}_1^n(TP, TP')\begin{bmatrix} TP' & 1 & TP \\ 0 & t & t \\ TP' & t & T \end{bmatrix}\eta[1s\,N|H_{\pi N}|N']\delta_{jj'}.$$

$\pi_1^2 N/\pi_1 N'$, fig. 7.14:

$$M = \sqrt{2}\,\frac{1}{\hat{j}'\hat{t}}\begin{bmatrix} l_1 & l_1 & LP \\ 0 & j & j \\ l_1 & j' & I \end{bmatrix}\begin{bmatrix} 1 & 1 & TP \\ 0 & t & t \\ 1 & t & T \end{bmatrix}\eta[\pi_1 N|H_{\pi N}|N'].$$

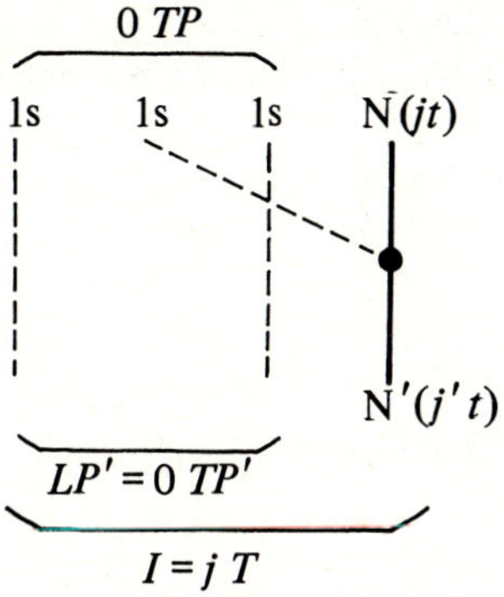

Fig. 7.13.

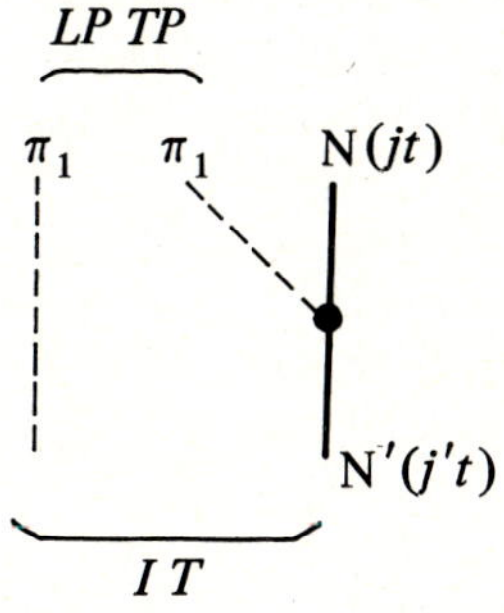

Fig. 7.14.

(ii) Two pion groups in the final state.

$\pi_1^n \pi_2 \mathrm{N}/\pi_1'^n \mathrm{N}'$, fig. 7.15 case A,

$\pi_1 \pi_2^n \mathrm{N}/\pi_1'^n \mathrm{N}'$, fig. 7.15 case B.

These two cases differ only in the order of the pion groups which is imposed by the standard order.

Case A:

$$M=\frac{1}{\hat{j}'\hat{t}}\begin{bmatrix} LP_1 & l_2 & LP \\ 0 & j & j \\ LP_1 & j' & I \end{bmatrix}\begin{bmatrix} TP_1 & 1 & TP \\ 0 & t & t \\ TP_1 & t & T \end{bmatrix}\eta[\pi_2\mathrm{N}|H_{\pi\mathrm{N}}|\mathrm{N}']\delta_{\{\pi_1\}\{\pi_1'\}}.$$

Case B:

$$M=\frac{1}{\hat{j}'\hat{t}}(-)^{l_1+LP_2-LP+1+TP_2-TP}\begin{bmatrix} LP_2 & l_1 & LP \\ 0 & j & j \\ LP_2 & j' & I \end{bmatrix}$$

$$\times\begin{bmatrix} TP_2 & 1 & TP \\ 0 & t & t \\ TP_2 & t & T \end{bmatrix}\eta[\pi_1\mathrm{N}|H_{\pi\mathrm{N}}|\mathrm{N}']\delta_{\{\pi_2\}\{\pi_1'\}}.$$

Here $\{\pi_i\}$ denotes all the quantum numbers needed to specify the pion cloud $\pi_i^{n_i}$. The $\delta_{\{\pi_i\}\{\pi_j\}}$ denote Kronecker deltas over all the quantum numbers of the pion clouds $\{\pi_i\}$ and $\{\pi_j\}$. If the spectator pion cloud is $1s^n$, then because of the standard

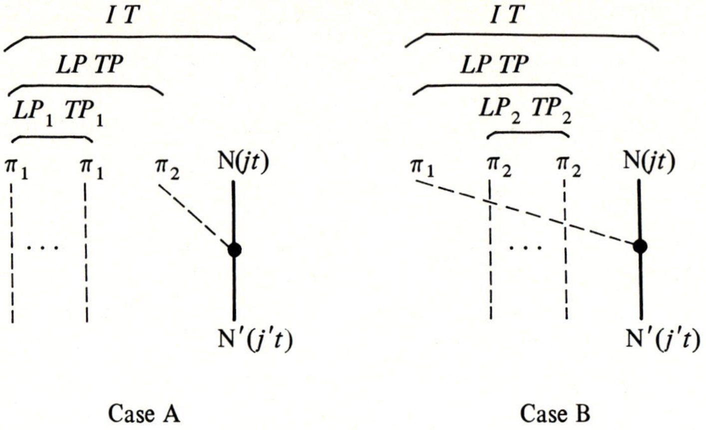

Fig. 7.15.

order, we have only case A which now is reduced to the simpler form:

$$M = \frac{1}{\hat{j}'\hat{t}} \begin{bmatrix} TP_1 & 1 & TP \\ 0 & t & t \\ TP_1 & t & T \end{bmatrix} \eta[\pi_2 \mathrm{N} | H_{\pi\mathrm{N}} | \mathrm{N}'] .$$

$1s^{n_1}\pi_2^{n_2}\mathrm{N}/1s^{n_1-1}\pi_2'^{n_2}\mathrm{N}'$, fig. 7.16.
Creation of a pion in a 1s state when $\pi_2^{n_2}$ is spectator:

$$M = \sqrt{n_1}\, \frac{1}{\hat{j}'\hat{t}} \mathrm{CFP}_1^{n_1}(TP_1, TP_1') \begin{bmatrix} TP_1' & 1 & TP_1 \\ TP_2 & 0 & TP_2 \\ TP' & 1 & TP \end{bmatrix}$$

$$\times \begin{bmatrix} TP' & 1 & TP \\ 0 & t & t \\ TP' & t & T \end{bmatrix} \eta[1s\mathrm{N} | H_{\pi\mathrm{N}} | \mathrm{N}'] \delta_{\{\pi_2\}\{\pi_2'\}} \delta_{jj'} .$$

(iii) Three pion groups in the final state.
The tables of Chapter 5 contain only cases where each of the three groups has only one pion. The three cases shown in fig. 7.17 differ only in the order of the pion labels which is imposed by the standard order of the list.

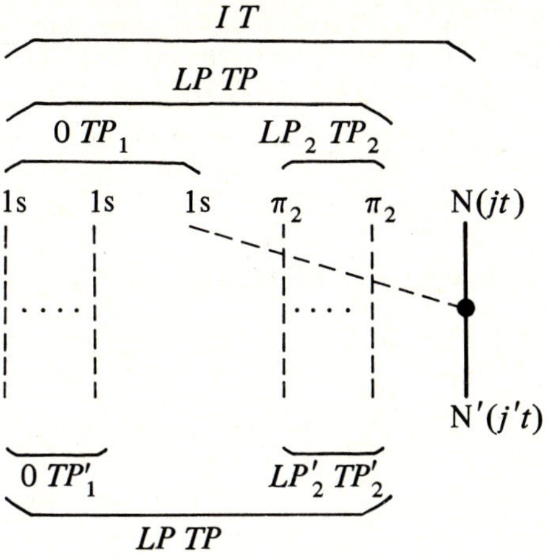

Fig. 7.16.

$\pi_1\pi_2\pi_3\mathrm{N}/\pi_1'\pi_2'\mathrm{N}'$, fig. 7.17.

Case A:

$$M=\frac{1}{\hat{j}'\hat{t}}\begin{bmatrix} LP_{12} & l_3 & LP \\ 0 & j & j \\ LP_{12} & j' & I \end{bmatrix}$$

$$\times\begin{bmatrix} TP_{12} & 1 & TP \\ 0 & t & t \\ TP_{12} & t & T \end{bmatrix}\eta[\pi_3\mathrm{N}|H_{\pi\mathrm{N}}|\mathrm{N}']\delta_{\pi_1\pi_1'}\delta_{\pi_2\pi_2'}\delta_{LP_{12}LP'}\delta_{TP_{12}TP'}.$$

Case B:

$$M=\frac{1}{\hat{j}'\hat{t}}\begin{bmatrix} l_1 & l_2 & LP_{12} \\ l_3 & 0 & l_3 \\ LP' & l_2 & LP \end{bmatrix}\begin{bmatrix} 1 & 1 & TP_{12} \\ 1 & 0 & 1 \\ TP' & 1 & TP \end{bmatrix}$$

$$\times\begin{bmatrix} LP' & l_2 & LP \\ 0 & j & j \\ LP' & j' & I \end{bmatrix}\begin{bmatrix} TP' & 1 & TP \\ 0 & t & t \\ TP' & t & T \end{bmatrix}\eta[\pi_2\mathrm{N}|H_{\pi\mathrm{N}}|\mathrm{N}']$$

$$\times\delta_{\pi_1\pi_1'}\delta_{\pi_3\pi_2'}.$$

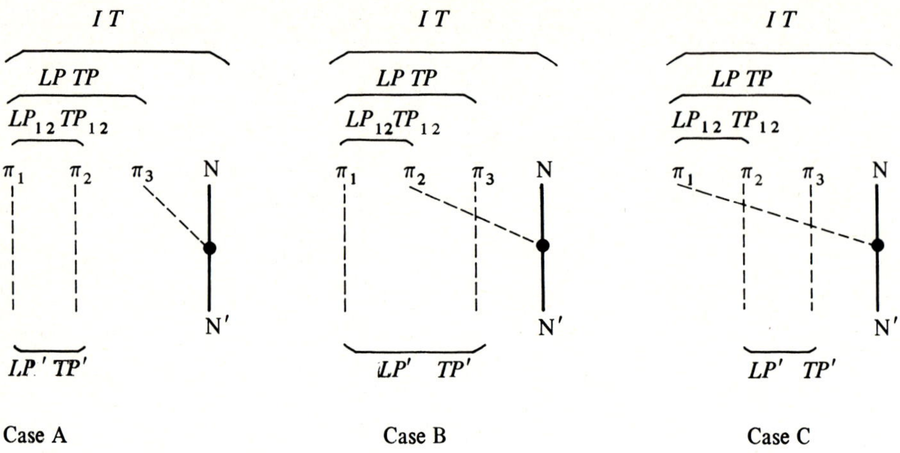

Fig. 7.17.

Case C:

$$M=(-)^{l_1+l_2-LP_{12}-TP_{12}}\frac{1}{\hat{j}'\hat{t}}\begin{bmatrix} l_2 & l_1 & LP_{12} \\ l_3 & 0 & l_3 \\ LP' & l_1 & LP \end{bmatrix}\begin{bmatrix} 1 & 1 & TP_{12} \\ 1 & 0 & 1 \\ TP' & 1 & TP \end{bmatrix}$$

$$\times\begin{bmatrix} LP' & l_1 & LP \\ 0 & j & j \\ LP' & j' & I \end{bmatrix}\begin{bmatrix} TP' & 1 & TP \\ 0 & t & t \\ TP' & t & T \end{bmatrix}\eta[\pi_1 \mathrm{N}|H_{\pi\mathrm{N}}|\mathrm{N}']\delta_{\pi_2\pi_1'}\delta_{\pi_3\pi_2'}.$$

7.3.2. Extension to the two-nucleon system

The general pion emission matrix elements for the two-nucleon system with a pion cloud are represented in fig. 7.18. The placement of the vertex point indicates that the pion can be emitted by either of the fermions. The expression for the matrix element will be written with the help of the matrix elements $\langle\{\pi\}\mathrm{N};\, YX|M|\{\pi\}\mathrm{N}';\, YX\rangle$ given in the previous section. Here Y and X are the total angular momentum and isospin respectively at the vertex point, while I and T refer to the total system. On the graph 7.19 the top end box of course represents the invariant forms of the above matrix elements. They are obtained by multiplying the full matrix elements of the previous section with the factor $\hat{X}\hat{Y}$. They also contain the phase η. Thus they are represented in the recoupling graphs by the usual end box modified by the label η. The antisymmetrization of the fermions yields four terms. Their recoupling graphs are very similar. One of them is shown in fig. 7.19. The final result is the real symmetric matrix element (the normalization factors $\mathcal{N}(\mathrm{NM})$ are given in sect. 7.2.6):

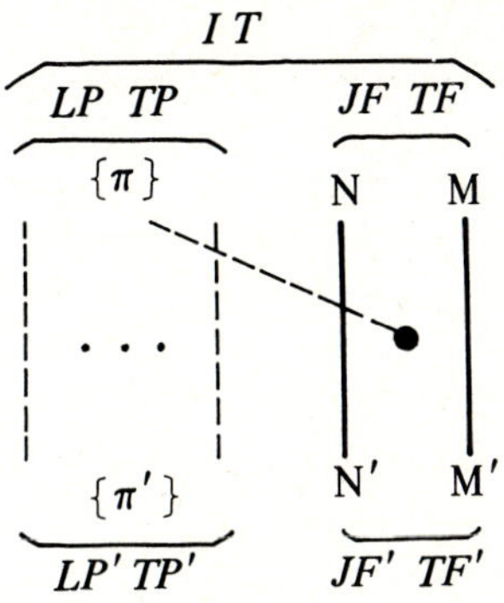

Fig. 7.18.

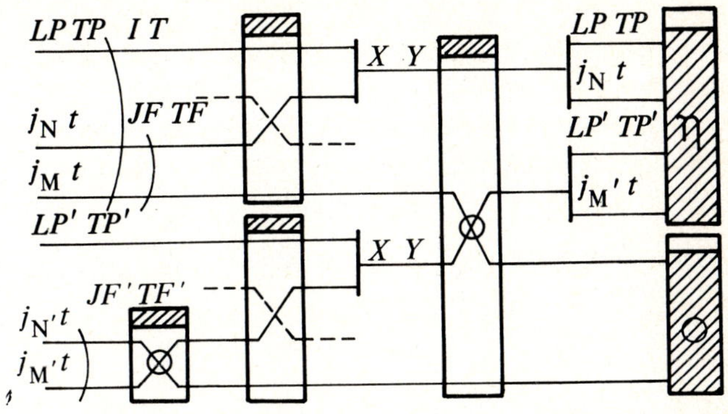

Fig. 7.19.

$\{\pi\}$NM/$\{\pi'\}$N′M′, fig. 7.18.

$$M=\sum_X\begin{bmatrix}TP & 0 & TP\\ t & t & TF\\ X & t & T\end{bmatrix}\begin{bmatrix}TP' & 0 & TP'\\ t & t & TF'\\ X & t & T\end{bmatrix}\frac{1}{\hat{X}}\,\mathfrak{N}(\mathrm{NM})\,\mathfrak{N}(\mathrm{N'M'})$$

$$\times\left(\delta_{\mathrm{MM'}}\sum_Y\begin{bmatrix}LP & 0 & LP\\ j_{\mathrm{N}} & j_{\mathrm{M}} & JF\\ Y & j_{\mathrm{M}} & I\end{bmatrix}\begin{bmatrix}LP' & 0 & LP'\\ j_{\mathrm{N'}} & j_{\mathrm{M'}} & JF'\\ Y & j_{\mathrm{M'}} & I\end{bmatrix}\frac{1}{\hat{Y}}[\{\pi\}\mathrm{N}|M|\{\pi'\}\mathrm{N'}]\right.$$

$$+(-)^{j_{\mathrm{N'}}+j_{\mathrm{M'}}-JF'-TF'}\delta_{\mathrm{MN'}}\sum_Y\begin{bmatrix}LP & 0 & LP\\ j_N & j_{\mathrm{M}} & JF\\ Y & j_{\mathrm{M}} & I\end{bmatrix}\begin{bmatrix}LP' & 0 & LP'\\ j_{\mathrm{M'}} & j_{\mathrm{N'}} & JF'\\ Y & j_{\mathrm{N'}} & I\end{bmatrix}\frac{1}{\hat{Y}}$$

$$\times[\{\pi\}\mathrm{N}|M|\{\pi'\}\mathrm{M'}]$$

$$+(-)^{j_{\mathrm{N}}+j_{\mathrm{M}}-JF-TF}\delta_{\mathrm{NM'}}\sum_Y\begin{bmatrix}LP & 0 & LP\\ j_{\mathrm{M}} & j_{\mathrm{N}} & JF\\ Y & j_{\mathrm{N}} & I\end{bmatrix}\begin{bmatrix}LP' & 0 & LP'\\ j_{\mathrm{N'}} & j_{\mathrm{M'}} & JF'\\ Y & j_{\mathrm{M'}} & I\end{bmatrix}\frac{1}{\hat{Y}}$$

$$\times[\{\pi\}\mathrm{M}|M|\{\pi'\}\mathrm{N'}]$$

$$+(-)^{j_{\mathrm{N}}+j_{\mathrm{M}}-JF-TF}(-)^{j_{\mathrm{N'}}+j_{\mathrm{M'}}-JF'-TF'}\delta_{\mathrm{NN'}}\sum_Y\begin{bmatrix}LP & 0 & LF\\ j_{\mathrm{M}} & j_{\mathrm{N}} & JF\\ Y & j_{\mathrm{N}} & I\end{bmatrix}$$

$$\left.\times\begin{bmatrix}LP' & 0 & LP'\\ j_{\mathrm{M'}} & j_{\mathrm{N'}} & JF'\\ Y & j_{\mathrm{N'}} & I\end{bmatrix}\frac{1}{\hat{Y}}[\{\pi\}\mathrm{M}|M|\{\pi'\}\mathrm{M'}]\right).$$

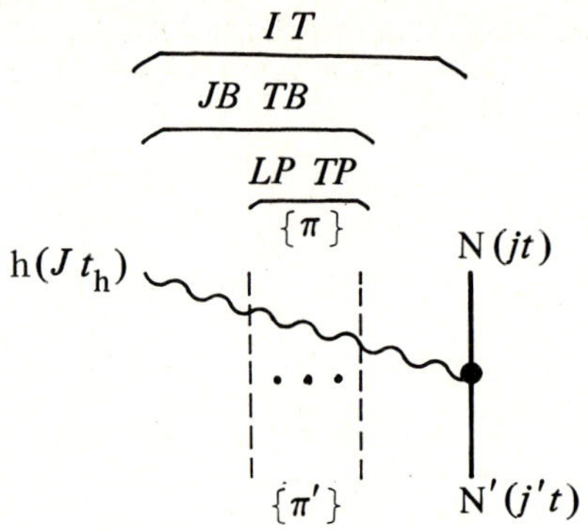

Fig. 7.20.

7.4. THE MANY-BODY MATRIX ELEMENTS WITH HEAVY BOSONS

In the system a heavy boson can interact either with the nucleon or with the pion cloud. We consider here configurations with at most one heavy boson, and thus only its interaction with the pion cloud poses a non-trivial many-body problem. In this section the set of quantum numbers for the heavy boson is denoted h (for heavy): $\mathrm{h} = (\nu_{\mathrm{h}} J)$ for the σ and $\mathrm{h} = (\kappa \nu_{\mathrm{h}} J t_{\mathrm{h}})$ for the ρ and ω.

7.4.1. The heavy boson-nucleon matrix elements

For the emission of a heavy boson by a nucleon the matrix element is calculated exactly as in sect. 7.3.1 for pion emission. This yields for the elementary process the real symmetric matrix

$$M = \frac{1}{\hat{j}'\hat{t}} \eta [\mathrm{hN} | H_{\mathrm{hN}} | \mathrm{N}']$$

with the phases η of sect. 7.2.2 and where the invariant matrix element is given in sect. 6.5.

For the emission of a heavy boson by a single nucleon with a pion cloud as a spectator we use the notation of fig. 7.20 which conforms to the coupling scheme eq. (7.4) adopted for the basis vectors. The matrix element includes a complete pion cloud overlap denoted $\delta_{\{\pi\}\{\pi'\}}$. We include the case of the σ-meson by putting $t_{\mathrm{h}} = 0$ and $H_{\mathrm{hN}} = H_{\sigma\mathrm{N}}$ of sect. 6.4. For the ρ and ω, $t_{\mathrm{h}} = 1$ or 0, and $H_{\mathrm{hN}} = H_{\mathrm{VN}}$ of sect. 6.5.:

$\mathrm{h}\{\pi\}\mathrm{N}/\{\pi'\}\mathrm{N}'$, fig. 7.20.

$$M = \frac{1}{\hat{j}'\hat{t}} (-)^{J+LP-JB+t_{\mathrm{h}}+TP-TB} \begin{bmatrix} LP & J & JB \\ 0 & j & j \\ LP & j' & I \end{bmatrix} \begin{bmatrix} TP & t_{\mathrm{h}} & TB \\ 0 & t & t \\ TP & t & T \end{bmatrix} \eta [\mathrm{hN} | H_{\mathrm{hN}} | \mathrm{N}'] \delta_{\{\pi\}\{\pi'\}}.$$

For configurations with two nucleons the notation is that of fig. 7.21. As in sect. 7.3.2 the antisymmetrization is performed explicitly and yields several recoupling graphs. One of these is given in fig. 7.22.

h $\{\pi\}$NM/$\{\pi'\}$N'M', fig. 7.21.

$$M=\frac{1}{\widehat{JF'}\widehat{TF'}}\begin{bmatrix} LP & J & JB \\ 0 & JF & JF \\ LP & JF' & I \end{bmatrix}\begin{bmatrix} TP & t_h & TB \\ 0 & TF & TF \\ TP & TF' & T \end{bmatrix}\begin{bmatrix} t & t & TF \\ t & t & TF' \\ t_h & 0 & t_h \end{bmatrix}$$

$$\times\hat{t}\,\mathcal{N}(\mathrm{NM})\,\mathcal{N}(\mathrm{N'M'})\,(-)^{J+LP-JB+t_h+TP-TB}$$

$$\times\delta_{\{\pi\}\{\pi'\}}\left(\delta_{\mathrm{MM'}}\begin{bmatrix} j_\mathrm{N} & j_\mathrm{M} & JF \\ j_\mathrm{N'} & j_\mathrm{M'} & JF' \\ J & 0 & J \end{bmatrix}\hat{j}_\mathrm{M}\eta[\mathrm{hN}|H_\mathrm{hN}|\mathrm{N'}]\right.$$

$$+(-)^{j_\mathrm{N'}+j_\mathrm{M'}-JF'-TF'}\delta_{\mathrm{MN'}}\begin{bmatrix} j_\mathrm{N} & j_\mathrm{M} & JF \\ j_\mathrm{M'} & j_\mathrm{N'} & JF' \\ J & 0 & J \end{bmatrix}\hat{j}_\mathrm{M}\eta[\mathrm{hN}|H_\mathrm{hN}|\mathrm{M'}]$$

$$+(-)^{j_\mathrm{N}+j_\mathrm{M}-JF-TF}\delta_{\mathrm{NM'}}\begin{bmatrix} j_\mathrm{M} & j_\mathrm{N} & JF \\ j_\mathrm{N'} & j_\mathrm{M'} & JF' \\ J & 0 & J \end{bmatrix}\hat{j}_\mathrm{N}\eta[\mathrm{hM}|H_\mathrm{hN}|\mathrm{N'}]$$

$$+(-)^{j_\mathrm{N}+j_\mathrm{M}-JF-TF}(-)^{j_\mathrm{N'}+j_\mathrm{M'}-JF'-TF'}\delta_{\mathrm{NN'}}\begin{bmatrix} j_\mathrm{M} & j_\mathrm{N} & JF \\ j_\mathrm{M'} & j_\mathrm{N'} & JF' \\ J & 0 & J \end{bmatrix}$$

$$\left.\times\hat{j}_\mathrm{N}\eta[\mathrm{hM}|H_\mathrm{hN}|\mathrm{M'}]\right)$$

7.4.2. The sigma- and rho-pion matrix elements

We now consider the cases where the sigma or the rho interacts with the pion cloud. As discussed in sect. 7.2.3, any number of spectator fermions can be present since they do not contribute to the expression of the matrix element in the adopted coupling scheme, eq. (7.4). We give only the cases which arise in the configuration spaces of sect. 5.4. They are listed respectively for two-pion annihilation, pion

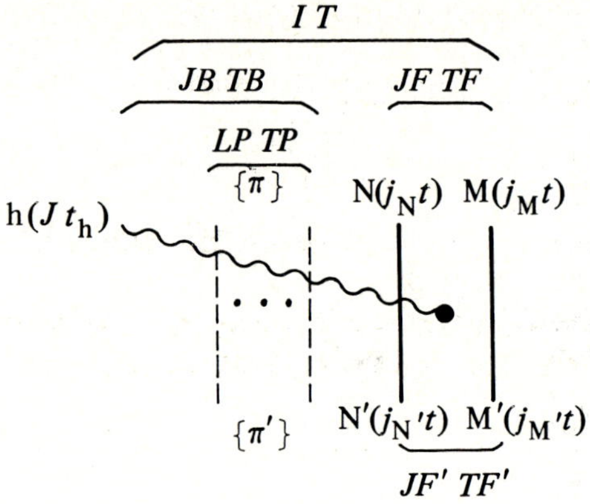

Fig. 7.21.

scattering, two-pion creation processes. The many-body matrix elements are written in the case of the ρ-2π interaction whose elementary matrix element is given in sect. 6.7. There the index ρ denotes the quantum numbers of the ρ state and the index P distinguishes between the different processes, eq. (6.9). The σ-2π many-body matrix elements are obtained from those of the ρ by setting $t_h = 0$ and replacing the elementary matrix element of $H_{\rho\pi}$ by the elementary matrix element of the σ-2π vertex of sect. 6.9.

The many-body matrix elements are classified according to the number of pions in the initial state, n'.

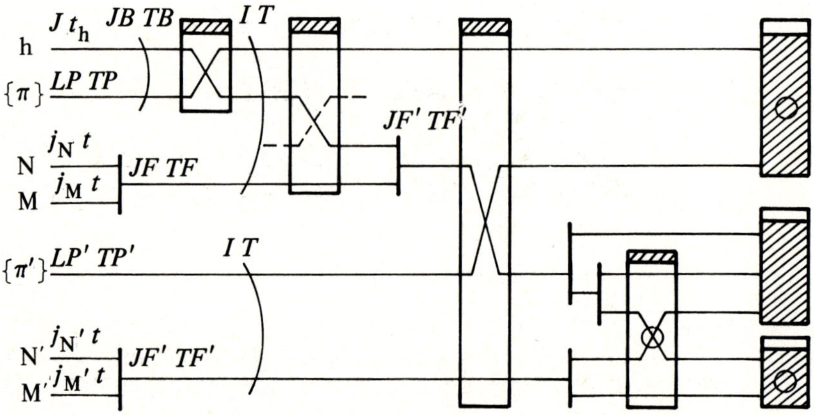

Fig. 7.22.

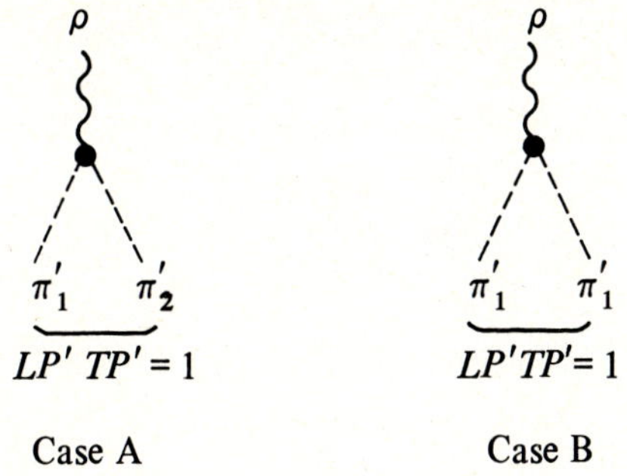

Fig. 7.23.

(i) Two-pion annihilation (P = 1).

$n' = 2$

$\rho/\pi_1'\pi_2'$ (case A) or $\rho/\pi_1'^2$ (case B), fig. 7.23.

$$\textit{Case A}: \quad M = \sqrt{4}\,\frac{1}{\hat{J}\hat{1}}\eta\left[\rho|H_{\rho\pi}|\pi_1'\pi_2'\right].$$

$$\textit{Case B}: \quad M = \sqrt{2}\,\frac{1}{\hat{J}\hat{1}}\eta\left[\rho|H_{\rho\pi}|\pi_1'^2\right].$$

$n' = 3$

$\rho\pi_1/\pi_1'^2\pi_2'$ with $\pi_1 \equiv \pi_2'$ (case A) or $\pi_1 \equiv \pi_1'$ (case B), fig. 7.24.

$$\textit{Case A}: \quad M = \sqrt{2}\,\frac{1}{\hat{J}\hat{1}}\eta\left[\rho|H_{\rho\pi}|\pi_1'^2\right]\delta_{LP_1'J}\delta_{TP_1'1}\delta_{\pi_1\pi_2'}.$$

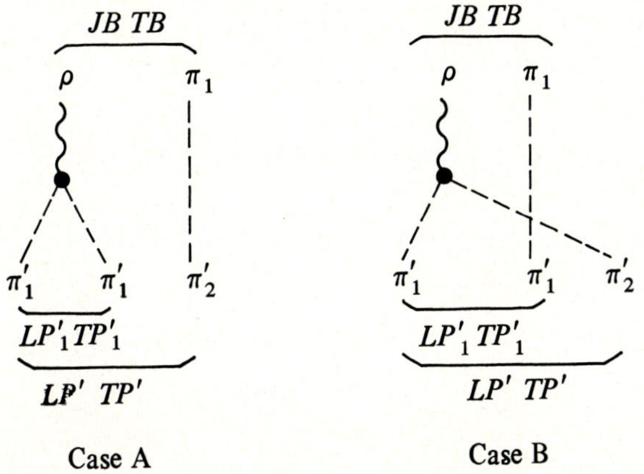

Fig. 7.24.

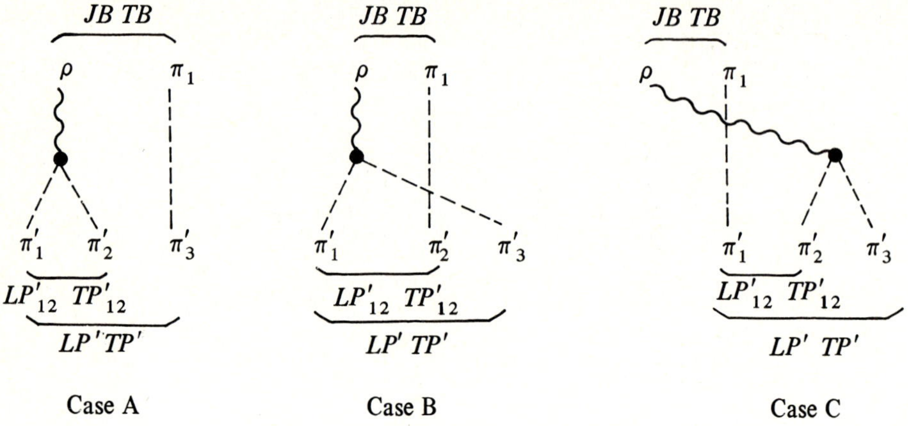

Fig. 7.25.

$$Case\ B:\quad M=\sqrt{8}\,\frac{1}{\hat{J}\hat{1}}\begin{bmatrix} l'_1 & l'_1 & LP'_1 \\ l'_2 & 0 & l'_2 \\ J & l'_1 & LP' \end{bmatrix}\begin{bmatrix} 1 & 1 & TP'_1 \\ 1 & 0 & 1 \\ 1 & 1 & TP' \end{bmatrix}\eta\left[\rho|H_{\rho\pi}|\pi'_1\pi'_2\right]\delta_{\pi_1\pi'_1}\,.$$

$\rho\pi_1/\pi'_1\pi'_2\pi'_3$ with $\pi_1\equiv\pi'_3$ (case A) or $\pi_1\equiv\pi'_2$ (case B) or $\pi_1\equiv\pi'_1$ (case C), fig. 7.25.

$$Case\ A:\quad M=\sqrt{4}\,\frac{1}{\hat{J}\hat{1}}\eta\left[\rho|H_{\rho\pi}|\pi'_1\pi'_2\right]\delta_{LP'_{12}J}\delta_{TP'_{12}1}\delta_{\pi_1\pi'_3}\,.$$

$$Case\ B:\quad M=\sqrt{4}\,\frac{1}{\hat{J}\hat{1}}\begin{bmatrix} l'_1 & l'_2 & LP'_{12} \\ l'_3 & 0 & l'_3 \\ J & l'_2 & LP' \end{bmatrix}\begin{bmatrix} 1 & 1 & TP'_{12} \\ 1 & 0 & 1 \\ 1 & 1 & TP' \end{bmatrix}\eta\left[\rho|H_{\rho\pi}|\pi'_1\pi'_3\right]\delta_{\pi_1\pi'_2}\,.$$

$$Case\ C:\quad M=\sqrt{4}\,\frac{1}{\hat{J}\hat{1}}(-)^{l'_1+J-LP'-TP'}\begin{bmatrix} l'_1 & l'_2 & LP'_{12} \\ 0 & l'_3 & l'_3 \\ l'_1 & J & LP' \end{bmatrix}$$

$$\times\begin{bmatrix} 1 & 1 & TP'_{12} \\ 0 & 1 & 1 \\ 1 & 1 & TP' \end{bmatrix}\eta\left[\rho|H_{\rho\pi}|\pi'_2\pi'_3\right]\delta_{\pi_1\pi'_1}\,.$$

$\underline{n'=4}$

$\rho 1s^2/1s^3\nu s$ with $\nu>1$, fig. 7.26.

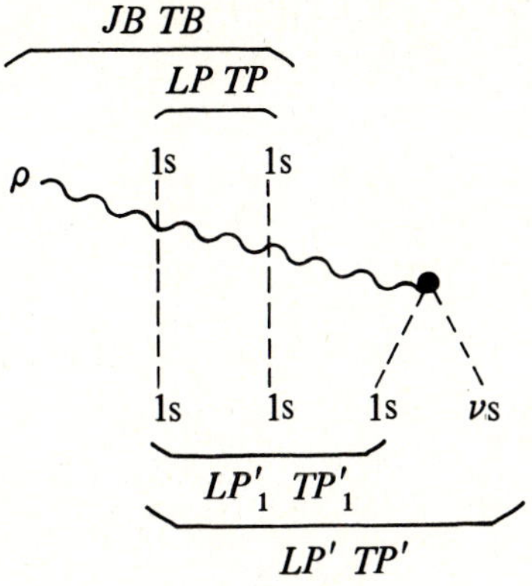

Fig. 7.26.

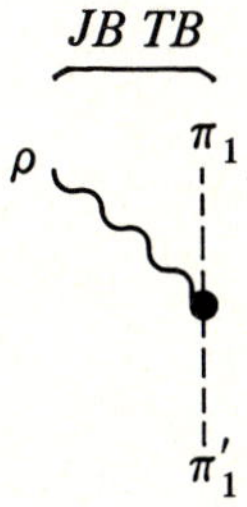

Fig. 7.27.

$$M = (-)^{TP+1-TP'} \sqrt{12}\, \frac{1}{\hat{1}} \mathrm{CFP}_1^3(TP_1', TP) \begin{bmatrix} TP & 1 & TP_1' \\ 0 & 1 & 1 \\ TP & 1 & TP' \end{bmatrix} \eta\left[\rho | H_{\rho\pi} [1s\, \nu s]\, \delta_{J0}\right].$$

(ii) Pion scattering (P = 2)

$n' = 1$

$\rho\pi_1/\pi_1'$, fig. 7.27.

$$M = \sqrt{4}\, \frac{1}{\hat{l}_1'\hat{1}} \eta\left[\rho\pi_1 | H_{\rho\pi} | \pi_1'\right].$$

$n' = 2$

$\rho 1s^2/1s^2$, fig. 7.28.

$$M = \sqrt{16}\, \frac{1}{\hat{1}} \begin{bmatrix} 1 & 0 & 1 \\ 1 & 1 & TP \\ 1 & 1 & TP' \end{bmatrix} \eta\left[\rho 1s | H_{\rho\pi} | 1s\right] \delta_{J0}.$$

$\rho 1s^2/1s\nu s$ with $\nu > 1$, fig. 7.29.

$$M = (-)^{TP'} \sqrt{8}\, \frac{1}{\hat{1}} \begin{bmatrix} 1 & 0 & 1 \\ 1 & 1 & TP \\ 1 & 1 & TP' \end{bmatrix} \eta\left[\rho 1s | H_{\rho\pi} | \nu s\right] \delta_{J0}.$$

(iii) Two-pion creation (P = 3)

$n' = 0$

$\rho\pi_1\pi_2$/vacuum (case A) or $\rho\pi_1^2$/vacuum (case B), fig. 7.30.

$$\textit{Case A}: \quad M = \sqrt{4}\, \eta\left[\rho\pi_1\pi_2 | H_{\rho\pi} |\right] \delta_{J0}.$$

$$\textit{Case B}: \quad M = 0.$$

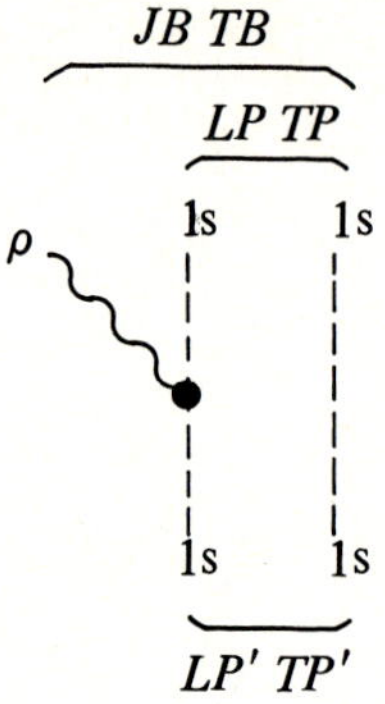

Fig. 7.28.

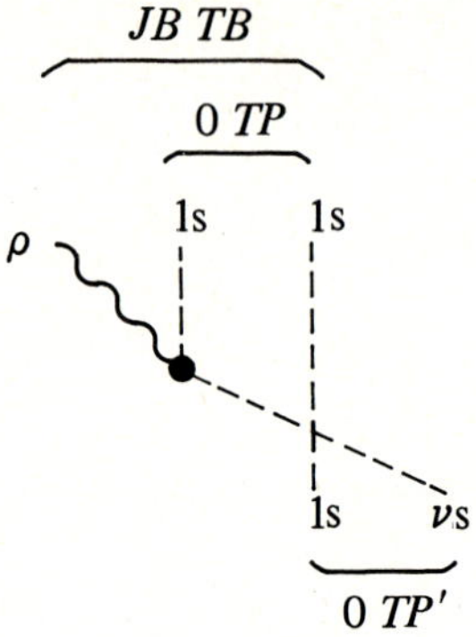

Fig. 7.29.

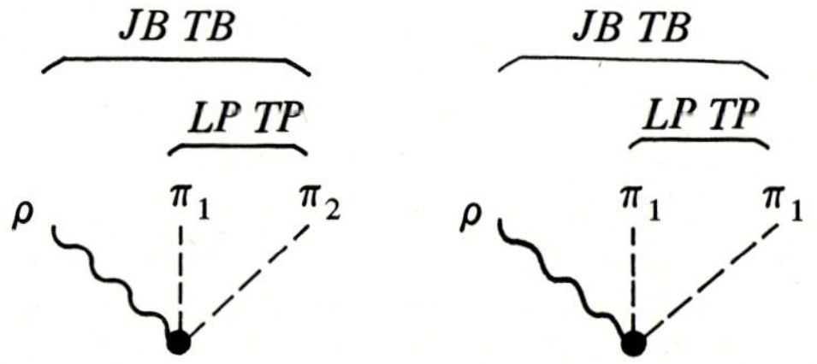

Case A Case B

Fig. 7.30.

7.4.3. The omega-pion matrix elements

The elementary invariant matrix element $H_{\omega\pi}$ is given in sect. 6.8. Again all fermion spectators factor out in the coupling scheme (7.4). All possible cases within the configurations of Chapter 5 lead to the four processes of fig. 7.31. They correspond to the elementary matrix element of eq. (6.11) with multiplicity factors M, recalling the selection rule $\pi_1 \neq \pi_2 \neq \pi_3$ or $\pi_1 \neq \pi_2 \neq \pi_3'$, etc. as explained in sect. 6.8. The corresponding real symmetric hamiltonian matrix for this elementary process is:

$$M = \begin{cases} \mathsf{M}\eta[\omega\pi_1\pi_2\pi_3|H_{\omega\pi}|], & \text{for case A} \\ \mathsf{M}\dfrac{1}{\hat{l}_1'\hat{1}}\eta[\omega\pi_1\pi_2|H_{\omega\pi}|\pi_1'], & \text{for case B} \\ \mathsf{M}\dfrac{1}{\widehat{LP}'\hat{1}}\eta[\omega\pi_1|H_{\omega\pi}|\pi_1'\pi_2'], & \text{for case C} \\ \mathsf{M}\dfrac{1}{\hat{J}}\eta[\omega|H_{\omega\pi}|\pi_1'\pi_2'\pi_3'], & \text{for case D.} \end{cases}$$

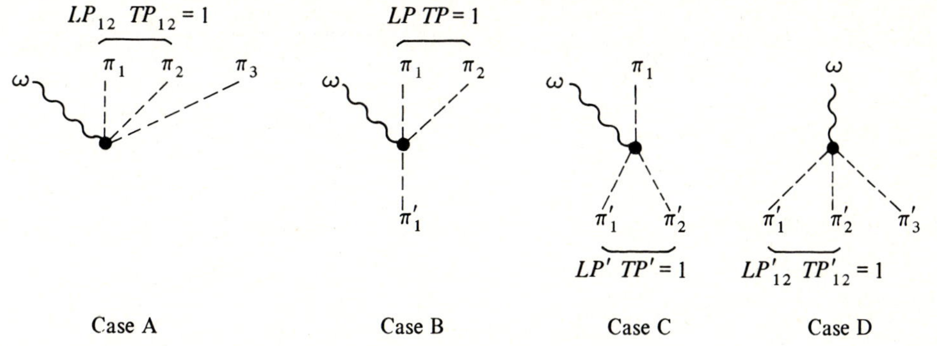

Fig. 7.31.

7.5. THE MANY-BODY MATRIX ELEMENTS OF THE PION-PION INTERACTION

The elementary matrix element of the pion-pion interaction $[\pi_i\pi_j|H_\pi|\pi_k\pi_l]$ given in eq. (6.6) has the symmetry properties of the invariant triple product, once the coupling $[\pi_i\pi_j]^{[LT]}$ has been specified. Hence all the processes have the same elementary matrix element. Of course the full matrix elements differ in recoupling and multiplicities **M**.

We treat here the pion-pion interaction in the pion cloud alone. The presence of spectators (heavy bosons, nucleons) is handled in the standard way described in sect. 7.2.3.

We now give all the examples which are contained in the configuration spaces of sect. 5.4. They are classified according to the number of pions created at the vertex (4, 3 or 2). In the drawings of the many-body vertices sometimes the selection rules are indicated.

7.5.1. *Creation of four pions*

$1s^4$/vacuum, fig. 7.32.

We note that the coupling in isospin space of the last pion is unique. Thus we need only the one-particle CFP_1^3 for analyzing the $1s^4$ configuration (we are using the simplified notation of sect. 4.6.4 for the CFP's):

$$M = \sqrt{24}\sum_T CFP_1^3(1, T)\,\eta\big[1s^4|H_\pi|\big].$$

Here, as for all π_i^2 cases, $T = 0$ or 2.

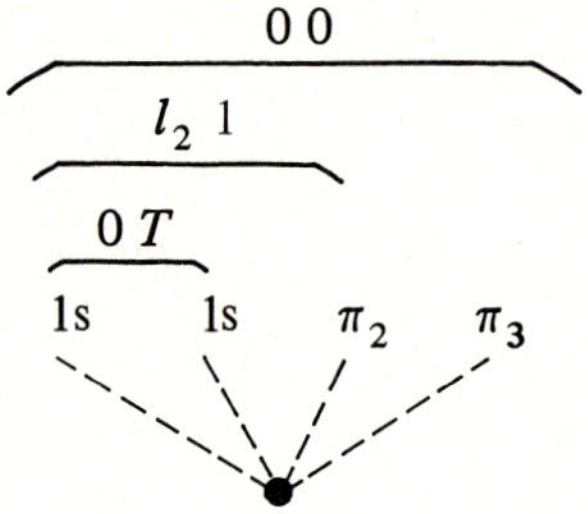

Fig. 7.32.

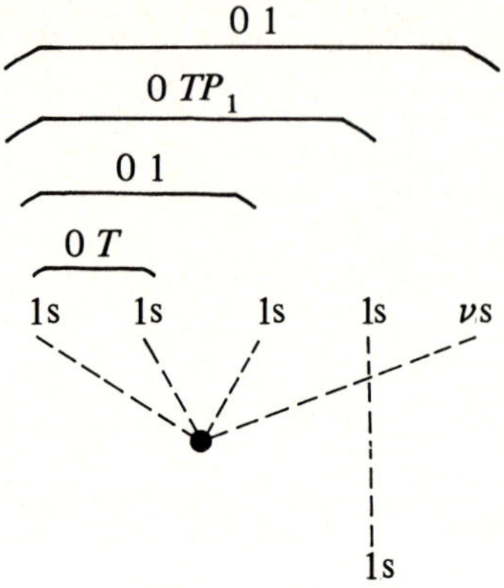

Fig. 7.33.

$1s^3\nu s$/vacuum with $\nu > 1$, fig. 7.32.
The state π_2 can only be a νs state with $\nu > 1$:

$$M = \sqrt{96}\sum_T \mathrm{CFP}_1^3(1, T)\eta\left[1s^3\nu s|H_\pi|\right]\delta_{TP_1 1}.$$

$1s^2\pi_2\pi_3$/vacuum, fig. 7.32.
This matrix element is different from zero only when $l_2 = l_3$:

$$M = \mathfrak{N}(\pi_2\pi_3)\sqrt{288}\,\eta\left[1s^2\pi_2\pi_3|H_\pi|\right]\delta_{l_2 l_3}.$$

$1s^4\nu s/1s$, fig. 7.33.

$$M = \sqrt{384}\,\frac{\widehat{TP_1}}{\hat{1}\hat{1}}\mathrm{CFP}_1^4(TP_1, 1)\sum_T \mathrm{CFP}_1^3(1, T)\eta\left[1s^3\nu s|H_\pi|\right].$$

$1s^4\pi_2/\pi_1'$, fig. 7.34.

$$M = \sqrt{24}\sum_T \mathrm{CFP}_1^3(1, T)\eta\left[1s^4|H_\pi|\right]\delta_{\pi_2\pi_1'}.$$

$1s^3 1p^2/1s$, fig. 7.35.

$$M = \sqrt{432}\,\frac{1}{\hat{1}}\,\frac{\widehat{TP_1}}{\widehat{TP_2}}\mathrm{CFP}_1^3(TP_1, TP_2)\eta\left[1s^2 1p^2|H_\pi|\right]\delta_{LP_2 0}.$$

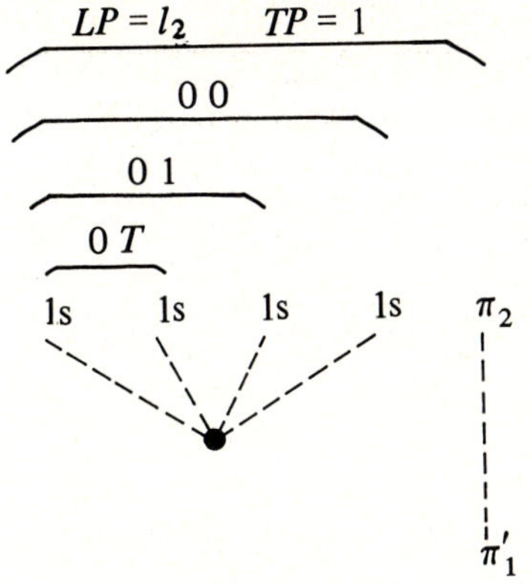

Fig. 7.34.

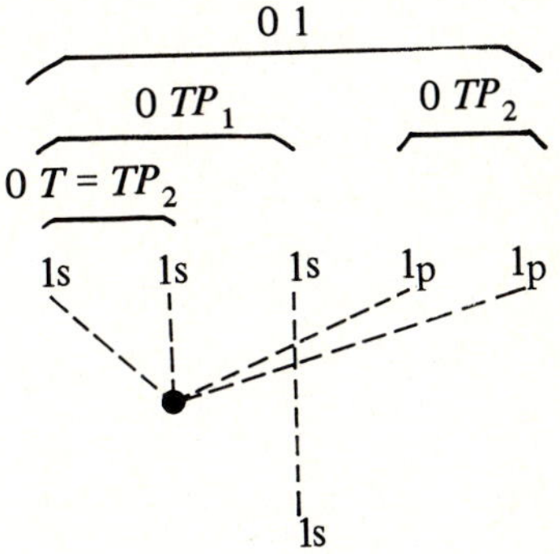

Fig. 7.35.

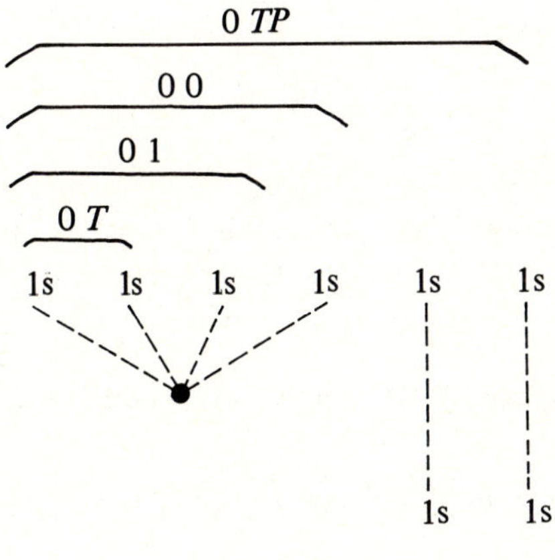

Fig. 7.36.

$1s^n/1s^{n-4}$, fig. 7.36.

$\underline{n=5}$

$$M = \sqrt{120}\,\mathrm{CFP}_1^5(1,0)\mathrm{CFP}_1^4(0,1)\sum_T \mathrm{CFP}_1^3(1,T)\eta\left[1s^4|H_\pi|\right].$$

$\underline{n=6}$

$$M = \sqrt{360}\,\mathrm{CFP}_2^6(TP,0,TP)\mathrm{CFP}_1^4(0,1)\sum_T \mathrm{CFP}_1^3(1,T)\eta\left[1s^4|H_\pi|\right].$$

Note the rapid increase of the multiplicity factor M as the number of equivalent pions increases in the pion cloud. This, of course, is the effect of stimulated emission of bosons.

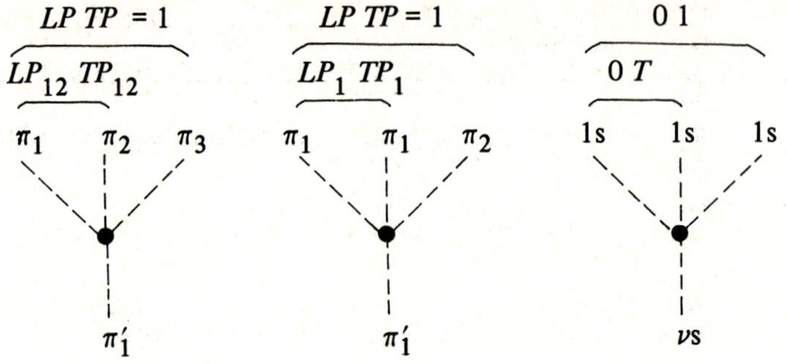

Case A Case B Case C

Fig. 7.37.

7.5.2. Creation of three pions

$\pi_1\pi_2\pi_3/\pi_1'$, fig. 7.37.
If $\pi_1 \neq \pi_2 \neq \pi_3$ we get, case A:

$$M = \sqrt{576}\,\frac{1}{\hat{l}_1'\hat{1}}\,\eta[\pi_1\pi_2\pi_3|H_\pi|\pi_1'].$$

$\pi_1^2\pi_2/\pi_1'$, case B:

$$M = \sqrt{288}\,\frac{1}{\hat{l}_1'\hat{1}}\,\eta[\pi_1^2\pi_2|H_\pi|\pi_1'].$$

$1s^3/\pi_1'$ with $\pi_1' = \nu s$ ($\nu \geqslant 1$), case C:

$$M = \sqrt{96}\,\frac{1}{\hat{1}}\sum_T \mathrm{CFP}_1^3(1,T)\,\eta[1s^3|H_\pi|\nu s]\,\delta_{l_1'0}.$$

$1s^2 1p^2/\pi_1'\pi_2'$, fig. 7.38.
$\pi_1' = 1s$, case A:

$$M = \frac{\sqrt{576}}{\mathfrak{N}(\pi_1'\pi_2')}\,\frac{1}{\hat{l}_2'\hat{1}}\begin{bmatrix} 1 & 1 & TP_1 \\ 0 & TP_2 & TP_2 \\ 1 & 1 & TP \end{bmatrix}\eta[1s1p^2|H_\pi|\pi_2'].$$

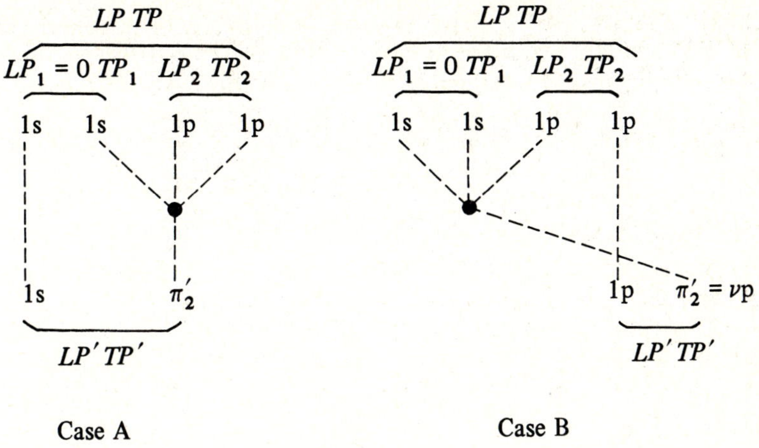

Fig. 7.38.

$\pi_1' = 1$p and $\pi_2' = \nu$p, $\nu \geqslant 1$, case B:

$$M = (-)^{TP+LP} \frac{\sqrt{576}}{\mathfrak{N}(\pi_1'\pi_2')} \frac{1}{\hat{l}_2'\hat{1}} \begin{bmatrix} TP_1 & 0 & TP_1 \\ 1 & 1 & TP_2 \\ 1 & 1 & TP \end{bmatrix} \eta\left[1s^2 1p|H_\pi|\pi_2'\right] \delta_{l_2' 1}.$$

$1s^3 1p^2/1s1p^2$, fig. 7.39.
Two terms contribute:

$$M = \sqrt{96}\, \frac{1}{\hat{TP}_1} \sum_T \mathrm{CFP}_1^3(TP_1, T)\, \eta\left[1s^3|H_\pi|1s\right] \delta_{TP_1,1} \delta_{\{\pi_2\}\{\pi_2'\}}$$

$$+ \sqrt{3456}\, \frac{1}{\hat{1}\hat{1}} \sum_{TT'} \mathrm{CFP}_1^3(TP_1, T) \begin{bmatrix} T & 1 & TP_1 \\ 1 & 1 & TP_2 \\ 1 & T' & TP \end{bmatrix} \begin{bmatrix} 0 & 1 & 1 \\ 1 & 1 & TP_2' \\ 1 & T' & TP \end{bmatrix} \eta\left[1s^2 1p|H_\pi|1p\right].$$

$1s^3 1p^2/1p^2\pi_2'$ with $\pi_2' = \nu$s, $\nu > 1$, fig. 7.40.

$$M = \sqrt{96}\, (-)^{TP_1' + 1 - TP} \frac{1}{\hat{1}} \sum_T \mathrm{CFP}_1^3(TP_1, T)\, \eta\left[1s^3|H_\pi|\nu s\right] \delta_{\{\pi_2\}\{\pi_1'\}}.$$

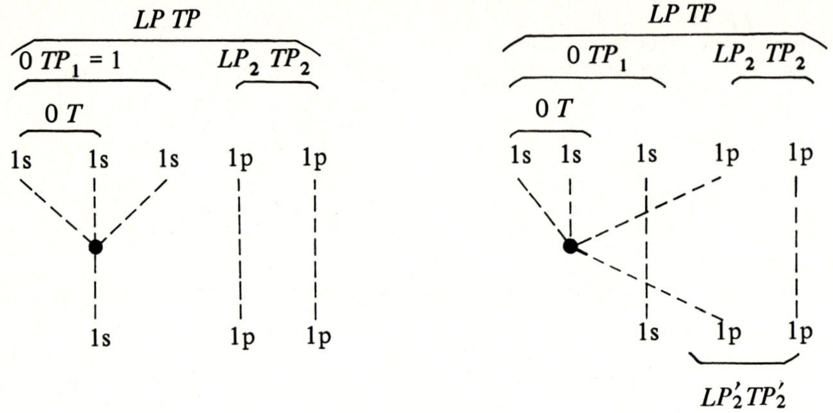

Fig. 7.39.

$1s^3 1p^2/1s1p2p$, fig. 7.41.

$$M = (-)^{TP'_{12}+1-TP-LP}\sqrt{1728}\,\frac{1}{\hat{1}\hat{1}}$$

$$\times \sum_T \mathrm{CFP}_1^3(TP_1, T)\begin{bmatrix} T & 1 & TP_1 \\ 1 & 1 & TP_2 \\ 1 & TP'_{12} & TP \end{bmatrix}\eta\left[1s^2 1p|H_\pi|2p\right].$$

$1s^n\pi_2/1s^{n-3}\nu s\pi'_3$ ($\nu > 1$ and $\pi_2 \equiv \pi'_3$), fig. 7.42.

$\underline{n = 3}$

$$M = \sqrt{96}\,\frac{1}{\hat{1}}\sum_T \mathrm{CFP}_1^3(1, T)\,\eta\left[1s^3|H_\pi|\nu s\right]\delta_{TP_1 1}.$$

$\underline{n = 4}$

$$M = (-)^{TP'_{12}}\sqrt{384}\,\frac{1}{\hat{1}}\mathrm{CFP}_1^4(TP_1, 1)$$

$$\times \sum_{TT''} \mathrm{CFP}_1^3(1, T)\begin{bmatrix} 1 & 1 & TP \\ 0 & 1 & 1 \\ 1 & T'' & TP \end{bmatrix}\begin{bmatrix} 1 & 1 & TP'_{12} \\ 0 & 1 & 1 \\ 1 & T'' & TP \end{bmatrix}\eta\left[1s^3|H_\pi|\nu s\right].$$

$1s^n\pi_2/1s^{n-3}\pi'_2\nu s$ ($\nu > 1$ and $\pi_2 \equiv \pi'_2$), fig. 7.43.

$\underline{n = 3}$

$$M = (-)^{\mathrm{TP}}\sqrt{96}\,\frac{1}{\hat{1}}\sum_T \mathrm{CFP}_1^3(1, T)\,\eta\left[1s^3|H_\pi|\nu s\right]\delta_{TP_1 1}.$$

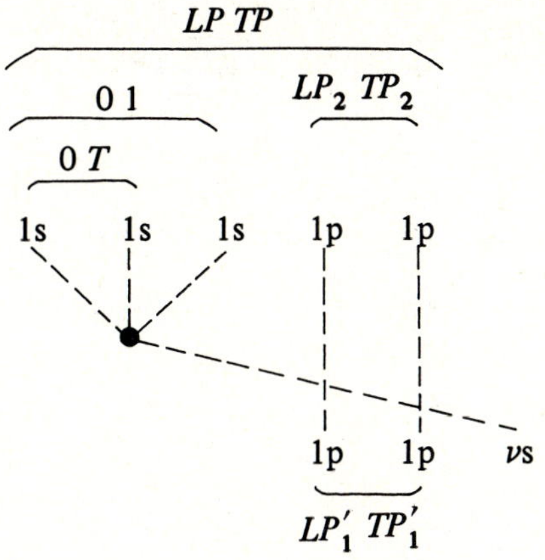

Fig. 7.40.

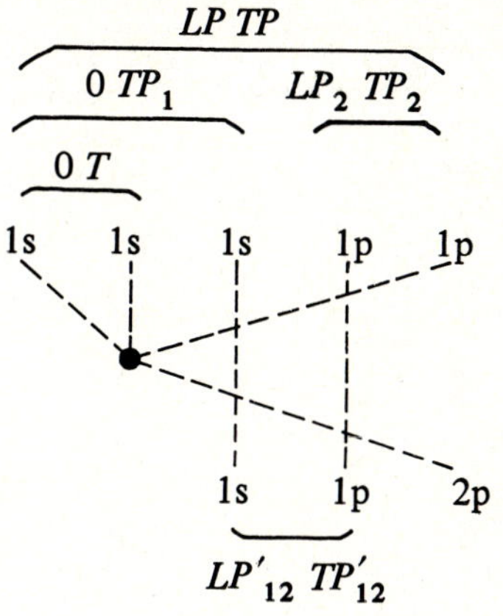

Fig. 7.41.

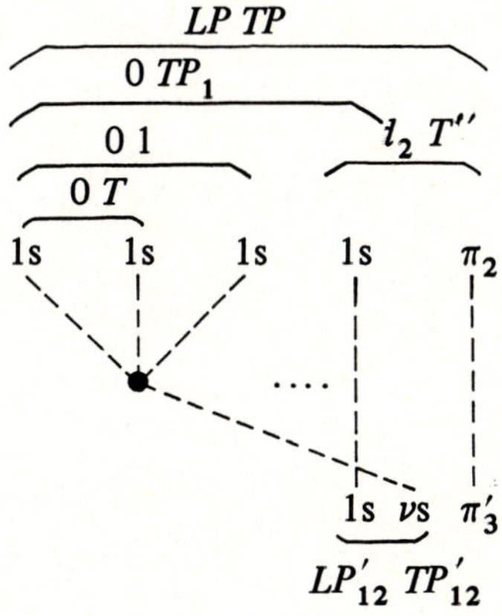

Fig. 7.42.

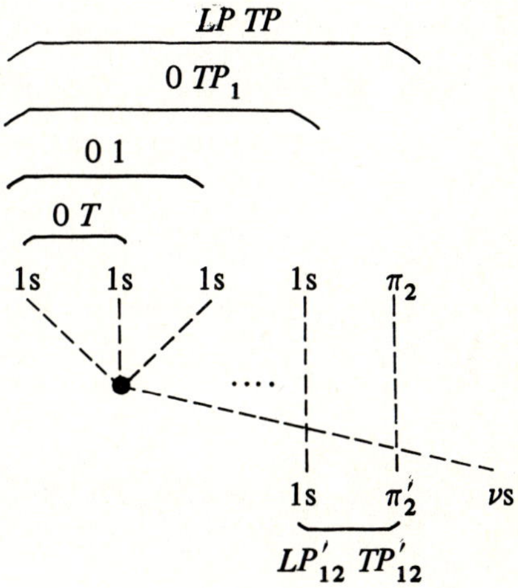

Fig. 7.43.

$\underline{n=4}$

$$M = (-)^{TP'_{12}+1+TP}\sqrt{384}\,\frac{1}{\hat{1}}\,\mathrm{CFP}_1^4(TP_1, 1)$$

$$\times \sum_T \mathrm{CFP}_1^3(1, T)\begin{bmatrix} 1 & 1 & TP_1 \\ 0 & 1 & 1 \\ 1 & TP'_{12} & TP \end{bmatrix} \eta\left[1\mathrm{s}^3|H_\pi|\nu\mathrm{s}\right].$$

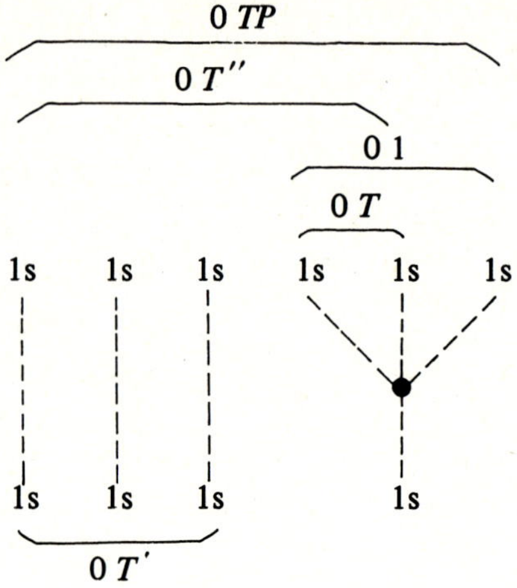

Fig. 7.44.

$1s^n/1s^{n-2}$, fig. 7.44.

$\underline{n=4}$

$$M=\sqrt{192}\,\frac{1}{\hat{1}}\mathrm{CFP}_1^4(TP,1)\sum_T \mathrm{CFP}_1^3(1,T)\,\eta\left[1s^3|H_\pi|1s\right].$$

$\underline{n=5}$

$$M=\sqrt{720}\,\frac{1}{\hat{1}}\sum_{T'}\mathrm{CFP}_2^5(TP,1,T')\mathrm{CFP}_1^3(TP,T')\sum_T \mathrm{CFP}_1^3(1,T)\,\eta\left[1s^3|H_\pi|1s\right].$$

$\underline{n=6}$

$$M=\sqrt{1920}\,\frac{1}{\hat{1}}\sum_{T''}\mathrm{CFP}_1^6(TP,T'')$$

$$\times\sum_{T'}\mathrm{CFP}_1^4(TP,T')\sum_T \mathrm{CFP}_2^5(T'',T',T)\begin{bmatrix}T' & T & T''\\ 0 & 1 & 1\\ T' & 1 & TP\end{bmatrix}\eta\left[1s^3|H_\pi|1s\right].$$

$1s^n/1s^{n-3}\pi_2'$ with $\pi_2'=\nu s$ ($\nu>1$), fig. 7.45.

$\underline{n=4}$

$$M=\sqrt{384}\,\frac{1}{\hat{1}}\mathrm{CFP}_1^4(TP,1)\sum_T \mathrm{CFP}_1^3(1,T)\,\eta\left[1s^3|H_\pi|\nu s\right].$$

$\underline{n=5}$

$$M=\sqrt{960}\,\frac{1}{\hat{1}}\mathrm{CFP}_2^5(TP,1,TP_1')\sum_T \mathrm{CFP}_1^3(1,T)\,\eta\left[1s^3|H_\pi|\nu s\right].$$

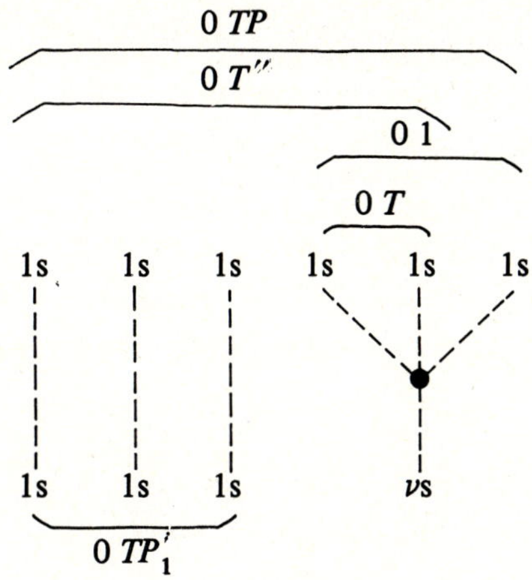

Fig. 7.45.

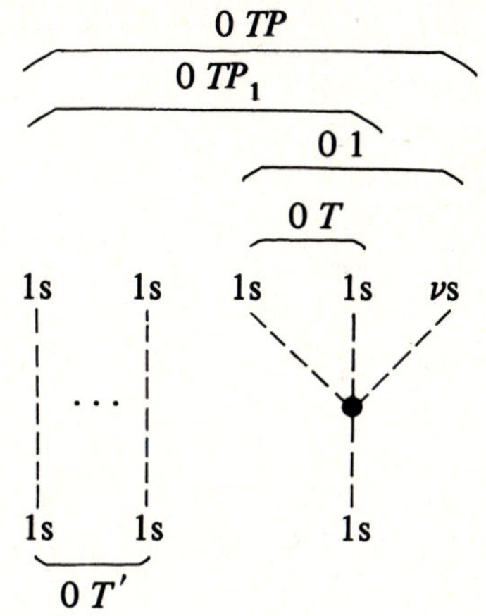

Fig. 7.46.

$\underline{n=6}$

$$M=\sqrt{1920}\,\frac{1}{\hat{1}}\sum_{T''}\mathrm{CFP}_1^6(TP,T'')\sum_{T}\mathrm{CFP}_2^5(T'',TP'_1,T)$$

$$\times\begin{bmatrix} TP'_1 & T & T'' \\ 0 & 1 & 1 \\ TP'_1 & 1 & TP \end{bmatrix}\eta\left[1s^3|H_\pi|\nu s\right].$$

$1s^n\pi_2/1s^{n-1}$ with $\pi_2=\nu s$ $(\nu>1)$, fig. 7.46.

$\underline{n=3}$

$$M=\sqrt{1728}\,\frac{1}{\hat{1}}\sum_{T}\mathrm{CFP}_1^3(TP_1,T)\begin{bmatrix} 1 & T & TP_1 \\ 0 & 1 & 1 \\ 1 & 1 & TP \end{bmatrix}\eta\left[1s^2\nu s|H_\pi|1s\right].$$

$\underline{n=4}$

$$M=\sqrt{5184}\,\frac{1}{\hat{1}}\sum_{T'}\mathrm{CFP}_1^3(TP,T')$$

$$\times\sum_{T}\mathrm{CFP}_2^4(TP_1,T',T)\begin{bmatrix} T' & T & TP_1 \\ 0 & 1 & 1 \\ T' & 1 & TP \end{bmatrix}\eta\left[1s^2\nu s|H_\pi|1s\right].$$

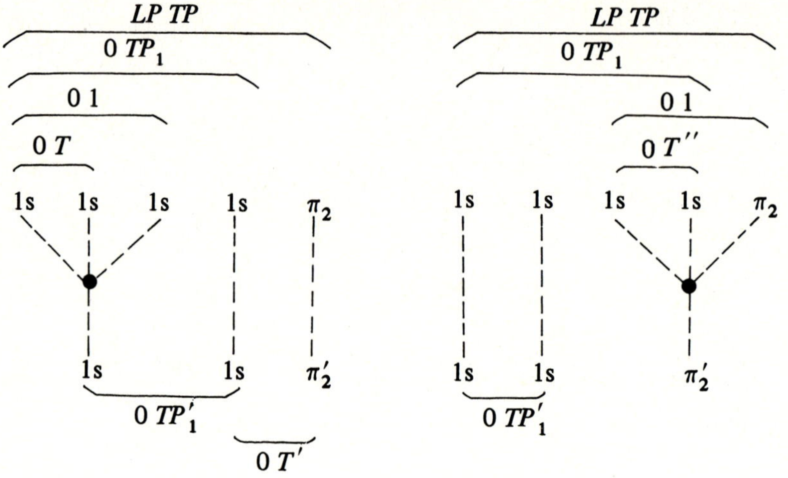

Fig. 7.47.

$1s^n\pi_2/1s^{n-2}\pi_2'$, fig. 7.47.

$\underline{n=3}$

$$M=\frac{1}{\hat{1}}\left(\sqrt{96}\sum_T \mathrm{CFP}_1^3(TP_1,T)\,\eta\left[1s^3|H_\pi|1s\right]\delta_{\pi_2\pi_2'}\,\delta_{TP_11}\right.$$

$$\left.+\sqrt{864}\,\frac{1}{\hat{l}_2'}\sum_{T''}\mathrm{CFP}_1^3(TP_1,T'')\begin{bmatrix}1 & T'' & TP_1\\ 0 & 1 & 1\\ 1 & 1 & TP\end{bmatrix}\eta\left[1s^2\pi_2|H_\pi|\pi_2'\right]\delta_{l_2l_2'}\right).$$

$\underline{n=4}$

$$M=\frac{1}{\hat{1}}\left(\sqrt{768}\sum_{TT'}\mathrm{CFP}_1^4(TP_1,1)\mathrm{CFP}_1^3(1,T)\begin{bmatrix}1 & 1 & TP_1\\ 0 & 1 & 1\\ 1 & T' & TP\end{bmatrix}\begin{bmatrix}1 & 1 & TP_1'\\ 0 & 1 & 1\\ 1 & T' & TP\end{bmatrix}\right.$$

$$\times\eta\left[1s^3|H_\pi|1s\right]\delta_{\pi_2\pi_2'}$$

$$\left.+\sqrt{1728}\,\frac{1}{\hat{l}_2'}\sum_{T''}\mathrm{CFP}_2^4(TP_1,TP_1',T'')\begin{bmatrix}TP_1' & T'' & TP_1\\ 0 & 1 & 1\\ TP_1' & 1 & TP\end{bmatrix}\eta\left[1s^2\pi_2|H_\pi|\pi_2'\right]\delta_{l_2l_2'}\right).$$

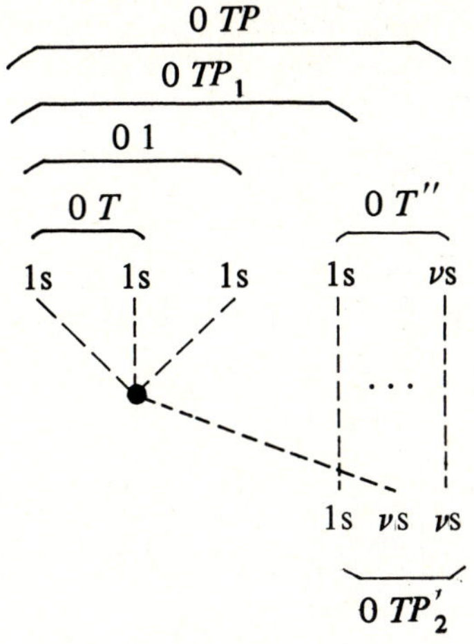

Fig. 7.48.

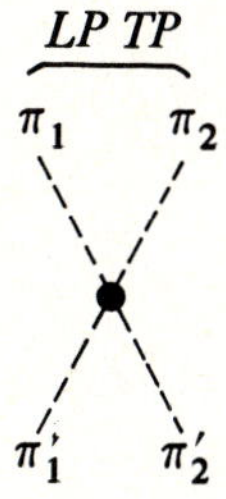

Fig. 7.49.

$1s^n \nu s / 1s^{n-3} \nu s^2$, fig. 7.48.

<u>$n=3$</u>

$$M = \sqrt{192}\,\frac{1}{\hat{1}} \sum_T \mathrm{CFP}_1^3(1,T)\,\eta\left[1s^3|H_\pi|\nu s\right]\delta_{TP_1,1}.$$

<u>$n=4$</u>

$$M = \sqrt{768}\,\frac{1}{\hat{1}}\,\mathrm{CFP}_1^4(TP_1,1)$$

$$\times \sum_{TT''} \mathrm{CFP}_1^3(1,T) \begin{bmatrix} 1 & 1 & TP_1 \\ 0 & 1 & 1 \\ 1 & T'' & TP \end{bmatrix} \begin{bmatrix} 0 & 1 & 1 \\ 1 & 1 & TP_2' \\ 1 & T'' & TP \end{bmatrix} \eta\left[1s^3|H_\pi|\nu s\right].$$

7.5.3. Creation of two pions

$\pi_1\pi_2/\pi_1'\pi_2'$, fig. 7.49.

$$M = \sqrt{576}\,\frac{1}{\widehat{LP}\,\widehat{TP}}\,\mathcal{N}(\pi_1\pi_2)\,\mathcal{N}(\pi_1'\pi_2')\,\eta\left[\pi_1\pi_2|H_\pi|\pi_1'\pi_2'\right].$$

$1s^3/1s\,\pi_2'\pi_3'$, fig. 7.50.

$$M = \sqrt{864} \sum_T \frac{1}{\hat{T}}\,\mathrm{CFP}_1^3(TP,T) \begin{bmatrix} 1 & 1 & TP_{12}' \\ 0 & 1 & 1 \\ 1 & T & TP \end{bmatrix} \eta\left[1s^2|H_\pi|\pi_2'\pi_3'\right]\delta_{l_2'l_3'}.$$

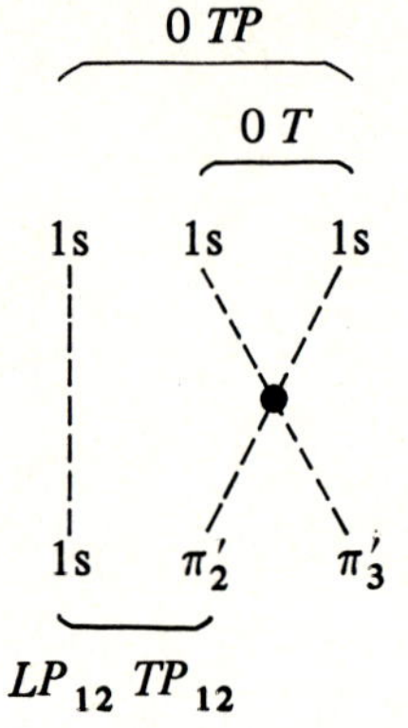

Fig. 7.50.

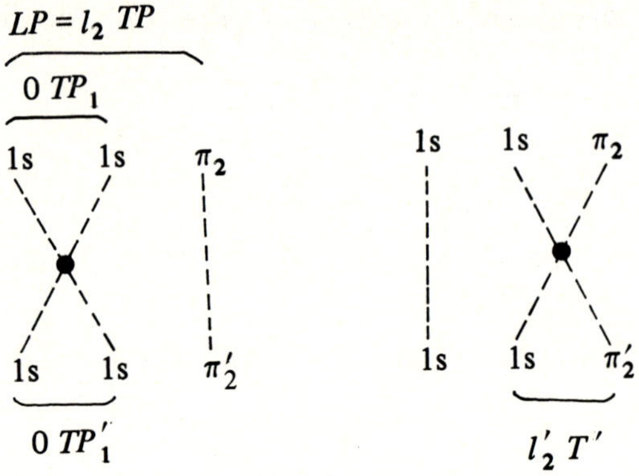

Fig. 7.51.

$1s^2\pi_2/1s^2\pi_2'$, fig. 7.51.

$$M=\sqrt{144}\,\frac{1}{\widehat{TP_1}}\eta\left[1s^2|H_\pi|1s^2\right]\delta_{\pi_2\pi_2'}\delta_{TP_1TP_1'}$$

$$+\sqrt{2304}\sum_{T'}\frac{1}{\hat{l}_2\hat{T}'}\begin{bmatrix}1 & 1 & TP_1\\ 0 & 1 & 1\\ 1 & T' & TP\end{bmatrix}\begin{bmatrix}1 & 1 & TP_1'\\ 0 & 1 & 1\\ 1 & T' & TP\end{bmatrix}\eta\left[1s\,\pi_2|H_\pi|1s\,\pi_2'\right]\delta_{l_2l_2'}\,.$$

$\pi_1\pi_2\pi_3/\pi_1'\pi_2'\pi_3'$, fig. 7.52.

We now consider the general case $n=n'=3$. The expression is written for the case where $\pi_2\neq\pi_3$ and $\pi_2'\neq\pi_3'$, but we may have $\pi_1=\pi_2$ and/or $\pi_1'=\pi_2'$. The various contributions are given by the nine graphs of fig. 7.52. For all these graphs resulting from the symmetrization operation, the remaining multiplicity factors are given by eq. (7.3).

We calculate here the two examples of the term (5), fig. 7.53 and of the term (7), fig. 7.54. The other terms are obtained by similar recoupling transformations.

$$M_{(5)}=\mathrm{M}\sum_{L,T}\frac{1}{\hat{L}\hat{T}}\begin{bmatrix}l_1 & l_2 & LP_{12}\\ l_3 & 0 & l_3\\ L & l_2 & LP\end{bmatrix}\begin{bmatrix}1 & 1 & TP_{12}\\ 1 & 0 & 1\\ T & 1 & TP\end{bmatrix}\begin{bmatrix}l_1' & l_2' & LP_{12}'\\ l_3' & 0 & l_3'\\ L & l_2' & LP\end{bmatrix}\begin{bmatrix}1 & 1 & TP_{12}'\\ 1 & 0 & 1\\ T & 1 & TP\end{bmatrix}$$

$$\times\eta\left[\pi_1\pi_3|H_\pi|\pi_1'\pi_3'\right]\delta_{\pi_2\pi_2'}\,,$$

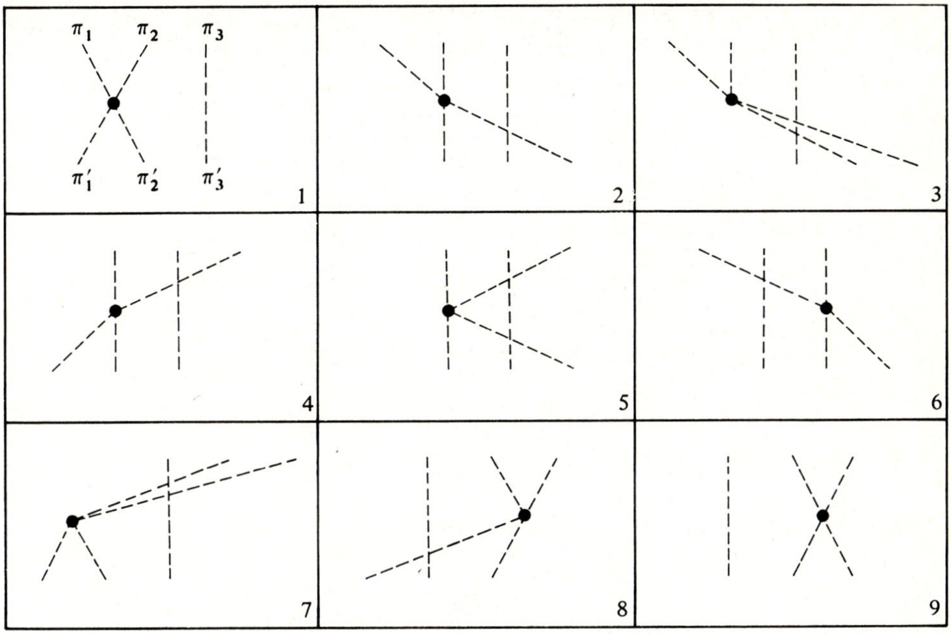

Fig. 7.52.

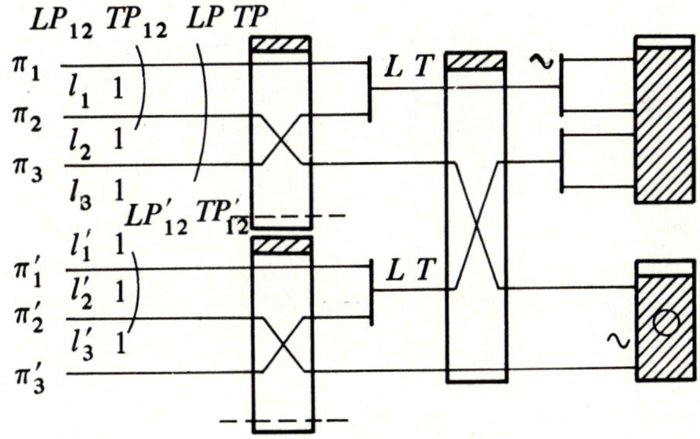

Fig. 7.53.

$$M_{(7)} = \mathsf{M}\frac{1}{\widehat{LP'_{12}}\widehat{TP'_{12}}}(-)^{LP'_{12}+l'_3-LP+TP'_{12}+1-TP}$$

$$\times\begin{bmatrix} l_1 & l_2 & LP_{12} \\ 0 & l_3 & l_3 \\ l_1 & LP'_{12} & LP \end{bmatrix}\begin{bmatrix} 1 & 1 & TP_{12} \\ 0 & 1 & 1 \\ 1 & TP'_{12} & TP \end{bmatrix}$$

$$\times\eta[\pi_2\pi_3|H_\pi|\pi'_1\pi'_2]\delta_{\pi_1\pi'_3}.$$

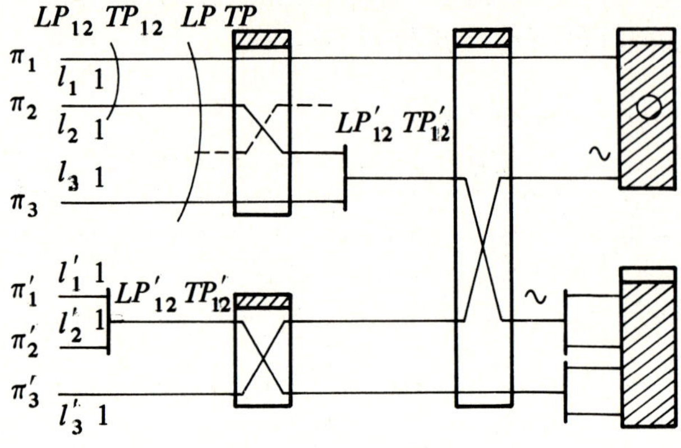

Fig. 7.54

$1s^21p^2/1s^21p^2$, fig. 7.55.

$$M = 12\frac{1}{\widehat{LP}}\left(\frac{1}{\widehat{TP}_2}\eta\left[1p^2|H_\pi|1p^2\right] + \frac{\widehat{LP}_2}{\widehat{TP}_1}\eta\left[1s^2|H_\pi|1s^2\right]\right)\delta_{TP_1TP_1'}\delta_{TP_2TP_2'}$$

$$+96\sum_{TT'}\frac{1}{\hat{1}\hat{T}}\begin{bmatrix}1 & 1 & TP_1\\ 1 & 1 & TP_2\\ T & T' & TP\end{bmatrix}\begin{bmatrix}1 & 1 & TP_1'\\ 1 & 1 & TP_2'\\ T & T' & TP\end{bmatrix}\eta\left[1s1p|H_\pi|1s1p\right].$$

$1s^3\pi_2/1s^3\pi_2'$, fig. 7.56.

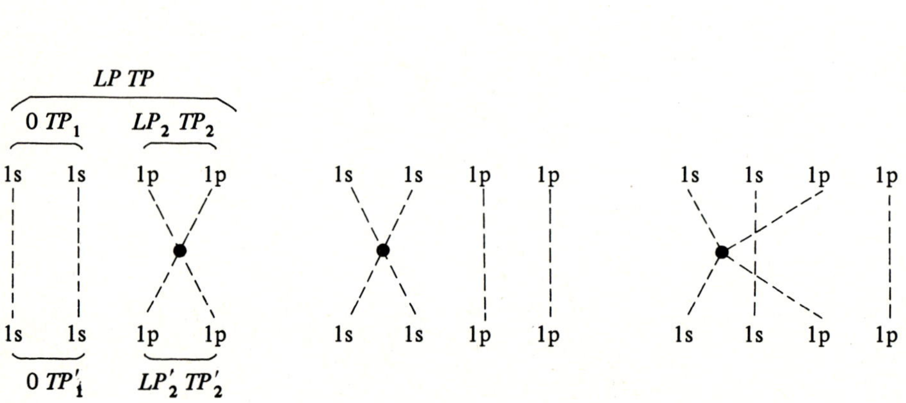

Fig. 7.55.

$$M = 36 \sum_{TT'} \frac{1}{\hat{T}} \mathrm{CFP}_1^3(TP_1, T)\mathrm{CFP}_1^3(TP_1', T)$$

$$\times \begin{bmatrix} T & 1 & TP_1 \\ 0 & 1 & 1 \\ T & T' & TP \end{bmatrix} \begin{bmatrix} T & 1 & TP_1' \\ 0 & 1 & 1 \\ T & T' & TP \end{bmatrix} \eta\left[1\mathrm{s}^2 | H_\pi | 1\mathrm{s}^2\right] \delta_{\pi_2 \pi_2'}$$

$$+ 72 \frac{1}{\hat{l}_2} \sum_{TT'} \frac{1}{\hat{T}'} \mathrm{CFP}_1^3(TP_1, T)\mathrm{CFP}_1^3(TP_1', T)$$

$$\times \begin{bmatrix} T & 1 & TP_1 \\ 0 & 1 & 1 \\ T & T' & TP \end{bmatrix} \begin{bmatrix} T & 1 & TP_1' \\ 0 & 1 & 1 \\ T & T' & TP \end{bmatrix} \eta[1\mathrm{s}\, \pi_2 | H_\pi | 1\mathrm{s}\, \pi_2'] \delta_{l_2 l_2'} .$$

$1s^3 1p^2/1s^3 1p^2$, fig. 7.57

$$M = 36 \sum_{TT''} \frac{1}{\hat{T}} \mathrm{CFP}_1^3(TP_1, T)\mathrm{CFP}_1^3(TP_1', T)$$

$$\times \begin{bmatrix} T & 1 & TP_1 \\ 0 & TP_2 & TP_2 \\ T & T'' & TP \end{bmatrix} \begin{bmatrix} T & 1 & TP_1' \\ 0 & TP_2' & TP_2' \\ T & T'' & TP \end{bmatrix} \eta\left[1\mathrm{s}^2 | H_\pi | 1\mathrm{s}^2\right]$$

$$+ 12 \frac{1}{\widehat{LP}\widehat{TP}_2} \eta\left[1\mathrm{p}^2 | H_\pi | 1\mathrm{p}^2\right] \delta_{TP_1 TP_1'} \delta_{TP_2 TP_2'} \delta_{LP_2 LP_2'}$$

$$+ 144 \frac{1}{\hat{1}} \sum_{TT'R} \frac{1}{\hat{T}'} \mathrm{CFP}_1^3(TP_1, T)\mathrm{CFP}_1^3(TP_1', T)$$

$$\times \begin{bmatrix} T & 1 & TP_1 \\ 1 & 1 & TP_2 \\ R & T' & TP \end{bmatrix} \begin{bmatrix} T & 1 & TP_1' \\ 1 & 1 & TP_2' \\ R & T' & TP \end{bmatrix} \eta[1\mathrm{s}1\mathrm{p} | H_\pi | 1\mathrm{s}1\mathrm{p}] .$$

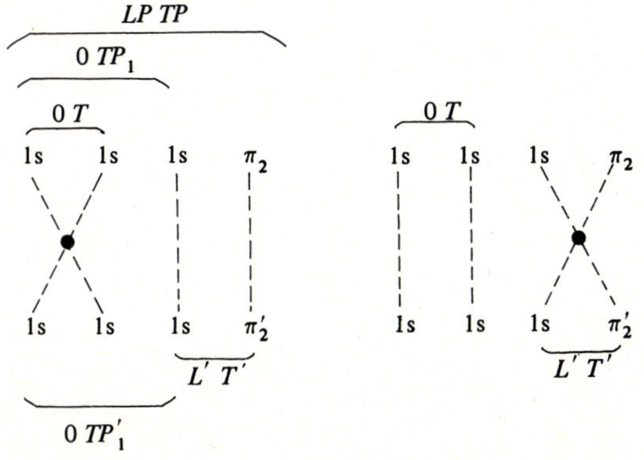

Fig. 7.56.

$1s^4\pi_2/1s^4\pi_2'$, fig. 7.58.

$$M = 72 \sum_{TRT'} \frac{1}{\hat{T}} \mathrm{CFP}_2^4(TP_1, R, T)\mathrm{CFP}_2^4(TP_1', R, T)$$

$$\times \begin{bmatrix} T & R & TP_1 \\ 0 & 1 & 1 \\ T & T' & TP \end{bmatrix} \begin{bmatrix} T & R & TP_1' \\ 0 & 1 & 1 \\ T & T' & TP \end{bmatrix} \eta\left[1s^2|H_\pi|1s^2\right]\delta_{\pi_2\pi_2'}$$

$$+96\frac{1}{\hat{l}_2} \sum_{TR} \frac{1}{\hat{T}} \mathrm{CFP}_1^4(TP_1, R)\mathrm{CFP}_1^4(TP_1', R)$$

$$\times \begin{bmatrix} R & 1 & TP_1 \\ 0 & 1 & 1 \\ R & T & TP \end{bmatrix} \begin{bmatrix} R & 1 & TP_1' \\ 0 & 1 & 1 \\ R & T & TP \end{bmatrix} \eta\left[1s\,\pi_2|H_\pi|1s\,\pi_2'\right]\delta_{l_2 l_2'}\,.$$

$1s^n/1s^n$, fig. 7.59.

<u>$n = 3$</u>

$$M = 36 \sum_T \frac{1}{\hat{T}} \left(\mathrm{CFP}_1^3(TP, T)\right)^2 \eta\left[1s^2|H_\pi|1s^2\right].$$

<u>$n \geqslant 4$</u>

$$M = 6n(n-1) \sum_{TR} \frac{1}{\hat{T}} \left(\mathrm{CFP}_2^n(TP, R, T)\right)^2 \eta\left[1s^2|H_\pi|1s^2\right].$$

$1s^n/1s^{n-1}\pi_2'$, fig. 7.60.

<u>$n = 3$</u>

$$M = 24\sqrt{3} \sum_T \frac{1}{\hat{T}} \mathrm{CFP}_1^3(TP, T) \begin{bmatrix} 1 & 1 & TP_1' \\ 0 & 1 & 1 \\ 1 & T & TP \end{bmatrix} \eta\left[1s^2|H_\pi|1s\pi_2'\right].$$

<u>$n \geqslant 4$</u>

$$M = 12(n-1)\sqrt{n} \sum_{TT'} \frac{1}{\hat{T}} \mathrm{CFP}_2^n(TP, T', T)\mathrm{CFP}_1^{n-1}(TP_1', T')$$

$$\times \begin{bmatrix} T' & 1 & TP_1' \\ 0 & 1 & 1 \\ T' & T & TP \end{bmatrix} \eta\left[1s^2|H_\pi|1s\pi_2'\right]\delta_{l_2'0}\,.$$

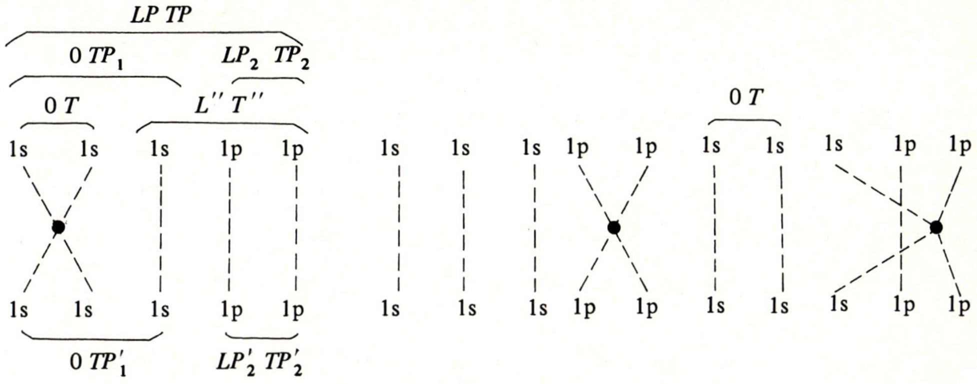

Fig. 7.57.

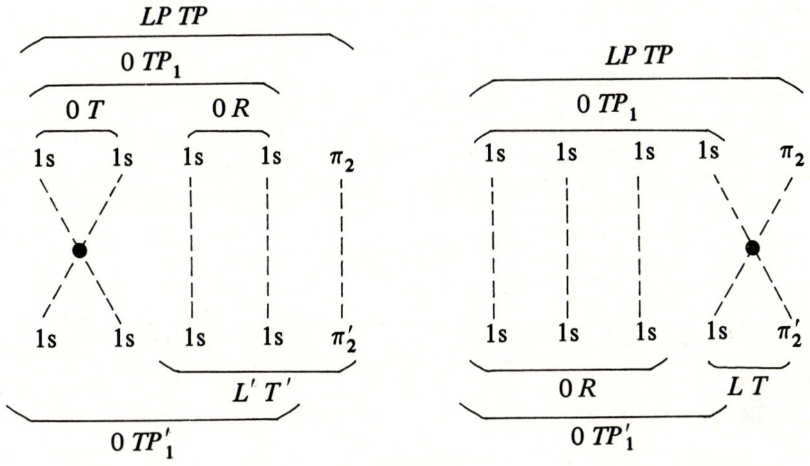

Fig. 7.58.

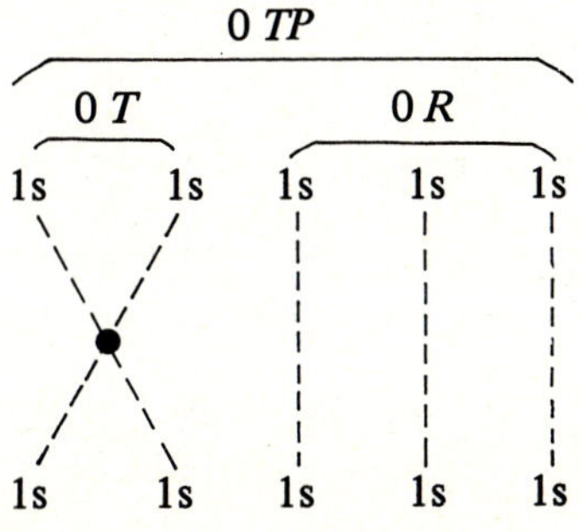

Fig. 7.59.

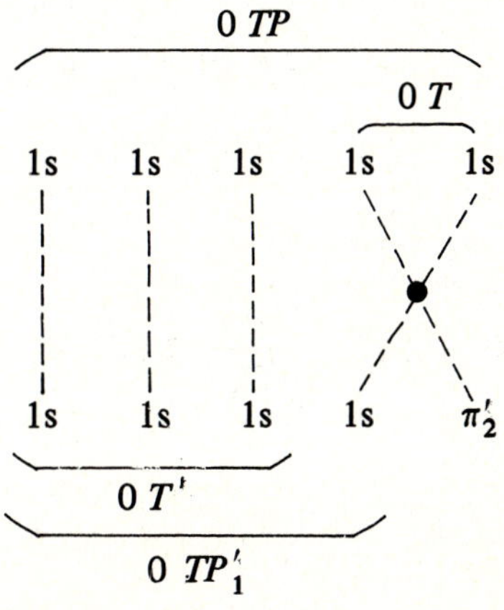

Fig. 7.60.

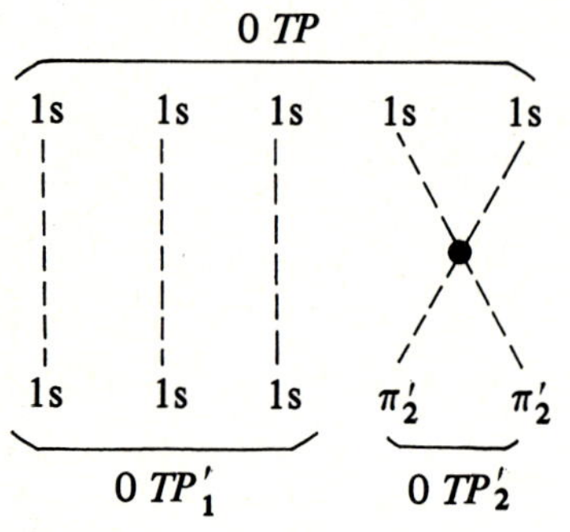

Fig. 7.61.

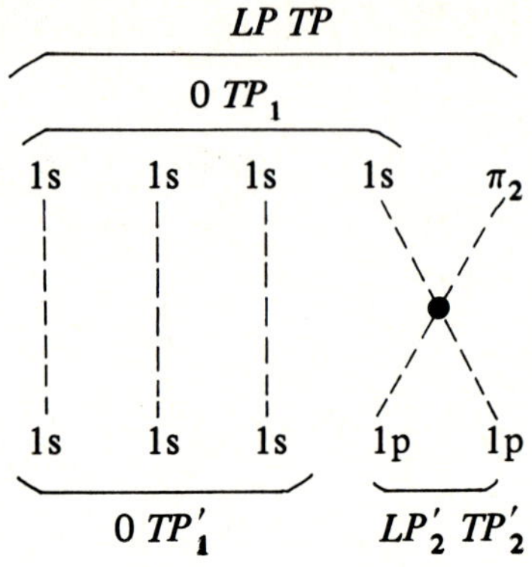

Fig. 7.62.

$1s^n/1s^{n-2}\pi_2'^2$, fig. 7.61.

$\underline{n=3}$

$$M = 12\sqrt{3}\,\frac{1}{\widehat{TP_2'}}\,\mathrm{CFP}_1^3(TP, TP_2')\,\eta\left[1s^2|H_\pi|\pi_2'^2\right].$$

$\underline{n \geqslant 4}$

$$M = 6\sqrt{2n(n-1)}\,\frac{1}{\widehat{TP_2'}}\,\mathrm{CFP}_2^n(TP, TP_1', TP_2')\,\eta\left[1s^2|H_\pi|\pi_2'^2\right].$$

$1s^n\pi_2/1s^{n-1}1p^2$, fig. 7.62.

$\underline{n=2}$

$$M = 24\,\frac{1}{\hat{l}_2\widehat{TP_2'}}\begin{bmatrix} 1 & 1 & TP_1 \\ 0 & 1 & 1 \\ 1 & TP_2' & TP \end{bmatrix}\eta\left[1s\,\pi_2|H_\pi|1p^2\right].$$

$\underline{n \geqslant 3}$

$$M = 12\sqrt{2n}\,\frac{1}{\hat{l}_2\widehat{TP_2'}}\,\mathrm{CFP}_1^n(TP_1, TP_1')\begin{bmatrix} TP_1' & 1 & TP_1 \\ 0 & 1 & 1 \\ TP_1' & TP_2' & TP \end{bmatrix}\eta\left[1s\,\pi_2|H_\pi|1p^2\right].$$

CHAPTER 8

Many-body center of mass

8.1. INTRODUCTION

The truncation of the Hilbert space yields solutions of the secular problem which are mixtures of states with different center-of-mass motion. This fact may be neglected when dealing with heavy systems. It must be accounted for in the case of light systems. In sect. 3.4.1 we have given a general method for extracting the c.m. motion energy from the solutions based on the extension to the relativistic case of the pseudo-hamiltonian method of non-relativistic shell-model calculations.

The many-body matrix elements of the c.m. pseudo-hamiltonian of sect. 3.4 are discussed in this chapter. The c.m. operators and the general treatment are presented in sect. 8.2. The c.m. operators involve the elementary matrix elements of the single-particle momentum and position coordinates, which are calculated for the spin 0, $\frac{1}{2}$ and 1 fields. The generalization to higher spin fields is straightforward.

The many-body aspect of the momentum operator is simple, and is treated in sect. 8.3. On the other hand the position operator shows the full complexity of a n-body operator. The method to factorize this many-body operator, which is presented in sect. 8.2.1, is applied to specific examples in sect. 8.4. Since the c.m. pseudo-hamiltonian conserves particle number and particle type one can conveniently work in either the Fock representation or the space representation of sect. 7.2.7. Here we give our examples using the space representation.

Finally let us recall that the actual application of the pseudo-hamiltonian method in the relativistic case requires care in fixing the values of the parameters ζ and Ω of the c.m. operators, eq. (3.12), so as to achieve a non-relativistic c.m. motion as discussed in sect. 3.4.1.

8.2. ELEMENTARY MOMENTUM AND POSITION OPERATORS

8.2.1. *Form of the operators*

The center-of-mass momentum and position operators, eq. (3.12), are calculated by expanding them into products of one-body operators. This is immediate for the c.m. momentum

$$P=\sum_i p_i,$$

$$P^2=\sum_i p_i^2+2\sum_{i<j} p_i\cdot p_j.$$

However, the form of the c.m. position operator (3.14) is

$$R=\sum_i \frac{E_i r_i}{\sum_j E_j},$$

where the $E_i=\sqrt{p_i^2+M_i^2}$ are the particle free energy operators. Thus, one sees that it is in fact a sum of many-body operators because of the denominator. It is non-separable in the individual particle variables. This difficulty arises from the relativistic kinematics. It will be handled by factorizing the N-body operator R^2 into products of one-body operators with the help of the Laplace transform

$$\frac{1}{\left(\sum_k E_k\right)}=\int_0^\infty \mathrm{d}z \exp\left(-z\left(\sum_k E_k\right)\right),$$

$$\frac{1}{\left(\sum_k E_k\right)^2}=\int_0^\infty \mathrm{d}z_1\,\mathrm{d}z_2 \exp\left(-(z_1+z_2)\left(\sum_k E_k\right)\right).$$

Thus with $z=z_1+z_2$ we have

$$R^2=\int_0^\infty z\,\mathrm{d}z\left(\sum_i\left(\prod_{k\neq i}\mathrm{e}^{-zE_k}\right)\left(E_i^2\mathrm{e}^{-zE_i}r_i^2\right)\right.$$

$$\left.+\sum_{i\neq j}\left(\prod_{k\neq i,j}\mathrm{e}^{-zE_k}\right)\left(E_i\mathrm{e}^{-zE_i}r_i\right)\cdot\left(E_j\mathrm{e}^{-zE_j}r_j\right)\right). \tag{8.1}$$

Since E and r are non-commuting variables we must use symmetrized operators in (8.1). To that end we define the symmetrized energy-weighted operators

$$X_i^2=\tfrac{1}{2}\left(E_i^2\mathrm{e}^{-zE_i}r_i^2+r_i^2E_i^2\mathrm{e}^{-zE_i}\right), \tag{8.2}$$

$$X_i=\tfrac{1}{2}\left(E_i\mathrm{e}^{-zE_i}r_i+r_iE_i\mathrm{e}^{-zE_i}\right). \tag{8.3}$$

Introducing also the abbreviation

$$Z_k=\mathrm{e}^{-zE_k}, \tag{8.4}$$

the expression (8.1) becomes

$$R^2 = \int_0^\infty z\,\mathrm{d}z \left(\sum_i \Big(\prod_{k \neq i} Z_k \Big) X_i^2 + \sum_{i \neq j} \Big(\prod_{k \neq i,\, j} Z_k \Big) \boldsymbol{X}_i \cdot \boldsymbol{X}_j \right).$$

Thus it is seen that the evaluation of the many-body c.m. matrix elements to begin with requires the knowledge of the elementary matrix elements of the single-particle operators:

$$\boldsymbol{p}_i,\, p_i^2,\, Z_i,\, \boldsymbol{X}_i \text{ and } X_i^2.$$

We now give the general expressions for the momentum and position operators in terms of the fields.

Consider a function $\mathcal{O}(\boldsymbol{p}, \boldsymbol{r})$ for a particle described by a field. Its operator expression in field theory is given in terms of the fourth component of the vector current, for example for spin 0,

$$\mathcal{O} = i \int \mathrm{d}^3 r : \left(\varphi^{(+)} \bar{\mathcal{O}} \pi^{(-)} - \pi^{(+)} \bar{\mathcal{O}} \varphi^{(-)} \right) :,$$

where $\bar{\mathcal{O}}$ is the symmetrized form of the function $\mathcal{O}(\boldsymbol{p}, \boldsymbol{r})$ in the non-commuting variables $\boldsymbol{p}$ and $\boldsymbol{r}$. In this definition the factor $\frac{1}{2}$ of the symmetrization has been taken into account explicitly by requiring that the annihilation parts of the field $\varphi^{(-)}$ and of the canonical conjugate field $\pi^{(-)}$ act first and the creation parts $\varphi^{(+)}$ and $\pi^{(+)}$ afterwards. One sees that this form gives indeed for the unit operator $\mathcal{O} = 1$ the number operator when substituting the definitions of the fields and carrying out the integration as shown below

$$\mathcal{N} = i \int \mathrm{d}^3 r : \left(\varphi^{(+)} \pi^{(-)} - \pi^{(+)} \varphi^{(-)} \right) := \sum_i : a_i^+ a_i :.$$

Substituting the expressions (2.7) and (2.19) for the conjugate fields we have for spin 0 and spin 1, respectively

$$\mathcal{O} = \int \mathrm{d}^3 r : \left(\varphi^{(+)} \bar{\mathcal{O}} i \vec{\partial}_t \varphi^{(-)} - \varphi^{(+)} i \overleftarrow{\partial}_t \bar{\mathcal{O}} \varphi^{(-)} \right) :, \tag{8.5}$$

$$\mathcal{O} = -\int \mathrm{d}^3 r : \left(\phi^{(+)} \bar{\mathcal{O}} \gamma_\mu \vec{\partial}_\mu \phi^{(-)} + \phi^{(+)} \gamma_\mu \overleftarrow{\partial}_\mu \bar{\mathcal{O}} \phi^{(-)} \right) :. \tag{8.6}$$

The fact that in (8.6) only the space components contribute has been utilized to simplify the expression. The same expression (8.6) holds for the spin-$\frac{3}{2}$ field upon the replacement $\phi \to \Phi$. For the spin-$\frac{1}{2}$ field, the corresponding expression is

$$\mathcal{O} = \int \mathrm{d}^3 r\, \psi^+ \bar{\mathcal{O}} \psi. \tag{8.7}$$

The evaluation of these matrix elements is carried through in the following sections. It requires various integrals over the position and momentum radial coordinates. The general form of these integrals is now derived using the example of a spin-0 field. Substituting in (8.5) the expansions (5.6) and (5.7) of the field we recognize that the radial integrals are of the form, for the spin-0 field

$$I=\int r^2\mathrm{d}r \sum_{\nu l\nu' l'}\left(\tfrac{1}{2}g_{\nu l}(r)i\frac{\overleftrightarrow{\partial}}{\partial t}\bar{\mathcal{O}}(r,p)g_{\nu' l'}(r)\right). \tag{8.8}$$

The function $\bar{\mathcal{O}}(\boldsymbol{r},\boldsymbol{p})$ has been factorized into the radial part $\bar{\mathcal{O}}(r,p)$ written here and an angular part. The angular parts will be evaluated in the following sections. They contribute angular momentum selection rules. In order to evaluate the radial integral the radial functions are expanded to yield

$$\int r^2\mathrm{d}r\,\tfrac{1}{2}g_{\nu l}(r)i\frac{\overleftrightarrow{\partial}}{\partial t}\bar{\mathcal{O}}(r,p)g_{\nu' l'}(r)=\int r^2\mathrm{d}r\,\frac{2}{\pi}\int p^2\mathrm{d}p\int p'^2\mathrm{d}p'\,\mathcal{P}_0(E,E')\bar{\mathcal{O}}(r,p)$$
$$\times f_{\nu l}(p)f_{\nu' l'}(p')j_l(pr)j_{l'}(p'r). \tag{8.9}$$

As always the time derivative is carried out in the Heisenberg picture before going into the Schrödinger picture by setting the time equal to zero. Thus,

$$\mathcal{P}_0(E,E')=\tfrac{1}{2}\left(\sqrt{\frac{E}{E'}}+\sqrt{\frac{E'}{E}}\right). \tag{8.10}$$

Note that the integral is both over position and momentum coordinates. Therefore one can use the operators $\boldsymbol{r}$ and $\boldsymbol{p}$ either in the position or the momentum representation. For convenience we choose to use the momentum space representation for both of them.

8.2.2. *The operator p^2*

The evaluation of the scalar operators p_i^2 presents no difficulty. Their expressions are for the spin 0, $\frac{1}{2}$ and 1 fields respectively:

$$p^2=\sum_{\nu\nu' l}\left[\nu lt|p^2|\nu' lt\right]:\left[\tilde{a}_{\nu}^{[lt]}a_{\nu'}^{[lt]}\right]^{[00]}:, \tag{8.11}$$

$$p^2=\sum_{\nu\nu' lj}\left[\nu ljt|p^2|\nu' ljt\right]:\left[\tilde{b}_{\nu l}^{[jt]}b_{\nu' l}^{[jt]}\right]^{[00]}:, \tag{8.12}$$

$$p^2=\sum_{\kappa\nu\nu' J}\left[\kappa\nu Jt|p^2|\kappa\nu' Jt\right]:\left[\tilde{A}_{\kappa\nu}^{[Jt]}A_{\kappa\nu'}^{[Jt]}\right]^{[00]}:. \tag{8.13}$$

Note that p^2 is diagonal also in κ for the spin-1 field.

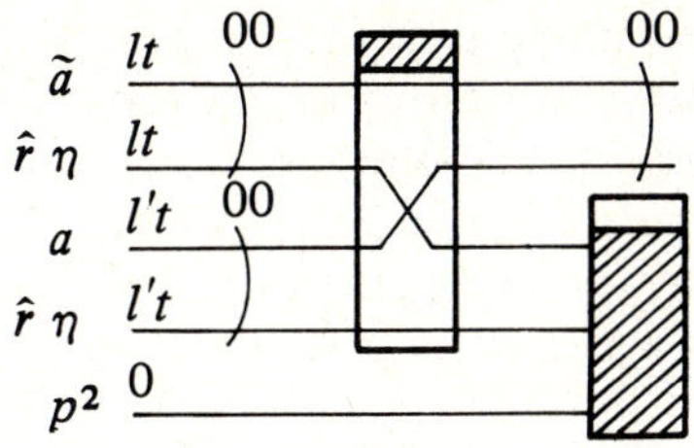

Fig. 8.1.

For spin 0, the invariant matrix element in (8.11) is according to fig. 8.1 and using the result of eq. (4.22):

$$[\nu lt|p^2|\nu' lt] = \hat{l}\hat{t}\int p^2\,\mathrm{d}p\, f_{\nu l}(p) f_{\nu' l}(p)\, p^2 .$$

For spin $\frac{1}{2}$, we obtain in detail from the expansion (5.8) together with (5.9)–(5.10), after completing the angular parts according to fig. 8.2:

$$p^2 = \sum_{\nu\nu' lj} \frac{2}{\pi}\int r^2\,\mathrm{d}r \int p^2\,\mathrm{d}p \int p'^2\,\mathrm{d}p'\, p^2 f_{\nu l}(p) f_{\nu' l}(p')$$

$$\times \left(\mathcal{P}^+_{1/2}(E, E')\, j_l(pr)\, j_l(p'r) + \mathcal{P}^-_{1/2}(E, E')\, j_\lambda(pr)\, j_\lambda(p'r)\right) \hat{j}\hat{t} : \left[\tilde{b}^{[jt]}_{\nu l} b^{[jt]}_{\nu' l}\right]^{[00]} : ,$$

with

$$\mathcal{P}^+_{1/2}(E, E') = \frac{1}{2}\sqrt{\frac{(E+M)(E'+M)}{EE'}} ,$$
$$\mathcal{P}^-_{1/2}(E, E') = \frac{1}{2}\sqrt{\frac{(E-M)(E'-M)}{EE'}} . \qquad (8.14)$$

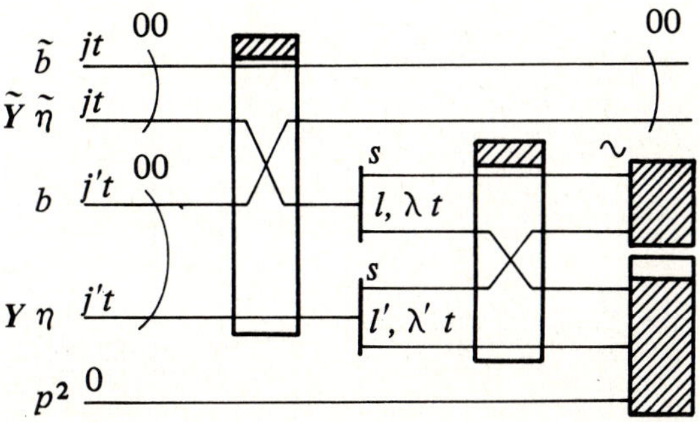

Fig. 8.2.

Carrying out the integration over r and p' we obtain the invariant matrix element

$$[\nu l j t|p^2|\nu' l j t] = \hat{j}\hat{t}\int p^2 \mathrm{d}p\, f_{\nu l}(p) f_{\nu' l}(p) p^2 .$$

The treatment for the vector mesons is similar. Introducing the expansions (5.11) through (5.13) in (8.6) and evaluating the angular part according to fig. 8.3 we obtain again for the invariant matrix element

$$[\kappa\nu J t|p^2|\kappa\nu' J t] = \hat{J}\hat{t}\int p^2 \mathrm{d}p\, f_{\nu J}(p) f_{\nu' J}(p) p^2$$

whatever the multipolarity κ.

The radial integrals entering these expressions are readily evaluated when using for the functions $f_{\nu l}(p)$ the harmonic oscillator basis (3.8) and yield

$$\int p^2 \mathrm{d}p\, f_{\nu l}(p) f_{\nu' l}(p) p^2 = -\frac{1}{\alpha^2}\left(\sqrt{\nu(\nu+l+\tfrac{1}{2})}\,\delta_{\nu'\nu+1} + \sqrt{(\nu-1)(\nu+l-\tfrac{1}{2})}\,\delta_{\nu'\nu-1}\right),$$

where α is the harmonic oscillator parameter.

Before leaving this section we note that the elementary matrix elements of the operators e^{-zE_i} which enter the expression of $\boldsymbol{R}^2$, eq. (8.1), are identical to those of p_i^2 as far as angular momentum is concerned. They are obtained by replacing p^2 by e^{-zE}. Thus for spin 0

$$\mathrm{e}^{-zE} = \sum_{\nu l} [\nu l t|\mathrm{e}^{-zE}|\nu' l t] : [\tilde{a}_\nu^{[lt]} a_{\nu'}^{[lt]}]^{[00]} : , \tag{8.15}$$

with

$$[\nu l t|\mathrm{e}^{-zE}|\nu' l t] = \hat{l}\hat{t}\int p^2 \mathrm{d}p f_{\nu l}(p) f_{\nu' l}(p) \mathrm{e}^{-zE} .$$

The forms for spin $\frac{1}{2}$ and spin 1 are given by the same replacement.

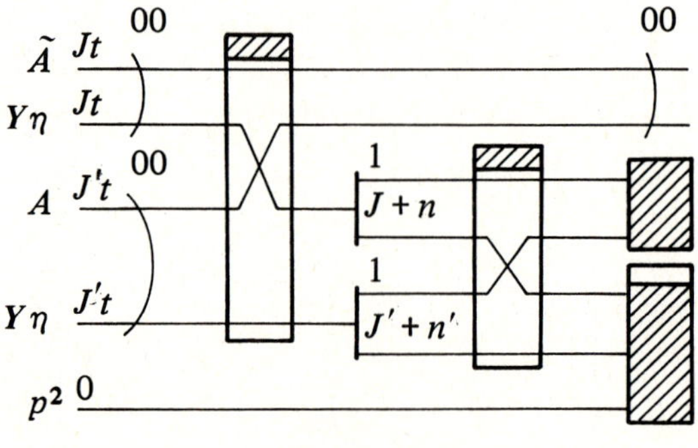

Fig. 8.3.

8.2.3 *The operator* $\boldsymbol{p}$

For the case of spin 0 the single-particle linear momentum operator $\boldsymbol{p}_i$ expressed in terms of the field operators is defined as in eq. (8.5)

$$\boldsymbol{p} = \int \mathrm{d}^3 r : \varphi^{(+)}(\boldsymbol{r}) i \overleftrightarrow{\partial}_t \varphi^{(-)}(\boldsymbol{r}) : \hat{1}[p^{[1]} e^{[1]}]^{[0]}$$

$$= \sum_{\nu l \nu' l'} [\nu l t | p | \nu' l' t] : [\tilde{a}_\nu^{[lt]} a_{\nu'}^{[l't]} e^{[1]}]^{[00]} : . \tag{8.16}$$

The invariant matrix element is calculated according to fig. 8.4. Here the placement of $\boldsymbol{p}$ is free since we work in momentum space. We obtain

$$[\nu l t | p | \nu' l' t] = -i \hat{t} \alpha_{ll'} \int p^2 \mathrm{d} p\, f_{\nu l}(p) f_{\nu' l'}(p) p .$$

Here the factor $\alpha_{ll'}$, eq. (4.24), imposes $l = l' \pm 1$.

For spin $\frac{1}{2}$, the linear momentum operator is,

$$\boldsymbol{p} = \int \mathrm{d}^3 r : \psi^+(\boldsymbol{r}) \psi(\boldsymbol{r}) : \hat{1}[p^{[1]} e^{[1]}]^{[0]} = \sum_{\nu l j \nu' l' j'} [\nu l j t | p | \nu' l' j' t] : [\tilde{b}_{\nu l}^{[jt]} b_{\nu' l'}^{[j't]} e^{[1]}]^{[00]} : .$$

The invariant matrix element is evaluated according to the graph of fig. 8.5 to yield

$$[\nu l j t | p | \nu' l' j' t] = -i \hat{t} \hat{s} \int p^2 \mathrm{d} p\, f_{\nu l}(p) f_{\nu' l'}(p) \frac{p}{2E}$$

$$\left((E+M) \begin{bmatrix} s & l & j \\ s & l' & j' \\ 0 & 1 & 1 \end{bmatrix} \alpha_{ll'} + (E-M) \begin{bmatrix} s & \lambda & j \\ s & \lambda' & j' \\ 0 & 1 & 1 \end{bmatrix} \alpha_{\lambda\lambda'} \right) .$$

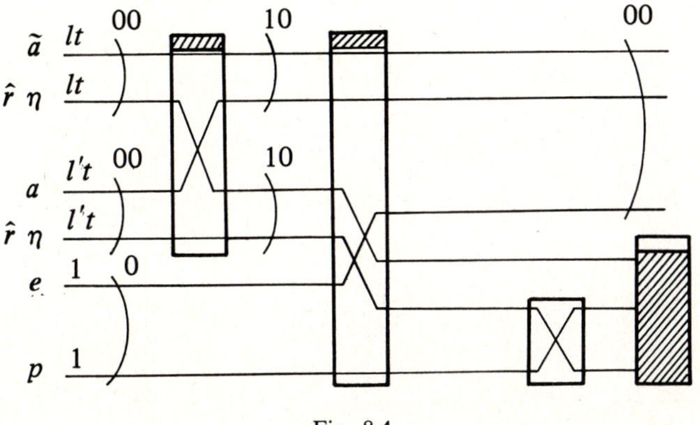

Fig. 8.4.

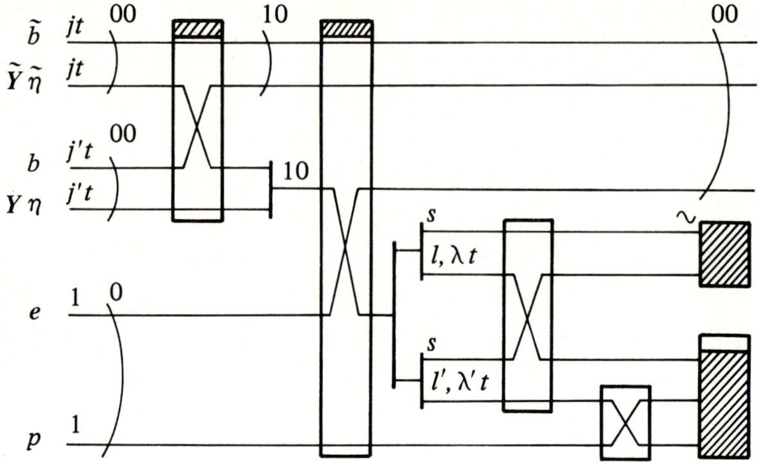

Fig. 8.5.

For the vector mesons, we have from (8.6)

$$p = -\int \mathrm{d}^3 r : \phi^{(+)}(\boldsymbol{r}) \left(\gamma_\mu \vec{\partial}_\mu + \gamma_\mu \overleftarrow{\partial}_\mu \right) \phi^{(-)}(\boldsymbol{r}) : \hat{1} \left[p^{[1]} e^{[1]} \right]^{[0]}$$

$$= \sum_{\kappa\nu J \kappa'\nu' J'} \left[\kappa\nu Jt | p | \kappa'\nu' J't \right] : \left[\tilde{A}^{[Jt]}_{\kappa\nu} A^{[J't]}_{\kappa'\nu'} e^{[1]} \right]^{[00]} : .$$

The elementary matrix elements in this expansion are not diagonal in the multipolarity κ. According to the graph of fig. 8.6 which contains the overlap box $[e^{[1]}|e^{[1]}] = \hat{1}$, we have

$$\left[\kappa\nu Jt | p | \kappa'\nu' J't \right] = -i\hat{t}\hat{1} \sum_{nn'} C_\kappa(n, J) C_{\kappa'}(n', J') \alpha_{J+n, J'+n'}$$

$$\times \begin{bmatrix} 1 & J+n & J \\ 1 & J'+n' & J' \\ 0 & 1 & 1 \end{bmatrix} \int p^2 \mathrm{d}p \, f_{\nu J}(p) \, p Q_{\kappa\kappa'} f_{\nu' J'}(p),$$

where the factor $Q_{\kappa\kappa'} = 1$ except for the cases where one only of the fields is longitudinal. Then,

$$Q_{\kappa\mathcal{L}} = Q_{\mathcal{L}\kappa} = \frac{E}{2M} + \frac{M}{2E}, \qquad \kappa \neq \mathcal{L}.$$

This factor arises from the normalizations of the fields. The unitary tranformation $C_\kappa(n, J)$ is given in eq. (5.27).

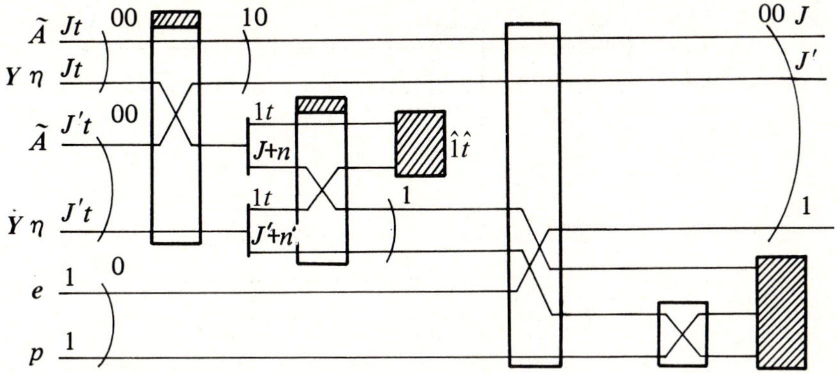

Fig. 8.6.

8.2.4. The operator r^2

Since r_i^2 is a scalar quantity the recoupling geometry is the same as that of p_i^2, and the angular parts of the invariant matrix elements have the same form. Thus the expressions for the symmetrized operator X^2 of eq. (8.2) are for the spin 0, $\frac{1}{2}$ and 1 fields, respectively

$$X^2 = \sum_{\nu\nu' l} [\nu lt|X^2|\nu' lt]:[\tilde{a}_\nu^{[lt]} a_{\nu'}^{[lt]}]^{[00]}:,$$

$$X^2 = \sum_{\nu\nu' lj} [\nu jt|X^2|\nu' jt]:[\tilde{b}_{\nu l}^{[jt]} b_{\nu' l}^{[jt]}]^{[00]}:,$$

$$X^2 = \sum_{\kappa\kappa'\nu\nu' J} [\kappa\nu Jt|X^2|\kappa\nu' Jt]:[\tilde{A}_{\kappa\nu}^{[Jt]} A_{\kappa\nu'}^{[Jt]}]^{[00]}:. \tag{8.17}$$

We will here and in the next section need the definition of the symmetrized energy weight factors, $i = 1$ and 2, i.e., linear or quadratic in the energy variable,

$$\mathcal{E}_i(E, E') = \tfrac{1}{2}(E^i \mathrm{e}^{-zE} + E'^i \mathrm{e}^{-zE'}).$$

For the spin-0 field the recoupling geometry is that of fig. 8.1 and we get

$$[\nu lt|X^2|\nu' lt] = \hat{l}\hat{t}\frac{2}{\pi}\int r^2 \mathrm{d}r\, r^2 \int p^2 \mathrm{d}p \int p'^2 \mathrm{d}p'\, j_l(pr)\, j_l(p'r)$$
$$\times \mathcal{E}_2(E, E')\mathcal{P}_0(E, E') f_{\nu l}(p) f_{\nu' l}(p'),$$

with the definition of eq. (8.10) for $\mathcal{P}_0$. This expression can be evaluated directly, or the integration over r may be performed analytically using the relations given at the end of sect. 4.2.4 and integrating by parts.

For the spin-$\frac{1}{2}$ field the recoupling geometry is that of fig. 8.2 and we have

$$[\nu ljt|X^2|\nu' ljt] = \hat{j}\hat{t}\frac{2}{\pi}\int r^2 \mathrm{d}r\, r^2 \int p^2 \mathrm{d}p \int p'^2 \mathrm{d}p'\, \mathcal{E}_2(E, E')$$

$$\times \{\mathcal{P}^+_{1/2}(E, E')\, j_l(pr)\, j_l(p'r) + \mathcal{P}^-_{1/2}(E, E')$$

$$\times j_\lambda(pr)\, j_\lambda(p'r)\} f_{\nu l}(p) f_{\nu' l}(p'),$$

with the definitions of eq. (8.14) for $\mathcal{P}_{1/2}$.

For the spin-1 field the form is, recognizing that the matrix element is not diagonal in the multipolarity κ in addition to ν,

$$[\kappa\nu Jt|X^2|\kappa'\nu' Jt] = \hat{J}\hat{t}\sum_n C_\kappa(n, J) C_{\kappa'}(n, J)\frac{2}{\pi}\int r^2 \mathrm{d}r\, r^2$$

$$\times \int p^2 \mathrm{d}p \int p'^2 \mathrm{d}p'\, j_{J+n}(pr)\, j_{J+n}(p'r)\, \mathcal{E}_2(E, E')$$

$$\times \mathcal{P}_{\kappa\kappa'}(E, E') f_{\nu J}(p) f_{\nu' J}(p'),$$

where the normalziation of the fields yields

$$\mathcal{P}_{\kappa\kappa'}(E, E') = \mathrm{n}_\kappa(E)\mathrm{m}_{\kappa'}(E') + \mathrm{m}_\kappa(E)\mathrm{n}_{\kappa'}(E')$$

with

$$\mathrm{n}_{\mathcal{M}} = \mathrm{n}_{\mathcal{E}} = \sqrt{\frac{1}{2E}}, \qquad \mathrm{n}_{\mathcal{L}} = \frac{1}{M}\sqrt{\frac{E}{2}},$$

$$\mathrm{m}_{\mathcal{M}} = \mathrm{m}_{\mathcal{E}} = \sqrt{\frac{E}{2}}, \qquad \mathrm{m}_{\mathcal{L}} = M\sqrt{\frac{1}{2E}}.$$

8.2.5. *The operator* $\boldsymbol{r}$

Again the symmetrized expression (8.3) must be used. The geometry is the same as that of the $\boldsymbol{p}$ operator. Thus the expression for the spin-0 field is with fig. 8.4

$$\boldsymbol{X} = \sum_{\nu\nu' l} [\nu lt|X|\nu' l't] : [\tilde{a}_\nu^{[lt]} a_{\nu'}^{[l't]} e^{[1]}]^{[00]} :,$$

with,

$$[\nu lt|X|\nu' l't] = \hat{t}|\alpha_{ll'}|\frac{2}{\pi}\int r^2 \mathrm{d}r\, r \int p^2 \mathrm{d}p \int p'^2 \mathrm{d}p'$$

$$\times j_l(pr)\, j_{l'}(p'r)\, \mathcal{E}_1(E, E')\, \mathcal{P}_0(E, E') f_{\nu l}(p) f_{\nu' l'}(p'). \quad (8.18)$$

The absolute values $|\alpha_{ll'}|$ arise from the definition (4.24) together with (4.17) and (4.18).

For spin $\frac{1}{2}$ the operator is

$$X = \sum_{\nu\nu' lj} [\nu ljt|X|\nu' l' j' t] : \left[\tilde{b}_{\nu l}^{[jt]} b_{\nu' l'}^{[j't]} e^{[1]}\right]^{[00]} : ,$$

where the invariant matrix element is calculated with a recoupling graph similar to fig. 8.5. Its expression is

$$[\nu ljt|X|\nu' l' j' t] = \hat{t}\hat{s}\frac{2}{\pi}\int r^2 \mathrm{d}r\, r \int p^2 \mathrm{d}p \int p'^2 \mathrm{d}p'$$

$$\times \mathcal{E}_1(E, E')\left(|\alpha_{ll'}| \mathcal{P}^+_{1/2}(E, E') \begin{bmatrix} s & l & j \\ s & l' & j' \\ 0 & 1 & 1 \end{bmatrix} j_l(pr)\, j_{l'}(p'r) \right.$$

$$\left. + |\alpha_{\lambda\lambda'}| \mathcal{P}^-_{1/2}(E, E') \begin{bmatrix} s & \lambda & j \\ s & \lambda' & j' \\ 0 & 1 & 1 \end{bmatrix} j_\lambda(pr)\, j_{\lambda'}(p'r) \right) f_{\nu l}(p) f_{\nu' l'}(p') . \tag{8.19}$$

For the spin-1 field the position operator is

$$X = \sum_{\kappa\nu J\kappa'\nu' J'} [\kappa\nu Jt|X|\kappa'\nu' J't] : \left[\tilde{A}_{\kappa\nu}^{[Jt]} A_{\kappa'\nu'}^{[J't]} e^{[1]}\right]^{[00]} :$$

with

$$[\kappa\nu Jt|X|\kappa'\nu' J't] = \hat{t}\hat{1}\sum_{nn'} C_\kappa(n, J) C_{\kappa'}(n', J')$$

$$\times |\alpha_{J+n, J'+n'}| \begin{bmatrix} 1 & J+n & J \\ 1 & J'+n' & J' \\ 0 & 1 & 1 \end{bmatrix} \frac{2}{\pi}\int r^2 \mathrm{d}r\, r \int p^2 \mathrm{d}p \int p'^2 \mathrm{d}p'$$

$$\times \mathcal{E}_1(E, E') \mathcal{P}_{\kappa\kappa'}(E, E')\, j_{J+n}(pr)\, j_{J'+n'}(p'r) f_{\nu J}(p) f_{\nu' J'}(p') .$$

8.3. CM MOMENTUM MANY-BODY MATRIX ELEMENTS

The evaluation of the many-body matrix elements of the c.m. pseudo-hamiltonian follows the lines of Chapter 7, i.e. it consists of the steps of separation of types, separation of groups within a type, and evaluation within a group. We consider first the c.m. momentum operator P^2

$$P^2 = \sum_i p_i^2 + \sum_{i \neq j} \boldsymbol{p}_i \cdot \boldsymbol{p}_j .$$

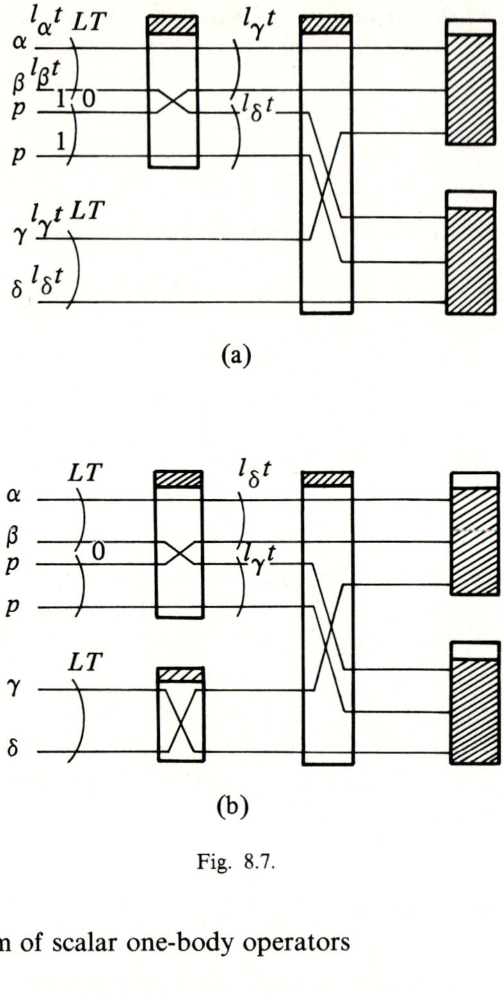

Fig. 8.7.

The first sum is a sum of scalar one-body operators

$$\sum_i p_i^2 = \sum_{\alpha\beta} [\alpha|p^2|\beta] : [\tilde{a}_\alpha a_\beta]^{[0]} :,$$

where the invariant matrix elements are given in the previous section. The second sum is a sum of scalar two-body operators of the general form

$$\sum_{i\neq j} \boldsymbol{p}_i \cdot \boldsymbol{p}_j = \sum_{\alpha\beta\gamma\delta} [\alpha\beta|\hat{1}[p^{[1]}p^{[1]}]^{[0]}|\gamma\delta] : [[\tilde{a}_\alpha \tilde{a}_\beta][a_\gamma a_\delta]]^{[0]} :, \tag{8.20}$$

where the invariant two-body matrix elements are calculated from the graph of fig. 8.7 in terms of the invariant elementary matrix elements of $\boldsymbol{p}_i$ for the different particle types. Note that with our definitions of the state vectors and operators the order of the indices in the invariant matrix element is here the same as that of the Fock operators. If the two particles belong in two different types, x and y, only the graph 8.7a exists. If both particles are bosons or fermions the exchange contribution 8.7b must be added. All the cases can be summarized as

$$\left[[\alpha\beta]^{[LT]}\|\hat{1}[p^{[1]}p^{[1]}]^{[0]}\|[\gamma\delta]^{[LT]}\right]$$

$$= \hat{1}\hat{L}\hat{T}\mathcal{N}(\alpha\beta)\mathcal{N}(\gamma\delta)\left(\begin{bmatrix} l_\alpha & l_\beta & L \\ 1 & 1 & 0 \\ l_\gamma & l_\delta & L \end{bmatrix}[\alpha|p|\gamma][\beta|p|\delta]\right.$$

$$\left.+\varepsilon(-)^{l_\gamma+l_\delta-L+2t-T}\begin{bmatrix} l_\alpha & l_\beta & L \\ 1 & 1 & 0 \\ l_\delta & l_\gamma & L \end{bmatrix}[\alpha|p|\delta][\beta|p|\gamma]\delta_{xy}\right). \qquad (8.21)$$

Here ε is the crossing phase associated with the statistics of the particles ($\varepsilon = -1$ for two fermions, $\varepsilon = +1$ for two bosons or one fermion and one boson). The norms are $\mathcal{N}(\alpha\beta) = 1$ for non-identical particles and $\mathcal{N}(\alpha\beta) = 1/\sqrt{1+\delta_{\alpha\beta}}$ for particles of a same type.

The many-body matrix elements of these scalar operators are directly given by the techniques of Chapter 7. Owing to the strong selection rules, viz. conservation of particle number and of particle type in addition to the selection rules of a scalar operator, here the phases η, sect. 7.2.2, for all matrix elements are simply, whatever the phase choice for the basis state vectors,

$$\eta_{\alpha\beta} = 1,$$

and the many-body matrix is of course real and symmetric. Since the many-body matrix elements of P^2 can be derived immediately from the examples of Chapter 7 we shall not list any of them. On the other hand the c.m. position operator R^2 is a new type of operator and its many-body aspects are now treated in detail.

8.4. CM POSITION MANY-BODY MATRIX ELEMENTS

8.4.1. Structure and multiplicities of the c.m. position matrix elements

As already mentioned, the c.m. position operator $\boldsymbol{R}$ is a full many-body operator. For a system of n particles the operators $\boldsymbol{R}$ and R^2 are n-body operators. In order to calculate the matrix elements of the n-body operator R^2, we factorize this operator into products of n single-particle operators according to the method of eq. (8.1)

$$R^2=\int_0^{\infty} z\,\mathrm{d}z\left(\sum_{i=1}^{n} X_i^2\Big(\prod_{k\neq i} Z_k\Big)+\sum_{i\neq j}^{n} \boldsymbol{X}_i\cdot\boldsymbol{X}_j\Big(\prod_{k\neq i,\,j} Z_k\Big)\right), \tag{8.22}$$

where the symmetrized position operators (8.2) and (8.3) together with the abbreviation (8.4) are introduced. The matrix elements of R^2 therefore will be computed as a function of the parameter z. At the end the integration over z must be carried out. The many-body character leads to two further aspects, viz. multiplicity and geometry. We consider first the question of structure and multiplicity and then give in the next section the matrix elements of R^2 with their full geometry for some cases of interest.

In order to study the structure of the many-body matrix element of R^2 we consider its form within a given particle type, which we write for compactness

$$R^2=\int_0^{\infty} z\,\mathrm{d}z\left(Z^n+X^2Z^{n-1}+\boldsymbol{X}\cdot\boldsymbol{X}Z^{n-2}\right). \tag{8.23}$$

The terms Z^n arise within a group when both operators $\boldsymbol{X}$ act on particles of other groups. The terms where the two operators $\boldsymbol{X}$ act each on a different type will be treated separately below in sect. 8.4.2 with reference to fig. 8.10.

In order to discuss the general structure and the multiplicities of each one of the products of n one-body operators in (8.23), we restrict ourselves to examples which cover the configuration spaces of Chapter 5. We consider a type made of groups denoted by a, b, c, in the lexicographic order. The various multiplicities and combinations of elementary matrix elements given below are most easily obtained in the space representation of sect. 7.2.7. The many-body states are fully symmetrized and normalized but, for simplicity, we drop the double underlining of the bra and ket used in that section. Since the geometry is not yet included, we use round brackets $(|\ \ |)$ for denoting the matrix elements. Table 8.1 gives a few specific examples.

These examples show how one goes from the expressions for Z^n to the cases X^2Z^{n-1} and $\boldsymbol{X}\cdot\boldsymbol{X}Z^{n-2}$. Formally, one has the general relations

$$\left(a_1^{n_1}a_2^{n_2}\ldots\,|X^2Z^{n-1}|\,b_1^{m_1}b_2^{m_2}\ldots\right)$$
$$=\sum_{ij}\left(a_i|X^2|b_j\right)\frac{\partial\left(a_1^{n_1}a_2^{n_2}\ldots\,|Z^n|\,b_1^{m_1}b_2^{m_2}\ldots\right)}{\partial\left(a_i|Z|b_j\right)},$$

$$\left(a_1^{n_1}a_2^{n_2}\ldots\,|\boldsymbol{X}\cdot\boldsymbol{X}Z^{n-2}|\,b_1^{m_1}b_2^{m_2}\ldots\right)$$
$$=\sum_{ijkl}\left(a_i|\boldsymbol{X}|b_j\right)\left(a_k|\boldsymbol{X}|b_l\right)\frac{\partial^2\left(a_1^{n_1}a_2^{n_2}\ldots\,|Z^n|\,b_1^{m_1}b_2^{m_2}\ldots\right)}{\partial\left(a_i|Z|b_j\right)\partial\left(a_k|Z|b_l\right)}.$$

The expression for the terms $\boldsymbol{X}Z^{n-1}$ is obtained from the first of these forms by the

TABLE 8.1
Structure and multiplicity of the matrix elements of some n-body operators

a^n/a^n

$$(a^n|Z^n|a'^n)=(a|Z|a')^n$$
$$(a^n|X^2Z^{n-1}|a'^n)=n(a|X^2|a')(a|Z|a')^{n-1}$$
$$(a^n|XXZ^{n-2}|a'^n)=n(n-1)(a|X|a')^2(a|Z|a')^{n-2}$$

$a^n/a'^{n-1}b'$

$$(a^n|Z^n|a'^{n-1}b')=\sqrt{n}\,(a|Z|a')^{n-1}(a|Z|b')$$
$$(a^n|X^2Z^{n-1}|a'^{n-1}b')=\sqrt{n}\,(a|Z|a')^{n-1}(a|X^2|b') +\sqrt{n}\,(n-1)(a|X^2|a')(a|Z|a')^{n-2}(a|Z|b')$$
$$(a^n|XXZ^{n-2}|a'^{n-1}b')=\sqrt{n}\,(n-1)(a|X|a')(a|Z|a')^{n-2}(a|X|b') +\sqrt{n}\,(n-1)(n-2)(a|X|a')^2(a|Z|a')^{n-3}(a|Z|b')$$

$a^{n-1}b/a'^{n-1}b'$

$$(a^{n-1}b|Z^n|a'^{n-1}b')=(a|Z|a')^{n-1}(b|Z|b') +(n-1)(a|Z|a')^{n-2}(a|Z|b')(b|Z|a')$$
$$\begin{aligned}(a^{n-1}b|X^2Z^{n-1}|a'^{n-1}b')=&(n-1)(a|X^2|a')(a|Z|a')^{n-2}(b|Z|b')\\ &+(a|Z|a')^{n-1}(b|X^2|b')\\ &+(n-1)(n-2)(a|X^2|a')(a|Z|a')^{n-3}(a|Z|b')(b|Z|a')\\ &+(n-1)(a|Z|a')^{n-2}(a|X^2|b')(b|Z|a')\\ &+(n-1)(a|Z|a')^{n-2}(a|Z|b')(b|X^2|a')\end{aligned}$$

replacement

$$\left(a_i|X^2|b_j\right)\rightarrow\left(a_i|\mathbf{X}|b_j\right).$$

Although the evaluation of these expressions presents no difficulty the full developments are cumbersome. Thus for the remaining configurations of the model spaces of Chapter 5 we give only in table 8.2 the expressions for the operator Z^n from which the matrix elements of the other operators are obtained from the above formulae.

8.4.2. *Complete c.m. position matrix elements*

With these relations and the elementary matrix elements of sect. 8.2 the many-body matrix elements of the c.m. position operator are readily calculated with the techniques of Chapter 7. The procedure is as follows. For a particular bra and ket, one first determines the structures and multiplicities of the products of one-body matrix elements according to the formulae of the previous sections. Then, for each of these products, one draws and evaluates the corresponding recoupling graph. For purposes of illustration we give here a few specific examples. Here again, as for the case of P^2,

$$\eta_{\alpha\beta}=1\,,$$

and of course the many-body matrix is real and symmetric.

TABLE 8.2
Structure and multiplicity of the matrix elements of Z^n

$$(a^{n-1}b|Z^n|a'^{n-2}b'^2) = \sqrt{2}\,(n-2)(a|Z|a')^{n-3}(a|Z|b')^2(b|Z|a') + \sqrt{2}\,2(a|Z|a')^{n-2}(b|Z|b')(a|Z|b')$$

$$(a^{n-2}b^2|Z^n|a'^{n-2}b'^2) = (a|Z|a')^{n-2}(b|Z|b')^2 + 2(n-2)(a|Z|b')(a|Z|a')^{n-3}(b|Z|a')(b|Z|b') + \tfrac{1}{2}(n-2)(n-3)\,(a|Z|b')^2(a|Z|a')^{n-4}(b|Z|a')^2$$

$$(a_1a_2a_3|Z^3|b_1b_2b_3) = \sum_{i\neq j\neq k=1}^{3} (a_1|Z|b_i)(a_2|Z|b_j)(a_3|Z|b_k)$$

$$(a_1^2a_2|Z^3|b_1b_2b_3) = \sqrt{2} \sum_{i\neq j\neq k=1}^{3} (a_1|Z|b_i)(a_1|Z|b_j)(a_2|Z|b_k)$$

$$(a^3|Z^3|b_1b_2b_3) = \sqrt{6} \sum_{i\neq j\neq k=1}^{3} (a|Z|b_i)(a|Z|b_j)(a|Z|b_k)$$

First consider the operator X^2Z^{n-1}. When there are several types with $n_1, n_2 \ldots$ particles, they are first separated according to the geometry given in sect. 7.2.3. Then the operator $X^2Z^{n_i-1}$ is applied to each type i in turn while the operators Z^{n_j} are applied to the remaining types j. Note furthermore that here each one-body operator (X^2 or Z) is a scalar and consequently the geometry is trivial. Within a type we get for example in the important case of the pion cloud and for some selected configurations

$$\left[[1s^n]^{[0T]}|Z^n|[1s^n]^{[0T]}\right] = n\hat{T}[1s|Z|1s]^n,$$

$$\left[[1s^n]^{[0T]}|X^2Z^{n-1}|[1s^n]^{[0T]}\right] = n\hat{T}\left[1s|X^2|1s\right][1s|Z|1s]^{n-1},$$

where the invariant elementary matrix elements of X^2 are given by (8.17) and of Z by (8.15). Likewise for $\nu \neq 1$

$$\left[[1s^n]^{[0T]}|Z^n|\left[[1s^{n-1}]^{[0R]}\nu s\right]^{[0T]}\right] = \hat{T}\sqrt{n}\,\mathrm{CFP}_1^n(T,R)[1s|Z|\nu s][1s|Z|1s]^{n-1},$$

$$\left[[1s^n]^{[0T]}|X^2Z^{n-1}|\left[[1s^{n-1}]^{[0R]}\nu s\right]^{[0T]}\right] = \hat{T}\,\mathrm{CFP}_1^n(T,R)$$

$$\times\left(\sqrt{n}\,[1s|X^2|\nu s][1s|Z|1s]^{n-1} + \sqrt{n}\,(n-1)[1s|X^2|1s][1s|Z|1s]^{n-2}[1s|Z|\nu s]\right).$$

When the type is made of several groups, the separation of the groups yields geometries similar to the ones of sect. 7.2.4. For example, for pions again, setting

$\pi_i \equiv i$ in the matrix elements, the geometry of fig. 8.8 yields for the following case

$$\left[[[12]^{[L_{12}T_{12}]}3]^{[LT]} \middle| Z^3 \middle| [[1'2']^{[L'_{12}T'_{12}]}3']^{[LT]} \right] = \frac{\hat{L}\hat{T}}{\hat{l}_1 \hat{l}_2 \hat{l}_3}$$

$$\times \Bigg([1|Z|1'][2|Z|2'][3|Z|3'] + (-)^{l'_1+l'_2+L'_{12}+T'_{12}} [1|Z|2'][2|Z|1'][3|Z|3']$$

$$+ \begin{bmatrix} l_1 & l_2 & L_{12} \\ l_3 & 0 & l_3 \\ L'_{12} & l_2 & L \end{bmatrix} \begin{bmatrix} 1 & 1 & T_{12} \\ 1 & 0 & 1 \\ T'_{12} & 1 & T \end{bmatrix} \big([1|Z|1'][3|Z|2'][2|Z|3']$$

$$+ (-)^{l'_1+l'_2+L'_{12}+T'_{12}} [1|Z|2'][3|Z|1'][2|Z|3'] \big)$$

$$+ (-)^{l_1+l_2+L_{12}+T_{12}} \begin{bmatrix} l_2 & l_1 & L_{12} \\ l_3 & 0 & l_3 \\ L'_{12} & l_1 & L \end{bmatrix} \begin{bmatrix} 1 & 1 & T_{12} \\ 1 & 0 & 1 \\ T'_{12} & 1 & T \end{bmatrix}$$

$$\times \big([2|Z|1'][3|Z|2'][1|Z|3'] + (-)^{l'_1+l'_2+L'_{12}+T'_{12}} [2|Z|2'][3|Z|1'][1|Z|3'] \big) \Bigg) .$$

We now turn to the operator $\boldsymbol{X} \cdot \boldsymbol{X} Z^{n-2}$. The elementary matrix elements of $\boldsymbol{X}$ are given in sect. 8.2.5. They are those of a vector operator. Note that in the case of the momentum operator $\boldsymbol{p} \cdot \boldsymbol{p}$ we constructed a scalar two-body invariant, eq. (8.21), so as to apply directly the rules of Chapter 7. Here it is more convenient to leave the operators in a factorized form so as to be able to use the structures and multiplicities given above. Simplifications arise because of the selection rules associated with the vector operator $\boldsymbol{X}$. For example the formula for the case $a^{n-1}b/a'^{n-1}b'$ for the pion cloud with a group $a^{n-1} = a'^{n-1} = 1s^{n-1}$ has the structure

$$(a^{n-1}b|\boldsymbol{X} \cdot \boldsymbol{X} Z^{n-2}|a^{n-1}b') = (a|\boldsymbol{X}|b') \cdot (b|\boldsymbol{X}|a') \frac{\partial^2(Z^n)}{\partial(a|Z|b')\, \partial(b|Z|a')}$$

$$= (n-1)(a|Z|a')^{n-2}(a|\boldsymbol{X}|b') \cdot (b|\boldsymbol{X}|a') ,$$

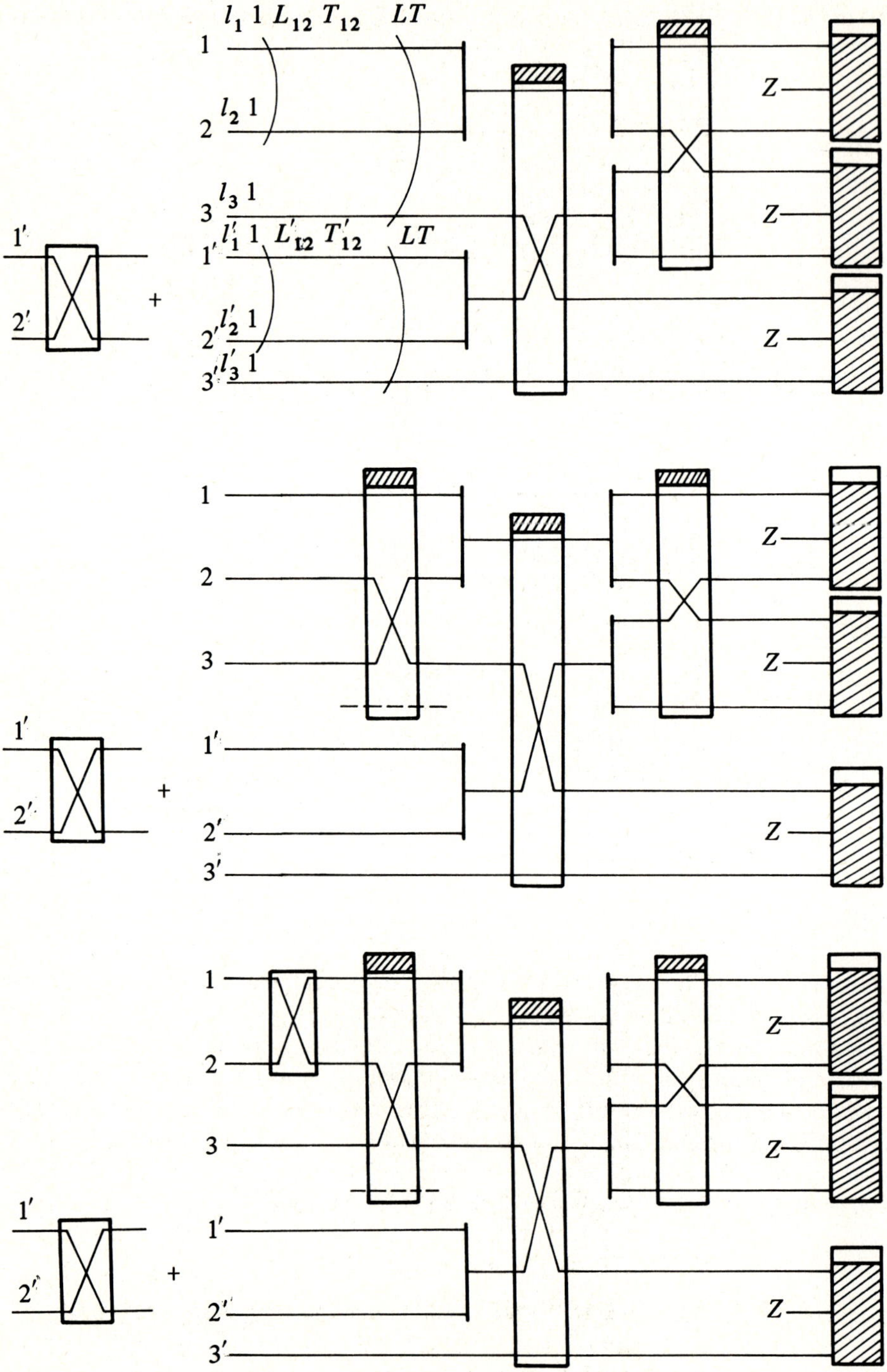

l_1 1 L_{12} T_{12} LT
1
l_2 1
2
l_3 1
3
l'_1 1 L'_{12} T'_{12} LT
1'
l'_2 1
2'
l'_3 1
3'
1'
2'
+
Z
Z
Z
1
2
3
1'
2'
3'
1'
2'
+
Z
Z
Z
1
2
3
1'
2'
3'
1'
2'
+
Z
Z
Z

Fig. 8.8.

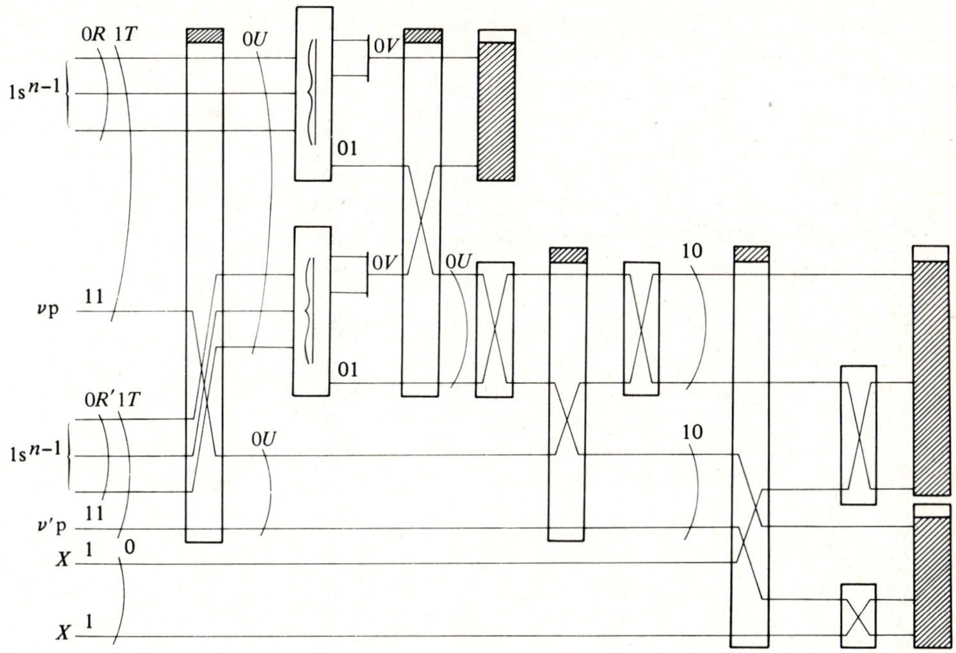

Fig. 8.9.

where it has been taken into account that $\boldsymbol{X}$ has only non-diagonal matrix elements. The geometry is calculated according to fig. 8.9 and the result is, taking into account the selection rule $\Delta l = 1$,

$$\left[[[1\mathrm{s}^{n-1}]^{[0R]}\nu\mathrm{p}]^{[1T]}\middle|\hat{1}[X^{[1]}X^{[1]}]^{[0]}Z^{n-2}\middle|[[1\mathrm{s}^{n-1}]^{[0R']}\nu'\mathrm{p}]^{[1T]}\right]$$

$$=\frac{n-1}{\hat{1}}\sum_V \hat{V}\,\mathrm{CFP}_1^{n-1}(R,V)\mathrm{CFP}_1^{n-1}(R',V)$$

$$\times\sum_U(-)^U\hat{U}\begin{bmatrix} R & 1 & T \\ R' & 1 & T \\ U & U & 0\end{bmatrix}\begin{bmatrix} V & 1 & R \\ V & 1 & R' \\ 0 & U & U\end{bmatrix}[1\mathrm{s}^{n-2}|Z|1\mathrm{s}^{n-2}][\nu\mathrm{p}|X|1\mathrm{s}][1\mathrm{s}|X|\nu'\mathrm{p}].$$

The elementary invariant matrix elements of $\boldsymbol{X}$ are given in eq. (8.18).

We now come to the case where the operator $\boldsymbol{X}\cdot\boldsymbol{X}Z^{n-2}$ distributes its vector operators $\boldsymbol{X}$ between two types. The calculation requires then the usual geometry for type separation. For example for a single nucleon N with a pion cloud, the

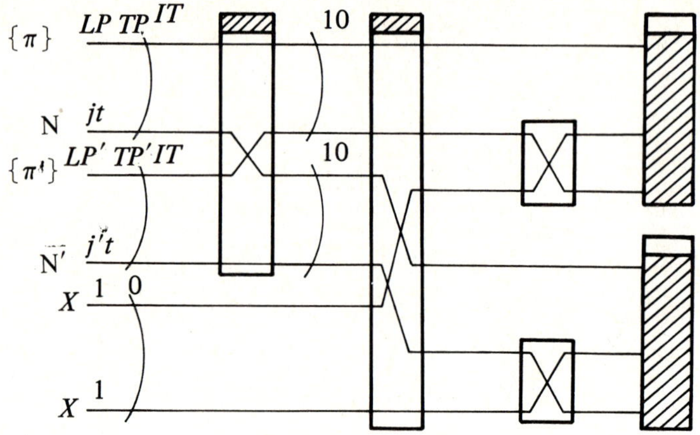

Fig. 8.10.

many-body matrix elements are inserted in the coupling diagram of fig. 8.10 to yield

$(n=n_1+n_2+\cdots=m_1+m_2+\cdots)$:

$$\left[\left[\left[\pi_1^{n_1}\pi_2^{n_2}\ldots\right]^{[LP\,TP]}\mathrm{N}^{[jt]}\right]^{[IT]}\middle|\hat{1}\left[X^{[1]}X^{[1]}\right]^{[0]}Z^{n-2}\middle|\left[\left[\pi_1'^{m_1}\pi_2'^{m_2}\ldots\right]^{[LP'TP']}\mathrm{N}'^{[j't]}\right]^{[IT]}\right]$$

$$=\frac{\hat{T}}{\widehat{TP}\hat{1}}\delta_{TP\,TP'}\begin{bmatrix} LP & j & I \\ LP' & j' & I \\ 1 & 1 & 0 \end{bmatrix}\left[\left[\pi_1^{n_1}\pi_2^{n_2}\ldots\right]^{[LP\,TP]}\middle|X^{[1]}Z^{n-2}\middle|\left[\pi_1'^{m_1}\pi_2'^{m_2}\ldots\right]^{[LP'TP']}\right]$$

$$\times\left[\mathrm{N}\middle|X^{[1]}\middle|\mathrm{N}'\right].$$

The elementary invariant matrix element for the fermion is given in eq. (8.19), while the structure and geometry of the boson many-body matrix element $\boldsymbol{X}Z^{n-2}$ is computed as that of the operator X^2Z^{n-1} treated above, except for the change of geometry required when going from the scalar operator X^2 to the vector operator $\boldsymbol{X}$.

These examples cover all the cases in the Hilbert spaces of Chapter 5 and they suffice to show how to generate the many-body matrix of the c.m. pseudo-hamiltonian.

CHAPTER 9

Electromagnetic interactions

9.1. INTRODUCTION

This chapter is devoted to the description of the interaction of the photon field with a relativistic many-body system treated according to the previous chapters. More particularly, we consider the important cases of the form factors for electron scattering and of the static electric quadrupole and magnetic dipole moments.

As discussed in sect. 2.5.1 we mainly restrict ourselves to those current interactions which arise from the minimal substitution into the free hamiltonian, but we also consider an interaction arising from the vector dominance model discussed in sect. 2.5.3, and the anomalous moment interaction of sect. 2.5.4. The invariant elementary matrix elements are derived in sect. 9.2 both for real photon transitions and for electron scattering form factors. The special case of the static moments is treated in sect. 9.3.

The many-body aspects are then carried out in sect. 9.4. In Chapter 7 these many-body aspects were discussed by giving in detail the expressions of the matrix elements between specific configurations. Here instead we use the occasion to describe a method with which one can automatically generate the many-body matrix elements of any operator.

9.2. ELECTRON SCATTERING FORM FACTORS

9.2.1. Form of the operators

The electron scattering form factor is defined for the vector current interaction as

$$G_\mu(\boldsymbol{q}) = \int \mathrm{d}^3 r J_\mu(\boldsymbol{r}) \mathrm{e}^{i\boldsymbol{q}\cdot\boldsymbol{r}},$$

where J_μ is the current four-vector and $\boldsymbol{q}$ is the momentum transfer.

The forms of the current J_μ derived from minimal coupling are given in sect. 2.5.1 for the different fields. In addition to the vector current interaction we also will take into account the anomalous moment interaction and the vector dominance interaction. These will be treated below.

It will be useful to introduce explicitly the polarization vector of the photon

$$d_\mu = \left(d_m^{(1)}, d_4\right).$$

Here $d_m^{(1)}$ are the amplitudes of the polarization vector along the components of the spherical unit vector $e_m^{[1]}$. This way, according to sect. 4.4, the space polarization of the photon field is given by the invariant form

$$\boldsymbol{d} = \hat{1}\left[d^{[1]}e^{[1]}\right]^{[0]}.$$

The amplitude d_4 is the amplitude of the time part of the polarization unit four-vector of the field.

Writing

$$\boldsymbol{J} = \hat{1}\left[J^{[1]}e^{[1]}\right]^{[0]},$$

$$\boldsymbol{A} = \hat{1}\left[d^{[1]}e^{[1]}\right]^{[0]}\mathrm{e}^{i\boldsymbol{q}\cdot\boldsymbol{r}},$$

we obtain the invariant form of the electron scattering form factor

$$\boldsymbol{G}(\boldsymbol{q}) = \hat{1}\int \mathrm{d}^3r\left[d^{[1]}J^{[1]}\right]^{[0]}\mathrm{e}^{i\boldsymbol{q}\cdot\boldsymbol{r}},$$

$$G_4(\boldsymbol{q}) = \int \mathrm{d}^3r\, d_4 J_4 \mathrm{e}^{i\boldsymbol{q}\cdot\boldsymbol{r}},$$

where we see that we have the convention $d_4 = 1$. One must not confuse the polarization vector component d_4 with the time-like unit vector $e_4 = i$ in our convention.

Considering the electron scattering form factor, it is well known that the real photon interaction is given by setting $|q| = \omega$, i.e. the photon energy transfer. Thus all the formulae of this section encompass the real photon interaction for which of course the longitudinal terms are absent and for which one must supply the appropriate photon normalization factors and creation-annihilation operators. More explicitly, for real photons one has the condition on the polarization vector

$$\boldsymbol{d}\cdot\boldsymbol{q} = 0,$$

$$d_4 = 0,$$

and comparison with the plane wave spin-1 field expansion, eq. (2.21), shows that the real photon absorption operator is obtained by augmenting the transversal part of the above form factor by the factor

$$\frac{1}{(2\pi)^{3/2}}\int \mathrm{d}^3q\,\frac{1}{\sqrt{2|q|}}A_{qd}\mathrm{e}^{-i|q|t}.$$

For photon emission the expression is obtained by hermitian conjugation.

Continuing the development of the electron scattering form factor the interaction is expanded into multipoles using

$$e^{i\boldsymbol{q}\cdot\boldsymbol{r}} = 4\pi \sum_{\Lambda} (i)^{\Lambda} \hat{\Lambda} \left[\hat{q}^{[\Lambda]} \hat{r}^{[\Lambda]}\right]^{[0]} j_{\Lambda}(qr).$$

Thus we define the discretized operator expansion in terms of invariant matrix elements denoted here for simplicity $G(q, \Lambda, L)$ and calculated in the following sections,

$$\boldsymbol{G}(\boldsymbol{q}) = \sum_{\Lambda L} G(q, \Lambda, L) \left[V^{[L]} \left[d^{[1]} \hat{q}^{[\Lambda]}\right]^{[L]}\right]^{[0]},$$

$$G_4(\boldsymbol{q}) = \sum_{L} G^4(q, L) d_4 \left[V^{[L]} \hat{q}^{[L]}\right]^{[0]}.$$

The tensorial quantity $V^{[L]}$ stands for the normal product of the Fock-space operators of the interacting particles. One also may introduce the expansion of the photon field in orthogonal multipolarity components, thus

$$\boldsymbol{G}(\boldsymbol{q}) = \boldsymbol{F}_{\mathfrak{M}}(\boldsymbol{q}) + \boldsymbol{F}_{\mathfrak{E}}(\boldsymbol{q}) + \boldsymbol{F}_{\mathfrak{L}}(\boldsymbol{q}),$$

$$G_4(\boldsymbol{q}) = F_4(\boldsymbol{q}).$$

In terms of the invariant matrix elements $G(q, \Lambda, L)$ we have for the magnetic multipolarity according to the unitary transformation eq. (5.27),

$$\boldsymbol{F}_{\mathfrak{M}}(\boldsymbol{q}) = \sum_{\Lambda L} G(q, \Lambda, L) \left[V^{[L]} \left[d^{[1]} \hat{q}^{[\Lambda]}\right]^{[L]}\right]^{[0]} \delta_{\Lambda L}. \tag{9.1}$$

Likewise for the electric multipolarity

$$\boldsymbol{F}_{\mathfrak{E}}(\boldsymbol{q}) = \sum_{\Lambda L} \frac{1}{\hat{L}} \left(\sqrt{L+1}\, \delta_{\Lambda, L-1} + \sqrt{L}\, \delta_{\Lambda, L+1}\right) G(q, \Lambda, L) \left[V^{[L]} \left[d^{[1]} \hat{q}^{[\Lambda]}\right]^{[L]}\right]^{[0]} \tag{9.2}$$

and for the longitudinal multipolarity, decomposed into its space-like and time-like parts, absent for real photons,

$$\boldsymbol{F}_{\mathfrak{L}}(\boldsymbol{q}) = \sum_{\Lambda L} \frac{1}{\hat{L}} \left(\sqrt{L}\, \delta_{\Lambda, L-1} - \sqrt{L+1}\, \delta_{\Lambda, L+1}\right) G(q, \Lambda, L) \left[V^{[L]} \left[d^{[1]} \hat{q}^{[\Lambda]}\right]^{[L]}\right]^{[0]},$$

$$F_4(\boldsymbol{q}) = \sum_{L} G^4(q, L) d_4 \left[V^{[L]} \hat{q}^{[L]}\right]^{[0]}. \tag{9.3}$$

Writing the expressions in these forms exhibits clearly the angular variable $\hat{q}$ which will enter the expressions for the angular distributions. The amplitudes $d_m^{[1]}$

contain the polarization information of the experiment. The radial q dependence is contained in the matrix element. Note also that the space-like matrix element is a function of both L and Λ, while in the time-like element $L=\Lambda$.

In electron scattering one frequently defines the transverse form factor

$$\boldsymbol{F}(\boldsymbol{q})_{\text{trans.}}=\boldsymbol{F}_{\mathfrak{M}}(\boldsymbol{q})+\boldsymbol{F}_{\mathcal{E}}(\boldsymbol{q}),$$

and the longitudinal form factor four-vector

$$F(\boldsymbol{q})_{\text{long.}}=(\boldsymbol{F}_{\mathfrak{L}}(\boldsymbol{q}),F_4(\boldsymbol{q})).$$

9.2.2. Treatment of the recoil and the relativistic boost

The recoil associated with the momentum transferred to the system in the interaction process must be accounted for. We must however distinguish the pseudo-center-of-mass motion introduced into the solutions by the use of a truncated basis, sect. 3.4., from this recoil of the target. We shall discuss in succession the extraction of the pseudo-c.m. motion and the corrections required to account for the physical recoil in a relativistic framework.

We first discuss the removal of the effect of the pseudo-hamiltonian. We assume that the pseudo-hamiltonian, eq. (3.12),

$$\mathcal{H}=H+\tfrac{1}{2}\zeta(P^2+\Omega^2R^2)$$

has solutions approximately of the form (3.16)

$$\Phi\approx\phi_0(\boldsymbol{R})\chi(\boldsymbol{\xi}_i),$$

where $\phi_0(\boldsymbol{R})$ is the normalized 1s c.m. wave function and $\chi(\boldsymbol{\xi}_i)$ is the physical wave function of the relative intrinsic particle coordinates $\boldsymbol{\xi}_i$. From the relation

$$\langle\Phi|e^{i\boldsymbol{q}\cdot\boldsymbol{r}}|\Phi\rangle=\langle\phi_0|e^{i\boldsymbol{q}\cdot\boldsymbol{R}}|\phi_0\rangle\langle\chi|e^{i\boldsymbol{q}\cdot\boldsymbol{\xi}}|\chi\rangle,$$

we see that the physical form factor is of the form

$$F_\mu(\boldsymbol{q})_{\text{corr.}}=e^{q^2/4\Omega^2}F_\mu(\boldsymbol{q})_{\text{calc.}},\tag{9.4}$$

where $F_\mu(\boldsymbol{q})_{\text{calc.}}$ is the form factor calculated from the solutions Φ with the laboratory coordinates $\boldsymbol{r}_i$. Note that the resulting correction can be important even for small values of the transferred momentum $\boldsymbol{q}$ depending on the magnitude of the parameter Ω.

We now turn to the effect of the relativistic recoil of the target system at large momentum transfer. For computing matrix elements it is best to use the brick-wall coordinate system in which the recoiling composite object moves with velocity $-\boldsymbol{V}$ before the interaction and $+\boldsymbol{V}$ after the interaction. We consider first the case of elastic scattering. We have

$$\mp V = \mp \frac{P}{E} = \mp \frac{q}{2} \frac{1}{\left(M^2 + \frac{1}{4}q^2\right)^{1/2}},$$

where M is the rest mass of the system, P is the c.m. momentum, and q is the momentum transfer in the brick-wall system. The Lorentz factor is

$$\gamma = \left(1 + \left(\frac{q}{2M}\right)^2\right)^{1/2}.$$

In the brick-wall system the Lorentz contraction is the same in the initial and final states. We must, however, evaluate the effect of the spin operators contained in the boost. We begin by evaluating the effect of the boost on the value of the one-body matrix element involving the interacting constituent particles. The overlap matrix elements arise by replacing the interaction by the unit operator.

Instead of applying the general Lorentz transformation operator for going into the brick-wall system, we shall explicitly generate the effects of the relativistic boost upon the discretized states in a way which shows the physics contained in this operation. We begin with the example of a spin-0 constituent in the basis state $\varphi_{\nu l}$ of the solution Ψ calculated in the laboratory system. In the brick-wall system, assuming the transferred momentum to be along the z direction, we get for the elementary matrix element (recall that q here is given in the brick-wall system), omitting the factors of eq. (2.31),

$$F_\mu(\boldsymbol{q}) = \langle \varphi'_{\nu l}(\boldsymbol{V}) | \overleftrightarrow{\partial}_\mu \mathrm{e}^{iqz} | \varphi'_{\nu l}(-\boldsymbol{V}) \rangle .$$

Here $\varphi'_{\nu l}(\pm \boldsymbol{V})$ is the state of the particle boosted with velocity $\pm \boldsymbol{V}$. For the spin-0 case, the boosted functions $\varphi'_{\nu l}$ differ from the laboratory functions $\varphi_{\nu l}$ by a Lorentz contraction and a change of normalization

$$\varphi'_{\nu l}(x, y, z) = \sqrt{\gamma}\, \varphi_{\nu l}(x, y, \gamma z).$$

This form is of course the same for the initial and final states. Thus we have for the elementary matrix element in the brick-wall system

$$F_\mu(\boldsymbol{q}) = \int \mathrm{d}x\,\mathrm{d}y\,\mathrm{d}z\, \gamma \varphi^*(x, y, \gamma z)\, \overleftrightarrow{\partial}_\mu \mathrm{e}^{iqz} \varphi(x, y, \gamma z),$$

or, by a change of variable $Z = \gamma z$,

$$F_\mu(\boldsymbol{q}) = \int \mathrm{d}x\,\mathrm{d}y\,\mathrm{d}Z\, \varphi^*(x, y, Z)\, \overleftrightarrow{\partial}_\mu \mathrm{e}^{i(q/\gamma)Z} \varphi(x, y, Z).$$

Thus the effect of the relativistic recoil for a spin-0 constituent is fully taken into account in the form factor by a re-scaling of the momentum transfer in eq. (9.4)

$$F_\mu(\boldsymbol{q})_{\text{physical}} = F_\mu(\boldsymbol{q}/\gamma)_{\text{corrected}}\left(1 + \delta_{\mu z}(\gamma - 1)\right).$$

The last factor arises from the derivative $\overleftrightarrow{\partial}_z$. Finally, in this expression $\boldsymbol{q}$ is in the brick-wall system while the measured form factors usually are given in the laboratory system or the c.m. system.

For a constituent of non-zero spin, in addition to compensating for the Lorentz contraction, we must also evaluate the effect of the Lorentz transformation on the spinors. Thus we will have in the brick-wall system

$$\psi_{\nu l}^{[j]\prime}(x, y, z)=\sqrt{\gamma}\left(\psi_{\nu l}^{[j]}(x, y, \gamma z)+\sum_k \delta\psi_{\nu l k}^{[j]}(x, y, \gamma z)\right).$$

For example, for spin $\frac{1}{2}$ we have in the brick-wall system, omitting the factors in eq. (2.32),

$$F_\mu(\boldsymbol{q})=\frac{1}{N}\int \mathrm{d}x\,\mathrm{d}y\,\mathrm{d}Z(\bar{\psi}(x, y, Z)+\delta\bar{\psi}(x, y, Z))\mathrm{e}^{i(q/\gamma)Z}\gamma_\mu(\psi(x, y, Z)$$
$$+\delta\psi(x, y, Z))$$

with

$$N=\langle\psi+\delta\psi|\psi+\delta\psi\rangle .$$

We outline the calculation of the change $\delta\psi$ in the spin-$\frac{1}{2}$ wave function of the interacting constituent in the brick-wall system. There only the small components are modified by the boost through the operator $\boldsymbol{\sigma}\cdot\boldsymbol{p}$ which in the brick-wall system writes

$$\boldsymbol{\sigma}\cdot\boldsymbol{p}=-i\boldsymbol{\sigma}\cdot\nabla=-i\left(\sigma_x\partial_x+\sigma_y\partial_y+\gamma\sigma_z\partial_Z\right)=-i\left(\boldsymbol{\sigma}\cdot\nabla_\xi+(\gamma-1)\sigma_z\partial_Z\right),$$

where $\boldsymbol{\xi}=(x, y, Z)$ and $\nabla_\xi=(\partial_x, \partial_y, \partial_Z)$. In order to evaluate the correction $\delta\psi$ we introduce the invariant form ($\mp$ for the initial and final states respectively)

$$\mp(\gamma-1)\sigma_z\partial_Z=\hat{1}^2\left[h^{[1]}\sigma^{[1]}\right]^{[0]}\left[h^{[1]}\,\nabla_\xi^{[1]}\right]^{[0]},$$

with the amplitudes, according to sect. 4.2.3,

$$h_m^{[1]}=-i(\gamma-1)^{1/2}\delta_{m0}.$$

From fig. 9.1 we obtain for $\delta\psi(\boldsymbol{\xi})$ in the final state

$$\delta\psi_{\nu l k}^{[j]}(\boldsymbol{\xi})=\sqrt{\tfrac{1}{2}}\sum_K\alpha_{\lambda l}\frac{\hat{k}}{\hat{s}\hat{\lambda}}\begin{bmatrix}1 & 1 & K\\ s & l & j\\ s & \lambda & k\end{bmatrix}\sqrt{6}\,(-)^{j-k}$$
$$\times\left(\left[b_{\nu l}^{[j]}\begin{pmatrix}0\\ \left[\left[h^{[1]}h^{[1]}\right]^{[K]}Y_\lambda^{[k]}(\boldsymbol{\xi})\,\delta v_{\nu l\lambda}(\xi)\right]^{[j]}\end{pmatrix}\right]^{[0]}\right.$$
$$\left.+\left[\tilde{a}_{\nu l}^{[j]}\begin{pmatrix}\left[\left[h^{[1]}h^{[1]}\right]^{[K]}Y_\lambda^{[k]}(\boldsymbol{\xi})\,\delta v_{\nu l\lambda}(\xi)\right]^{[j]}\\ 0\end{pmatrix}\right]^{[0]}\right).$$

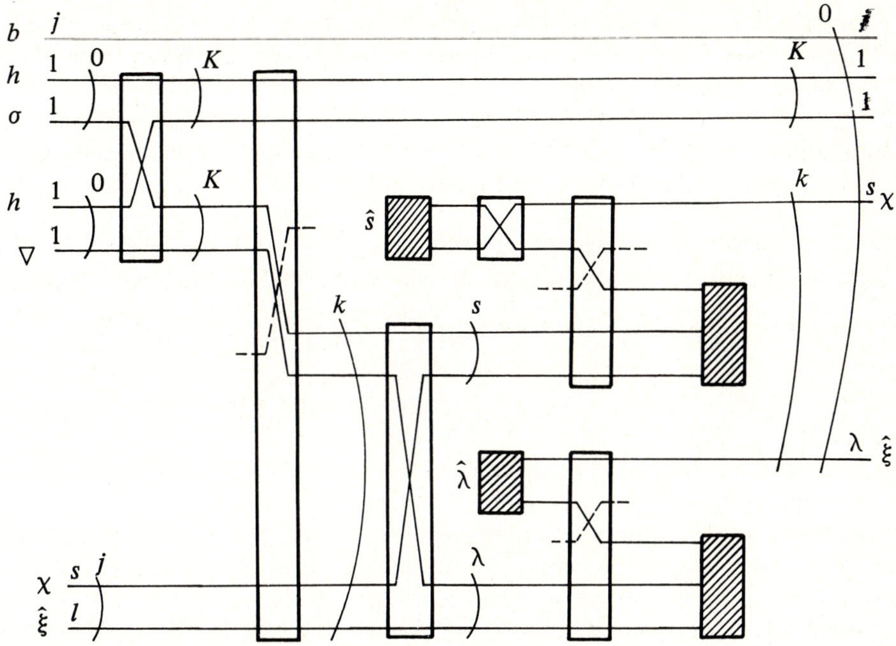

Fig. 9.1.

Here K is even owing to

$$[h^{[1]}h^{[1]}]^{[0]} = \frac{\gamma - 1}{\hat{1}},$$

$$[h^{[1]}h^{[1]}]_0^{[1]} = 0,$$

$$[h^{[1]}h^{[1]}]_0^{[2]} = \frac{2}{\sqrt{6}}(\gamma - 1).$$

The normalization N can now be computed, which is needed for large momentum transfer. However, for low momentum transfers we expect the relativistic recoil corrections to be small, in contrast to the corrections resulting from the extraction of the 1s c.m. motion discussed above, since $\delta\psi$ would be of the order of

$$\gamma - 1 \sim \frac{q^2}{8M^2}.$$

For the case of inelastic scattering or real photon absorption (emission) the calculation is most easily performed in a generalized brick-wall system which is defined such that in that system the γ-factors are the same in the initial and the final states. Then, considering the excitation of the system of rest mass M by the energy ΔE so that the rest mass of the final state is $M+\Delta E$, we have $\boldsymbol{P}_{\text{init.}} = -M\gamma\boldsymbol{v}$, $E_{\text{init.}} = M\gamma$ and $\boldsymbol{P}_{\text{fin.}} = (M+\Delta E)\gamma\boldsymbol{v}$, $E_{\text{fin.}} = (M+\Delta E)\gamma$. The momentum transfer four-vector is $q_\mu = ((2M+\Delta E)\gamma\boldsymbol{v},\ i\Delta E\gamma))$. With obvious modifications, the above formulae then hold also for inelastic form factors.

9.2.3. *The photon-pion interaction*

Before going in sect. 9.4 into the calculation of the many-body matrix elements of eqs. (9.1), (9.2) and (9.3) we need the elementary electromagnetic interaction operators in the Fock representation for the different fields. Many elementary processes are possible. We shall designate their operators by G. Here, as in all the following subsects. 9.2, we consider explicitly the important case of electron scattering form factors. The expressions for real photon processes can easily be derived along the lines of sect. 9.2.1.

For the spin-0 field three photon absorption processes exist as shown in fig. 9.2: pion scattering (a), pion pair production (b) and annihilation (c).

The photon field couples to charged particles. For isospin-1 particles, the current operator contains the charge operator τ_z which has eigenvalues ± 1 for the charged pions and 0 for the neutral pions. In other words, τ_z eliminates π_0 for all reactions and will insure charge conservation for all processes as shown below. The invariant form of τ_z is

$$\tau_z = \hat{1}[c^{[1]}\tau^{[1]}]^{[0]}, \tag{9.5}$$

with the definition of the isospin amplitudes

$$c_m^{[1]} = -i\delta_{m0}. \tag{9.6}$$

Explicitly, with these values of $c_m^{[1]}$ the only components remaining in the isospin invariants, which will be calculated below are

$$(\eta_M^{[1]}\tau_0^{[1]}\eta_{-M}^{[1]}).$$

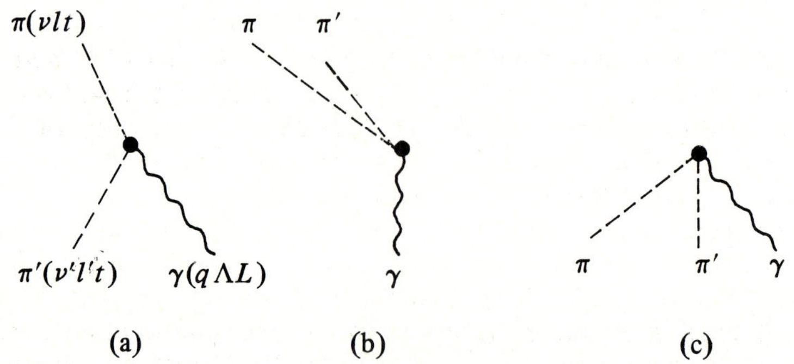

Fig. 9.2.

For pair production or annihilation opposite charges are implied by this form, since the corresponding wave functions, see (4.41) and (4.43), have the same tensorial character and opposite isospin projections. Thus only $(\pi^+\pi^-)$ pairs are created or absorbed. On the other hand, for scattering, the charge of the absorbed particle is given by the isospin projection of a contrastandard wave function while the charge of the emitted particle is given by the projection of a standard wave function. Thus for this process, in the above form, one of the wave functions must be converted to standard form to show the charge of the particle. For scattering the operator (9.5) indeed insures either $\pi^+ \to \pi^+$ or $\pi^- \to \pi^-$. Thus for the current eq. (2.31), we have for all cases

$$J_\mu = \tfrac{1}{2} ie\varphi \overleftrightarrow{\partial}_\mu \hat{1}[c^{[1]}\tau^{[1]}]^{[0]}\varphi .$$

In this expression the fields are the real fields of eq. (5.4). Introducing the polarization vector d_μ of the photon field as defined in sect. 9.2.1, we have for the space-like part (in the definition of the form factor, $\boldsymbol{A}$ is dimensionless)

$$\begin{aligned} \boldsymbol{G}_{\gamma\pi} &= \int \mathrm{d}^3 r : \boldsymbol{J}\cdot\boldsymbol{A} : \\ &= \tfrac{1}{2} ie\hat{1}^2 \int \mathrm{d}^3 r : \big(\varphi[c^{[1]}\tau^{[1]}]^{[0]}([d^{[1]}\nabla^{[1]}]^{[0]}\varphi) \\ &\quad - ([d^{[1]}\nabla^{[1]}]^{[0]}\varphi)[c^{[1]}\tau^{[1]}]^{[0]}\varphi\big)\mathrm{e}^{i\boldsymbol{q}\cdot\boldsymbol{r}} : , \end{aligned}$$

and for the time-like part

$$\begin{aligned} G^4_{\gamma\pi} &= \int \mathrm{d}^3 r : J_4 A_4 : \\ &= \tfrac{1}{2} ie\hat{1} \int \mathrm{d}^3 r : \big(\varphi[c^{[1]}\tau^{[1]}]^{[0]}(\partial_4\varphi) - (\partial_4\varphi)[c^{[1]}\tau^{[1]}]^{[0]}\varphi\big)\mathrm{e}^{i\boldsymbol{q}\cdot\boldsymbol{r}} : . \end{aligned}$$

In fact, the photon-pion interaction is analogous to the rho-pion interaction described in sect. 6.7 since charge conservation imposes an isospin transfer $T = 1$. The only difference is the replacement of the rho multipoles by the plane wave expansion of the photon. Hence the relations between the three processes of fig. 9.2 are the same as those of the rho-pion matrix elements.

The evaluation of their isospin part is given by the diagram of fig. 9.3, while the angular momentum recouplings are shown in fig. 9.4a and 9.4b for the two terms of $\boldsymbol{G}_{\gamma\pi}$ and in fig. 9.5 for $G^4_{\gamma\pi}$. The pion quantum numbers are $\pi(\nu l t)$ and $\pi'(\nu' l' t)$ while for the photon the multipoles are denoted L; the angular momentum arising from the expansion of the plane wave is denoted Λ. The photon thus is denoted by $\gamma(q\Lambda L)$. In the matrix elements we shall use the symbol γ to denote all the photon quantum numbers. We obtain the invariant matrix elements for the space-like part

$$\begin{aligned}
&\left[\pi|G_{\gamma\pi}|\gamma(q\Lambda L)\pi'\right] \\
&\quad= e\pi\sqrt{6}\, i^{\Lambda}\left(\sum_{\lambda}\frac{1}{\hat{\lambda}}\alpha_{\lambda l'}[l|\lambda|\Lambda]\begin{bmatrix} l' & 1 & \lambda \\ 0 & \Lambda & \Lambda \\ l' & L & l\end{bmatrix}\mathcal{F}_{\lambda}^{\nu\nu' ll'\Lambda}(q)\right. \\
&\qquad\left. -\sum_{\lambda'}\frac{1}{\hat{\lambda}'}\alpha_{\lambda' l}[\lambda'|l'|\Lambda]\begin{bmatrix} l & 1 & \lambda' \\ l' & \Lambda & \lambda' \\ L & L & 0\end{bmatrix}\mathcal{F}_{\lambda}^{\nu'\nu l' l\Lambda}(q)\right),
\end{aligned}$$

with the definition

$$\mathcal{F}_{\lambda}^{\nu_a\nu_b l_a l_b \Lambda}(q) = \int r^2\,\mathrm{d}r\, j_{\Lambda}(qr)\, g_{\nu_a l_a}(r)\, k_{\nu_b l_b \lambda}(r).$$

The time-like part is, performing as usual the time derivation in the Heisenberg picture before putting the time to zero, with fig. 9.5,

$$\left[\pi|G^4_{\gamma\pi}|\gamma(q,L)\pi'\right]^P = e\pi\sqrt{6}\, i^L[l|l'|L]\int r^2\,\mathrm{d}r\, j_L(qr)\,\Gamma^P_{\nu l\nu' l'}(r),$$

where the functions $\Gamma^P(r)$ have been given in sect. 6.7. The pion vertex operator for the form factor is thus given by the discretized operator expansion, writing $c^{[01]}$ for $c^{[1]}$ as in (4.28) and $\theta_\nu^{[l1]}$ as defined in (6.5),

$$\boldsymbol{G}_{\gamma\pi}(\boldsymbol{q}) = \sum\left[\pi|G_{\gamma\pi}|\gamma(q\Lambda L)\pi'\right]:\left[\left[\theta_\nu^{[l1]}\theta_{\nu'}^{[l'1]}c^{[01]}\right]^{[L0]}\left[d^{[1]}\hat{q}^{[\Lambda]}\right]^{[L]}\right]^{[0]}:,$$

$$G^4_{\gamma\pi}(\boldsymbol{q}) = \sum\left[\pi|G^4_{\gamma\pi}|\gamma(qL)\pi'\right]^P:\left[\theta_\nu^{[l1]}\theta_{\nu'}^{[l'1]}c^{[01]}\right]^{[L0]}\hat{q}^{[L]}]^{[0]}:.$$

The polarization and angular correlation information is contained in the factors $d^{[1]}$ and $\hat{q}^{[\Lambda]}$. From now on for compactness we will omit the explicit indices q, L, Λ as they are implied by the symbol γ in the notation for the invariant matrix element.

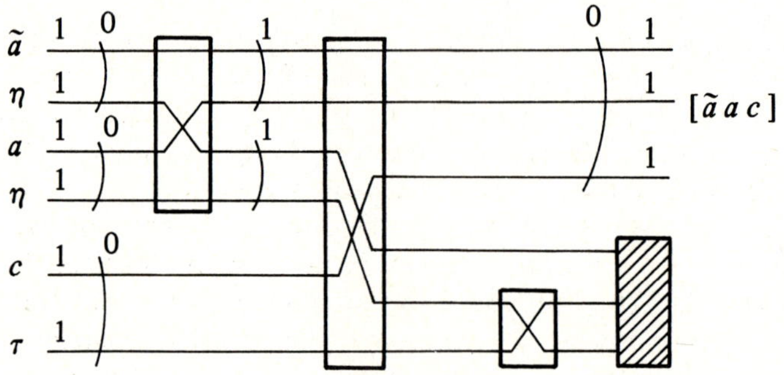

Fig. 9.3.

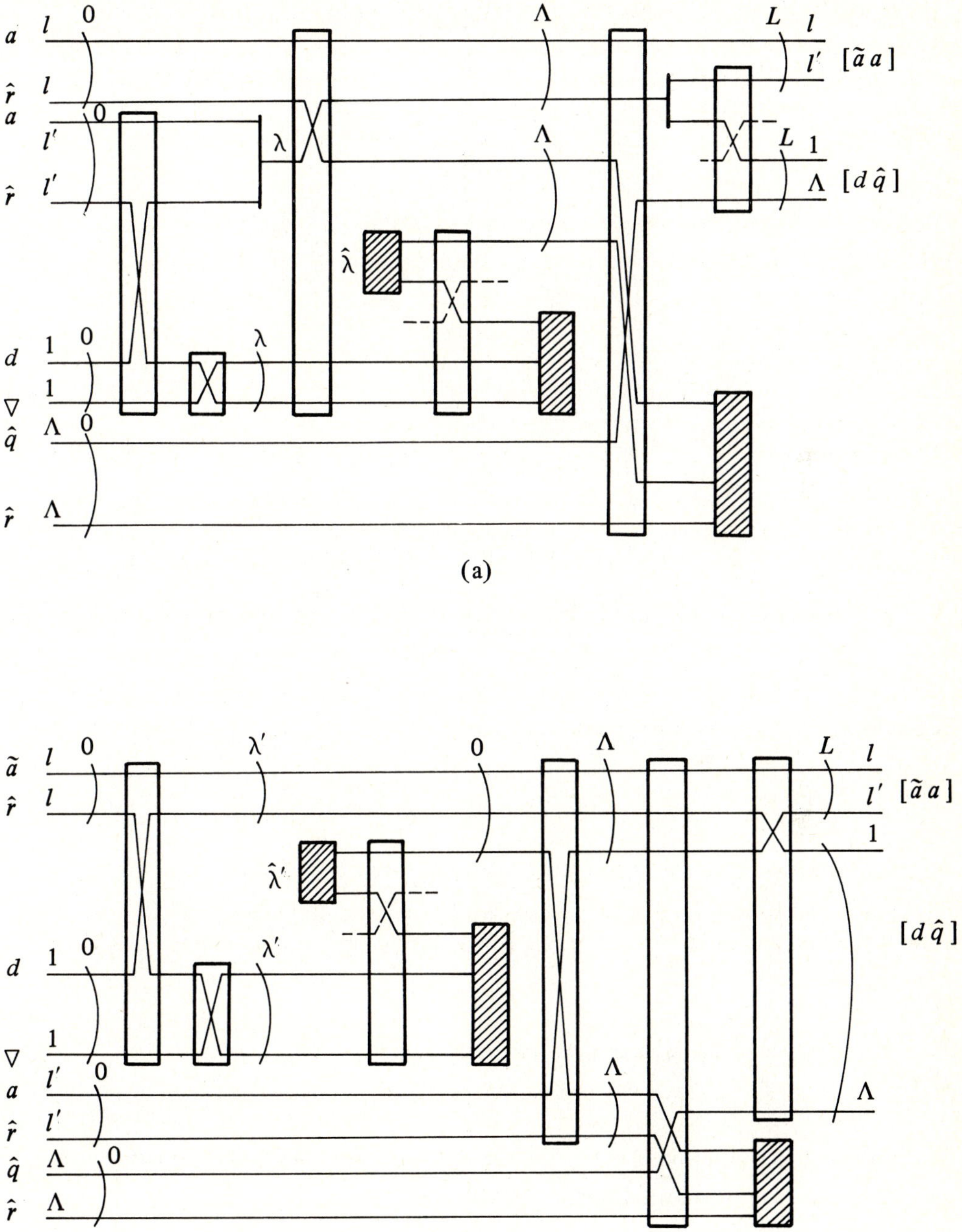

(a)

(b)

Fig. 9.4.

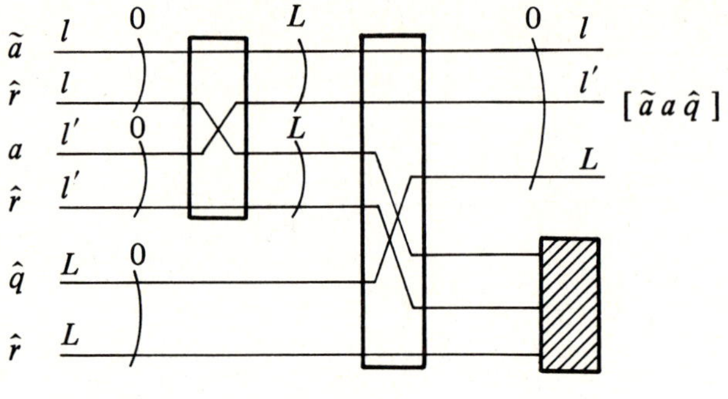

Fig. 9.5.

9.2.4. *The photon-nucleon current interaction*

The interaction of the photon with a spin-$\frac{1}{2}$ field comprises a current term treated below and an anomalous moment term calculated in the next section. The isospin charge projection operator for the nucleon is written as

$$\tfrac{1}{2}(1+\tau_z)=\tfrac{1}{2}\sum_{y=0,1}\hat{y}\left[c^{[y]}\tau^{[y]}\right]^{[0]},$$

where $c_m^{[1]}$ is defined in (9.6), while $c^{[0]}=1$ and $\tau^{[0]}=1$. We write with the spinor matrices γ_i, $i=x, y, z$, the invariant

$$\boldsymbol{d}\cdot\boldsymbol{\gamma}=\hat{1}\left[d^{[1]}\gamma^{[1]}\right]^{[0]}.$$

Then the operator of the form factor, with the current (2.32), is

$$\boldsymbol{G}_{\gamma\mathrm{N}}=\int \mathrm{d}^3r:\boldsymbol{J}\cdot\boldsymbol{A}:=\tfrac{1}{2}ie\hat{1}\sum_{y}\hat{y}\int \mathrm{d}^3r:\bar{\psi}\left[d^{[1]}\gamma^{[1]}\right]^{[0]}\left[c^{[y]}\tau^{[y]}\right]^{[0]}\psi\mathrm{e}^{i\boldsymbol{q}\cdot\boldsymbol{r}}:,$$

$$G_{\gamma\mathrm{N}}^4=\int \mathrm{d}^3r:J_4A_4:=\tfrac{1}{2}ie\sum_{y}\hat{y}\int \mathrm{d}^3r:\bar{\psi}\gamma_4\left[c^{[y]}\tau^{[y]}\right]^{[0]}\psi\mathrm{e}^{i\boldsymbol{q}\cdot\boldsymbol{r}}:.$$

Introducing the notation $\psi=\begin{pmatrix}L\\-S\end{pmatrix}$ for the large and small components of (5.8), with the fermion quantum numbers $\mathrm{N}(\nu ljt)$:

$$L=\sqrt{\tfrac{1}{2}}\,\hat{j}\left[b^{[j]}Y_l^{[j]}\right]^{[0]}u_{\nu l}(r),$$

$$S=\sqrt{\tfrac{1}{2}}\,\hat{j}\left[b^{[j]}Y_\lambda^{[j]}\right]^{[0]}v_{\nu l\lambda}(r),$$

the above expressions become

$$\boldsymbol{J}\cdot\boldsymbol{A} = -\tfrac{1}{2}e\hat{1}\sum_y \hat{y}\big(L^+[d^{[1]}\sigma^{[1]}]^{[0]}[c^{[y]}\tau^{[y]}]^{[0]}S$$

$$+S^+[d^{[1]}\sigma^{[1]}]^{[0]}[c^{[y]}\tau^{[y]}]^{[0]}L\big)\mathrm{e}^{i\boldsymbol{q}\cdot\boldsymbol{r}},$$

$$J_4A_4 = \tfrac{1}{2}ie\sum_y \hat{y}\big(L^+[c^{[y]}\tau^{[y]}]^{[0]}L + S^+[c^{[y]}\tau^{[y]}]^{[0]}S\big)\mathrm{e}^{i\boldsymbol{q}\cdot\boldsymbol{r}}.$$

The calculation of the isospin is given by fig. 9.3. The angular momentum recouplings are given by fig. 9.6 for the space-like part and by fig. 9.7 for the time-like part. Herewith we have, with $t=\frac{1}{2}$ and $s=\frac{1}{2}$

$$\big[\mathrm{N}|G_{\gamma\mathrm{N}}(y)|\gamma\mathrm{N}'\big] = ie\pi\hat{s}(-)^y[t|\tau^{[y]}|t]i^{\Lambda}\frac{\hat{L}}{\hat{\Lambda}}$$

$$\times\left(\begin{bmatrix} s & l & j \\ s & \lambda' & j' \\ 1 & \Lambda & L\end{bmatrix}[l|\lambda'|\Lambda]\int r^2\mathrm{d}r\, j_\Lambda(qr)u_{\nu l}(r)v_{\nu' l'\lambda'}(r)\right.$$

$$\left.+\begin{bmatrix} s & \lambda & j \\ s & l' & j' \\ 1 & \Lambda & L\end{bmatrix}[\lambda|l'|\Lambda]\int r^2\mathrm{d}r\, j_\Lambda(qr)v_{\nu l\lambda}(r)u_{\nu' l'}(r)\right),$$

$$\big[\mathrm{N}|G^4_{\gamma\mathrm{N}}(y)|\gamma\mathrm{N}'\big] = ie\pi\hat{s}(-)^y[t|\tau^{[y]}|t]i^{L}$$

$$\times\left(\begin{bmatrix} s & l & j \\ s & l' & j' \\ 0 & L & L\end{bmatrix}[l|l'|L]\int r^2\mathrm{d}r\, j_L(qr)u_{\nu l}(r)u_{\nu' l'}(r)\right.$$

$$\left.+\begin{bmatrix} s & \lambda & j \\ s & \lambda' & j' \\ 0 & L & L\end{bmatrix}[\lambda|\lambda'|L]\int r^2\mathrm{d}r\, j_L(qr)v_{\nu l\lambda}(r)v_{\nu' l'\lambda'}(r)\right).$$

Here

$$[\tfrac{1}{2}|\tau^{[0]}|\tfrac{1}{2}] = \hat{t} = \sqrt{2}\,,$$

$$[\tfrac{1}{2}|\tau^{[1]}|\tfrac{1}{2}] = i\sqrt{6}\,.$$

The photon-nucleon vertex operator for the form factor is thus, writing $c^{[0y]}=c^{[y]}$

$$G_{\gamma\mathrm{N}}(\boldsymbol{q}) = \sum\big[\mathrm{N}|G_{\gamma\mathrm{N}}(y)|\gamma\mathrm{N}'\big]:\big[[\tilde{b}^{[jt]}_{\nu l}b^{[j't]}_{\nu' l'}c^{[0y]}]^{[L0]}[d^{[1]}\hat{q}^{[\Lambda]}]^{[L]}\big]^{[0]}:,$$

$$G^4_{\gamma\mathrm{N}}(\boldsymbol{q}) = \sum\big[\mathrm{N}|G^4_{\gamma\mathrm{N}}(y)|\gamma\mathrm{N}'\big]:\big[[\tilde{b}^{[jt]}_{\nu l}b^{[j't]}_{\nu' l'}c^{[0y]}]^{[L0]}\hat{q}^{[L]}\big]^{[0]}:.$$

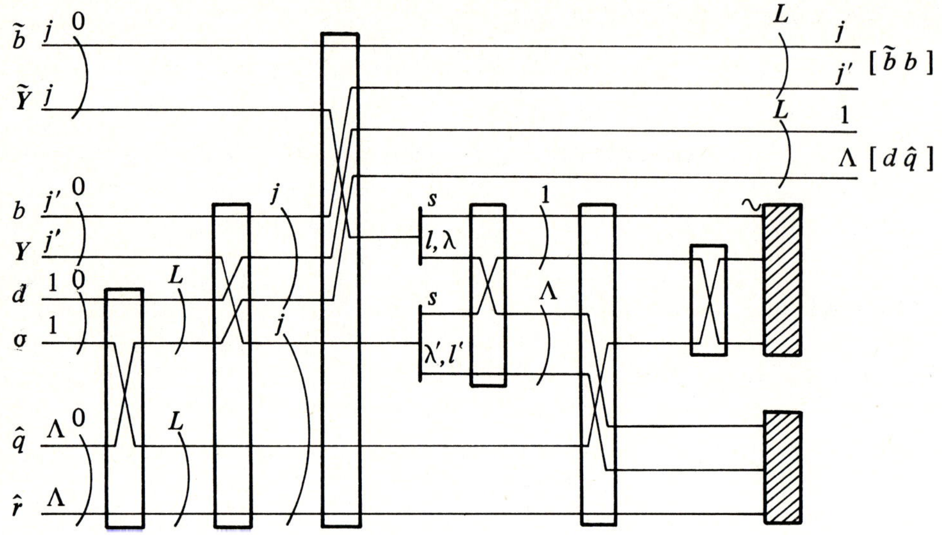

Fig. 9.6.

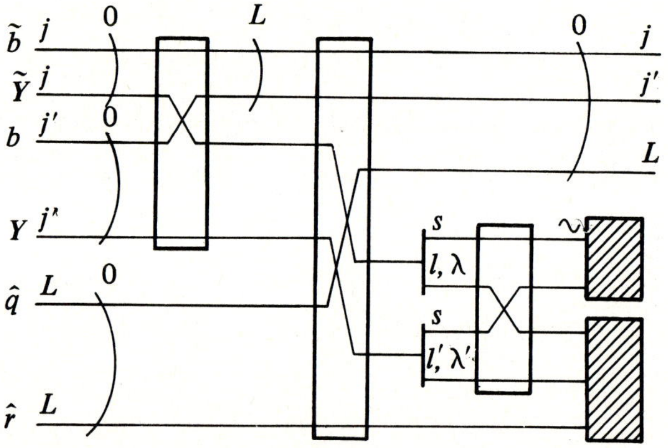

Fig. 9.7.

9.2.5. The photon-nucleon anomalous moment interaction

The partially dressed nature of the basic nucleon field leads to an electromagnetic interaction with the anomalous magnetic moments of the neutron and of the proton, which according to sect. 2.5.4, writes

$$G_{\gamma_a N} = -g_i \frac{1}{2M} \int d^3r \colon \bar{\psi} S_{\mu\nu} C_i \psi F_{\mu\nu} \colon .$$

Here g_p and g_n are the proton and neutron residual anomalous moments, i.e. that part of the moment which is not produced by the configuration space of the model.

The isospin operator is here a projector

$$C_i = \tfrac{1}{2}(1 \pm \tau_z)$$

for the proton and neutron states respectively ($i = \mathrm{p}$ or $i = \mathrm{n}$). Furthermore from sect. 2.3.2

$$S_{\mu\nu} = -\tfrac{1}{4}i[\gamma_\mu, \gamma_\nu]_-,$$

or, in detail,

$$S_{jk} = \tfrac{1}{2}\varepsilon_{ijk}\begin{bmatrix} \sigma_i & 0 \\ 0 & \sigma_i \end{bmatrix},$$

$$S_{k4} = -S_{4k} = \frac{1}{2}\begin{bmatrix} 0 & \sigma_k \\ \sigma_k & 0 \end{bmatrix}.$$

In invariant form the isospin operator is written as

$$\tfrac{1}{2}\sum_{y=0,1} g_y \hat{y}[c^{[y]}\tau^{[y]}]^{[0]},$$

where the $c^{[y]}$ and $\tau^{[y]}$ have been defined in the previous section and where

$$g_0 = g_\mathrm{p} + g_\mathrm{n},$$

$$g_1 = g_\mathrm{p} - g_\mathrm{n}.$$

Thus the isospin part contributes to the vertex the factor

$$\tfrac{1}{2}g_y(-)^y[t|\tau^{[y]}|t]:[\tilde{b}^{[t]}b^{[t]}c^{[y]}]^{[0]}:,$$

which can be checked along the lines of fig. 9.3.

The orbital part yields, writing $\psi = \begin{pmatrix} L \\ -S \end{pmatrix}$, a term of the form

$$-\int \mathrm{d}^3r:\psi^+\gamma_4\tfrac{1}{2}\left(\begin{pmatrix} \boldsymbol{\sigma} & 0 \\ 0 & \boldsymbol{\sigma} \end{pmatrix}\cdot \mathrm{rot}\,\boldsymbol{A} + \begin{pmatrix} 0 & \boldsymbol{\sigma} \\ \boldsymbol{\sigma} & 0 \end{pmatrix}\cdot(\nabla A_4 - \partial_4 \boldsymbol{A})\right)\psi:$$

$$= -\tfrac{1}{2}\int \mathrm{d}^3r:(L^+\boldsymbol{\sigma}\cdot \mathrm{rot}\,\boldsymbol{A}L - S^+\boldsymbol{\sigma}\cdot\mathrm{rot}\,\boldsymbol{A}S - L^+\boldsymbol{\sigma}\cdot\nabla A_4 S + S^+\boldsymbol{\sigma}\cdot\nabla A_4 L):.$$

In the last form the term $\partial_4 \boldsymbol{A}$ has been dropped in view of the fact that the energy transfer vanishes for elastic scattering, which is the case considered here.

The expression for rot $\boldsymbol{A}$ and ∇A_4 are obtained from sect. 4.5.2. Omitting this time the geometry graph the final matrix elements are, for the space-like part,

$$\left[\mathrm{N}|G_{\gamma_a\mathrm{N}}(y)|\gamma\mathrm{N}'\right] = i\pi(-)^y g_y \frac{1}{2M}\left[t|\tau^{[y]}|t\right] i^{\Lambda} q$$

$$\times \sum \frac{\hat{L}}{\hat{R}} \alpha_{R\Lambda} \begin{bmatrix} 1 & 1 & 1 \\ 0 & \Lambda & \Lambda \\ 1 & R & L \end{bmatrix}$$

$$\times \left(Q^R_{l'l} \begin{bmatrix} s & l & j \\ s & l' & j' \\ 1 & R & L \end{bmatrix} \int r^2 \mathrm{d}r\, u_{\nu l}(r) u_{\nu' l'}(r) j_R(qr) \right.$$

$$\left. - Q^R_{\lambda'\lambda} \begin{bmatrix} s & \lambda & j \\ s & \lambda' & j' \\ 1 & R & L \end{bmatrix} \int r^2 \mathrm{d}r\, v_{\nu l\lambda}(r) v_{\nu' l'\lambda'}(r) j_R(qr) \right),$$

and for the time-like part

$$\left[\mathrm{N}|G^4_{\gamma_a\mathrm{N}}(y)|\gamma\mathrm{N}'\right] = i\pi\sqrt{\tfrac{1}{2}}\,(-)^y g_y \frac{1}{2M}\left[t|\tau^{[y]}|t\right] i^{L} q$$

$$\times \sum \alpha_{RL} \left(- Q^R_{l\lambda'} \begin{bmatrix} s & l & j \\ s & \lambda' & j' \\ 1 & R & L \end{bmatrix} \int r^2 \mathrm{d}r\, u_{\nu l}(r) v_{\nu' l'\lambda'}(r) j_R(qr) \right.$$

$$\left. + Q^R_{\lambda l'} \begin{bmatrix} s & \lambda & j \\ s & l' & j' \\ 1 & R & L \end{bmatrix} \int r^2 \mathrm{d}r\, v_{\nu l\lambda}(r) u_{\nu' l'}(r) j_R(qr) \right).$$

We recall the definition of sect. 4.3.1 for the factors Q^R_{kl}.

Finally the photon-nucleon anomalous moment vertex operators for the elastic form factor are given by the discretized expansions

$$\boldsymbol{G}_{\gamma_a\mathrm{N}}(\boldsymbol{q}) = \sum \left[\mathrm{N}|G_{\gamma_a\mathrm{N}}(y)|\gamma\mathrm{N}'\right] : \left[\left[\tilde{b}^{[jt]}_{\nu l} b^{[j't]}_{\nu' l'} c^{[0y]}\right]^{[L0]} \left[d^{[1]} \hat{q}^{[\Lambda]}\right]^{[L]}\right]^{[0]} :,$$

$$G^4_{\gamma_a\mathrm{N}}(\boldsymbol{q}) = \sum \left[\mathrm{N}|G^4_{\gamma_a\mathrm{N}}(y)|\gamma\mathrm{N}'\right] : \left[\left[\tilde{b}^{[jt]}_{\nu l} b^{[j't]}_{\nu' l'} c^{[0y]}\right]^{[L0]} \hat{q}^{[L]}\right]^{[0]} : .$$

9.2.6. *The photon-vector meson interaction*

We consider only the case of the vector current interaction. Being neutral the ω-meson has no current. For the ρ-meson we here consider the scattering process of fig. 9.8. Vector meson pair production and annihilation can be easily derived following the same steps. The expressions for these processes are simply related by

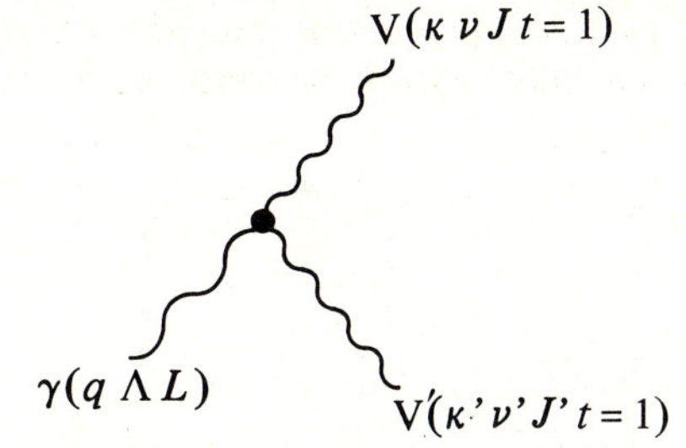

Fig. 9.8.

hermitian conjugation applied to one of the vector meson fields. Separating the space-like components $\boldsymbol{\phi}$, $\boldsymbol{\pi}$ and the time-like components ϕ_4, π_4 of the fields, the expression of the space-like part of the current (2.33) ($i = 1, 2, 3$) contains the terms

$$-\bar{\phi}\gamma_i\gamma_4\bar{\pi} = -\phi_4\bar{\boldsymbol{\pi}}, \qquad \pi\gamma_4\gamma_i\phi = \bar{\boldsymbol{\pi}}\phi_4,$$

while the time-like part ($\mu = 4$) contains the terms

$$-\bar{\phi}\bar{\pi} = -\boldsymbol{\phi}\cdot\bar{\boldsymbol{\pi}}, \qquad \pi\phi = \boldsymbol{\pi}\cdot\boldsymbol{\phi}.$$

The operator of the form factor thus is, introducing the charge projection operator τ_z as before

$$\boldsymbol{G}_{\gamma\mathrm{V}} = \int \mathrm{d}^3r : \boldsymbol{J}\cdot\boldsymbol{A} : = -ie\tfrac{1}{2}\int \mathrm{d}^3r : \left(\phi_4\hat{1}[c^{[1]}\tau^{[1]}]^{[0]}\boldsymbol{d}\cdot\boldsymbol{\pi} - \boldsymbol{\pi}\cdot\boldsymbol{d}\hat{1}[c^{[1]}\tau^{[1]}]^{[0]}\phi_4\right)e^{i\boldsymbol{q}\cdot\boldsymbol{r}} :,$$

$$G^4_{\gamma\mathrm{V}} = \int \mathrm{d}^3r : J_4A_4 : = -ie\tfrac{1}{2}\int \mathrm{d}^3r : \left(\boldsymbol{\phi}\hat{1}[c^{[1]}\tau^{[1]}]^{[0]}\boldsymbol{\pi} - \boldsymbol{\pi}\hat{1}[c^{[1]}\tau^{[1]}]^{[0]}\boldsymbol{\phi}\right)e^{i\boldsymbol{q}\cdot\boldsymbol{r}} :.$$

The evaluation of the matrix elements is given for the isospin in fig. 9.3 and for the angular recoupling in fig. 9.9 for the space-like part and fig. 9.10 for the time-like part respectively. Setting $\lambda = J + n$ and $V(\kappa, \nu, J, t)$ for the quantum numbers of the vector meson, we get for the space-like part

$$\left[\mathrm{V}|G_{\gamma\mathrm{V}}|\gamma\mathrm{V}'\right] = ie2\pi\sqrt{6}\,i^{\Lambda}\frac{\hat{L}^3}{\hat{\Lambda}}\sum_{nn'}\left((-)^{L+J+J'}[J|\Lambda|\lambda']\begin{bmatrix} J & 0 & J \\ \lambda' & 1 & J' \\ \Lambda & 1 & L \end{bmatrix}\int r^2\mathrm{d}r\right.$$

$$\times j_\Lambda(qr)\mathcal{X}_{\nu J}(r)C_{\kappa'}(n', J')\mathcal{Y}_{\kappa'\nu'J'n'}(r)\delta_{\kappa\mathfrak{L}} + [J'|\Lambda|\lambda]\begin{bmatrix} J' & 0 & J' \\ \lambda & 1 & J \\ \Lambda & 1 & L \end{bmatrix}\int r^2\mathrm{d}r$$

$$\left.\times j_\Lambda(qr)\mathcal{X}_{\nu'J'}(r)C_\kappa(n, J)\mathcal{Y}_{\kappa\nu Jn}(r)\delta_{\kappa'\mathfrak{L}}\right),$$

where in the second term the reordering of the operators brings in a phase. For the

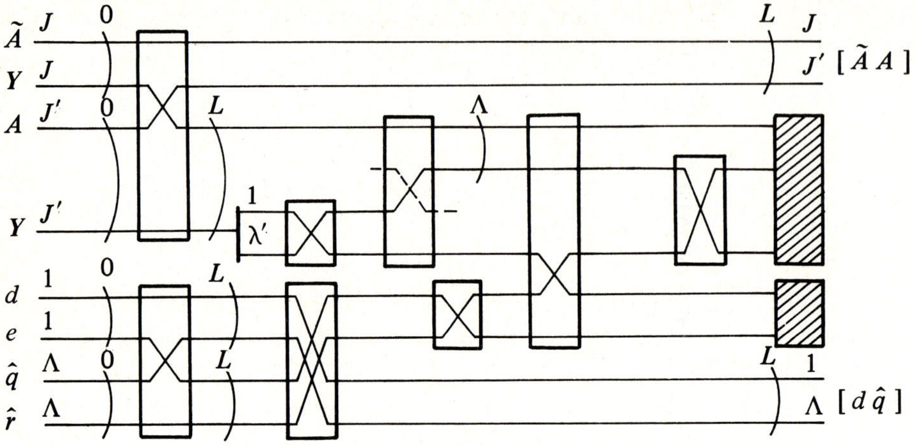

Fig. 9.9.

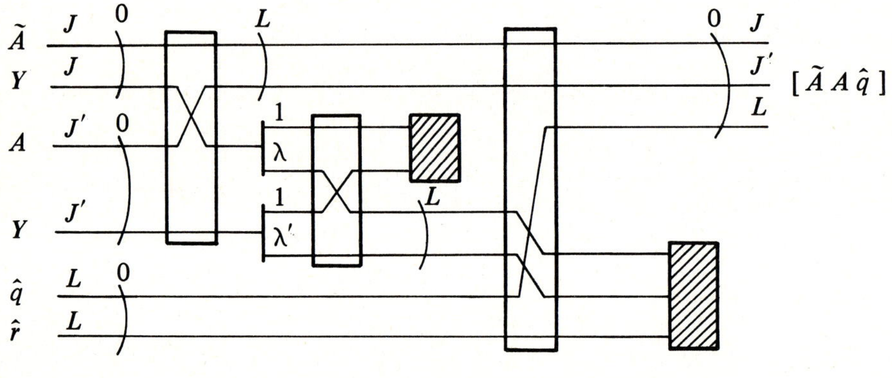

Fig. 9.10.

time-like part we have

$$\left[\mathrm{V}|G^4_{\gamma\mathrm{V}}|\gamma\mathrm{V}'\right] = ie2\pi\sqrt{6}\, i^L \hat{1} \sum_{nn'} [\lambda|\lambda'|L] \begin{bmatrix} 1 & \lambda & J \\ 1 & \lambda' & J' \\ 0 & L & L \end{bmatrix} C_\kappa(n, J) C_{\kappa'}(n', J')$$

$$\times \int r^2 \mathrm{d}r\, j_L(qr) \left(\mathcal{W}_{\kappa\nu Jn}(r) \mathcal{Y}_{\kappa'\nu'J'n'}(r) + \mathcal{Y}_{\kappa\nu Jn}(r) \mathcal{W}_{\kappa'\nu'J'n'}(r) \right).$$

We have written the vector meson fields keeping the notation of eqs. (5.18), (5.28), and (5.29). The vertex operator is then

$$G_{\gamma V}(q)=\sum[V|G_{\gamma V}|\gamma V']:\left[\left[\tilde{A}^{[J1]}_{\kappa\nu}A^{[J'1]}_{\kappa'\nu'}c^{[01]}\right]^{[L0]}\left[d^{[1]}\hat{q}^{[\Lambda]}\right]^{[L]}\right]^{[0]}:,$$

$$G^4_{\gamma V}(q)=\sum[V|G^4_{\gamma V}|\gamma V']:\left[\left[\tilde{A}^{[J1]}_{\kappa\nu}A^{[J'1]}_{\kappa'\nu'}c^{[01]}\right]^{[L0]}\hat{q}^{[L]}\right]^{[0]}:.$$

9.2.7. The photon-nucleon-delta current interaction

The current interaction of the delta photon-absorption process $\Delta \to N$ shown in fig. 9.11 is given by the replacement $\gamma_5\partial_\mu\varphi \to A_\mu$ in the $\pi N\Delta$ interaction, eq. (2.69), with the proper mass factor to make the coupling constant dimensionless,

$$G_{\gamma N\Delta} = -g_{\gamma N\Delta}\sqrt{2M_\Delta}\int d^3r:\bar{\psi}D^+_\mu C^+_p \Phi A_\mu:.$$

The inverse process $N \to \Delta$ is given by time reversal. In this expression the non-square four-vector matrices D_μ have been introduced in eq. (2.70), and the isospin operator C_p is a suitable projection operator. In invariant form it is written as

$$C_p = \sum_{y=1,2} a_y \hat{y}\left[c^{[y]}C^{[y]}\right]^{[0]},$$

where the triangularity between y, $t=\frac{1}{2}$ and $T=\frac{3}{2}$ requires the values $y=1,2$ and where the amplitudes are

$$c^{[1]}_m = -i\delta_{m0},$$

$$c^{[2]}_m = -\delta_{m0}.$$

Together with the value of $[t|\tilde{C}^{[1]}|T]=2i$ given in sect. 6.10, we see that the choice

$$a_1 = \sqrt{\tfrac{1}{8}},$$

$$a_2\left[t|\tilde{C}^{[2]}|T\right] = \sqrt{\tfrac{1}{2}},$$

would project on the proton and Δ^+ states. The isospin contributes to the vertex the factor

$$(-)^y a_y\left[t|\tilde{C}^{[y]}|T\right]:\left[\tilde{b}^{[t]}B^{[T]}c^{[y]}\right]^{[0]}:.$$

The remaining part of the current interaction has a structure similar to that of the $\pi N\Delta$ interaction of sect. 6.10. The integrand is of the form, using the notation of that section,

$$\begin{pmatrix} L \\ -S \end{pmatrix}^{+} \gamma_4 D_\mu^+ \begin{pmatrix} a \\ 0 \\ b \\ 0 \end{pmatrix} A_\mu = \begin{pmatrix} L \\ S \end{pmatrix}^{+} \begin{pmatrix} 0 & G_\mu^+ & -F_\mu^+ & 0 \\ F_\mu^+ & 0 & 0 & -G_\mu^+ \end{pmatrix} \begin{pmatrix} a \\ 0 \\ b \\ 0 \end{pmatrix} A_\mu$$

$$= (-L^+ \boldsymbol{F}^+ b + S^+ \boldsymbol{F}^+ a) \cdot \boldsymbol{A} .$$

As in sect. 6.10 the time-like component D_4^+ does not contribute owing to the transverse gauge adopted for the spin-$\frac{3}{2}$ field Φ. With a geometry similar to that of fig. 6.23 the complete matrix element, including isospin, is $(s = \frac{1}{2}, S = \frac{3}{2})$

$$[\mathrm{N}|G_{\gamma \mathrm{N}\Delta}(y)|\gamma\Delta] = -ig_{\gamma\mathrm{N}\Delta} 4\pi (-)^y a_y [t|\tilde{C}^{[y]}|T] \frac{\hat{j}_\mathrm{N}}{\hat{S}} i^{\Lambda} (-)^{L+j_\Delta+j_\mathrm{N}}$$

$$\times \sum \left(Q^{l_\mathrm{N}}_{\Lambda\lambda_\Delta} \begin{bmatrix} 1 & \Lambda & L \\ S & \lambda_\Delta & j_\Delta \\ s & l_\mathrm{N} & j_\mathrm{N} \end{bmatrix} \beta_{l_\Delta \lambda_\Delta j_\Delta} \int r^2 \mathrm{d}r\, u_{\nu_\mathrm{N} l_\mathrm{N}}(r)\, v_{\nu_\Delta l_\Delta \lambda_\Delta}(r)\, j_\Lambda(qr) \right.$$

$$\left. - Q^{\lambda_\mathrm{N}}_{\Lambda l_\Delta} \begin{bmatrix} 1 & \Lambda & L \\ S & l_\Delta & j_\Delta \\ s & \lambda_\mathrm{N} & j_\mathrm{N} \end{bmatrix} \int r^2 \mathrm{d}r\, v_{\nu_\mathrm{N} l_\mathrm{N} \lambda_\mathrm{N}}(r)\, u_{\nu_\Delta l_\Delta}(r)\, j_\Lambda(qr) \right),$$

and the vertex operator is

$$\boldsymbol{G}_{\gamma\mathrm{N}\Delta}(\boldsymbol{q}) = \sum [\mathrm{N}|G_{\gamma\mathrm{N}\Delta}(y)|\gamma\Delta] : \left[\left[\tilde{b}^{[j_\mathrm{N} t]}_{\nu_\mathrm{N} l_\mathrm{N}} B^{[j_\Delta T]}_{\nu_\Delta l_\Delta} c^{[0y]} \right]^{[L0]} [d^{[1]} \hat{q}^{[\Lambda]}]^{[L]} \right]^{[0]} : ,$$

$$G^4_{\gamma\mathrm{N}\Delta}(\boldsymbol{q}) = 0 .$$

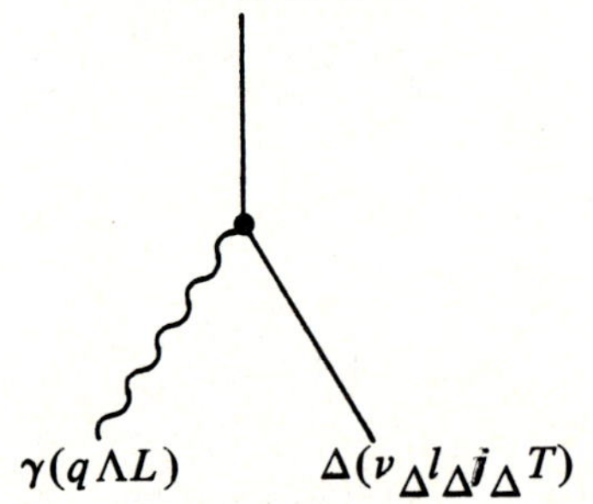

Fig. 9.11.

9.2.8. The photon-nucleon-delta anomalous moment interaction

The anomalous $\gamma N\Delta$ interaction contributing to the vertex of fig. 9.11 is of the form

$$G_{\gamma_a N\Delta} = -g_{\gamma_a N\Delta}\frac{1}{\sqrt{2M_\Delta}}\int d^3r : \bar{\psi} S^+_{\mu\nu} C_3 \Phi F_{\mu\nu} : ,$$

with

$$S^+_{\mu\nu} = \gamma_\mu D^+_\nu - \gamma_\nu D^+_\mu ,$$

where γ_μ are spin-$\frac{1}{2}$ matrices, and D_μ the non-square matrices of sect. 2.6.2. Furthermore,

$$F_{\mu\nu} = \partial_\mu A_\nu - \partial_\nu A_\mu .$$

Here C_3 is the charge conserving third component of the isospin-changing operator $\boldsymbol{C}$

$$C_3 = \hat{1}[c^{[1]} C^{[1]}]^{[0]}$$

with $c^{[1]}_m$ given by eq. (9.6).

Thus the interaction describes both $\Delta^0 \to \mathrm{n}$ and $\Delta^+ \to \mathrm{p}$ processes with the same coupling constant. If different coupling constants are to be used a form analogous to that of the previous section, i.e.

$$C_3 = \sum_{y=1,2} b_y \hat{y} [c^{[y]} C^{[y]}]^{[0]}$$

has to be employed adjusting the constants b_y, as in sect. 9.2.5.

The integrand is now developed as follows:

$$\bar{\psi}\,(\gamma_\mu D^+_\nu - \gamma_\nu D^+_\mu)\,\Phi(\partial_\mu A_\nu - \partial_\nu A_\mu)$$

$$= \begin{pmatrix} L \\ -S \end{pmatrix}^+ \gamma_4 \left(\gamma_\mu \begin{pmatrix} 0 & G^+_\nu & -F^+_\nu & 0 \\ F^+_\nu & 0 & 0 & -G^+_\nu \end{pmatrix} \right.$$

$$\left. - \gamma_\nu \begin{pmatrix} 0 & G^+_\mu & -F^+_\mu & 0 \\ F^+_\mu & 0 & 0 & -G^+_\mu \end{pmatrix} \right)$$

$$\times \begin{pmatrix} a \\ 0 \\ b \\ 0 \end{pmatrix} (\partial_\mu A_\nu - \partial_\nu A_\mu) = M_1 + M_2 .$$

One has

$$M_1 = -i(L^+\boldsymbol{\sigma}\times \boldsymbol{F}^+ a + S^+\boldsymbol{\sigma}\times\boldsymbol{F}^+ b)\cdot\nabla\times\boldsymbol{A},$$

$$M_2 = 2(L^+\boldsymbol{F}^+ b + S^+\boldsymbol{F}^+ a)\cdot(\nabla A_4 - \partial_4 \boldsymbol{A}).$$

Here the terms $D_4^+\Phi$ do not contribute owing to the transverse gauge of Φ and have been omitted. Furthermore $\partial_4\boldsymbol{A}$ vanishes in the case of elastic scattering.

For the calculation of the matrix element M_1 we note, from sects. 4.5.1 and 4.5.2, the invariant form

$$(\boldsymbol{\sigma}\times\boldsymbol{F})\cdot(\nabla\times\boldsymbol{A}) = -2\sqrt{3}\left[\left[\sigma^{[1]}F^{[1]}\right]^{[1]}\left[d^{[1]}\nabla^{[1]}\right]^{[1]}\right]^{[0]}\mathrm{e}^{i\boldsymbol{q}\cdot\boldsymbol{r}}.$$

We thus obtain, including isospin, the matrix element associated with the term M_1,

$$\left[\mathrm{N}|G_{\gamma_a\mathrm{N}\Delta}(y)|\gamma\Delta\right]_1 = ig_{\gamma_a\mathrm{N}\Delta}24\pi\frac{1}{2M_\Delta}(-)^y b_y\left[t|\tilde{C}^{[y]}|T\right]i^{\Lambda}q$$

$$\times\sum\frac{\hat{\Lambda}}{\hat{L}}\begin{bmatrix}1 & 1 & 1\\ \Lambda & \Lambda & 0\\ L & R & 1\end{bmatrix}\begin{bmatrix}s & S & 1\\ 1 & 1 & 1\\ s & s & 0\end{bmatrix}\alpha_{R\Lambda}$$

$$\times\left(Q^R_{l_\mathrm{N}l_\Delta}\begin{bmatrix}s & l_\mathrm{N} & j_\mathrm{N}\\ S & l_\Delta & j_\Delta\\ 1 & R & L\end{bmatrix}\int r^2\mathrm{d}r\, u_{\nu_\mathrm{N}l_\mathrm{N}}(r)u_{\nu_\Delta l_\Delta}(r)j_R(qr) + Q^R_{\lambda_\mathrm{N}\lambda_\Delta}\begin{bmatrix}s & \lambda_\mathrm{N} & j_\mathrm{N}\\ S & \lambda_\Delta & j_\Delta\\ 1 & R & L\end{bmatrix}\beta_{l_\Delta\lambda_\Delta j_\Delta}\right.$$

$$\left.\times\int r^2\mathrm{d}r\, v_{\nu_\mathrm{N}l_\mathrm{N}\lambda_\mathrm{N}}(r)v_{\nu_\Delta l_\Delta\lambda_\Delta}(r)j_R(qr)\right).$$

The matrix element associated with the term M_2 is

$$\left[\mathrm{N}|G_{\gamma_a\mathrm{N}\Delta}(y)|\gamma\Delta\right]_2 = -ig_{\gamma_a\mathrm{N}\Delta}\frac{8\pi}{\hat{1}}\frac{1}{2M_\Delta}(-)^y b_y\left[t|\tilde{C}^{[y]}|T\right]i^{L}q$$

$$\times\sum\alpha_{RL}\left(Q^R_{l_\mathrm{N}\lambda_\Delta}\begin{bmatrix}s & l_\mathrm{N} & j_\mathrm{N}\\ S & \lambda_\Delta & j_\Delta\\ 1 & R & L\end{bmatrix}\beta_{l_\Delta\lambda_\Delta j_\Delta}\right.$$

$$\times\int r^2\mathrm{d}r\, u_{\nu_\mathrm{N}l_\mathrm{N}}(r)v_{\nu_\Delta l_\Delta\lambda_\Delta}(r)j_R(qr) + Q^R_{\lambda_\mathrm{N}l_\Delta}\begin{bmatrix}s & \lambda_\mathrm{N} & j_\mathrm{N}\\ S & l_\Delta & j_\Delta\\ 1 & R & L\end{bmatrix}$$

$$\left.\times\int r^2\mathrm{d}r\, v_{\nu_\mathrm{N}l_\mathrm{N}\lambda_\mathrm{N}}(r)u_{\nu_\Delta l_\Delta}(r)j_R(qr)\right).$$

The full vertex operator is thus

$$\boldsymbol{G}_{\gamma_a \mathrm{N}\Delta}(\boldsymbol{q}) = \sum \Big(\big[\mathrm{N}|G_{\gamma_a \mathrm{N}\Delta}(y)|\gamma\Delta\big]_1 : \Big[\big[\tilde{b}^{[j_\mathrm{N} t]}_{\nu_\mathrm{N} l_\mathrm{N}} B^{[j_\Delta T]}_{\nu_\Delta l_\Delta} c^{[0y]}\big]^{[L0]} \big[d^{[1]} \hat{q}^{[\Lambda]}\big]^{[L]}\Big]^{[0]} :$$

$$+ \big[\mathrm{N}|G_{\gamma_a \mathrm{N}\Delta}(y)|\gamma\Delta\big]_2 : \Big[\big[\tilde{b}^{[j_\mathrm{N} t]}_{\nu_\mathrm{N} l_\mathrm{N}} B^{[j_\Delta T]}_{\nu_\Delta l_\Delta} c^{[0y]}\big]^{[L0]} \hat{q}^{[L]}\Big]^{[0]} : \Big),$$

$$G^4_{\gamma_a \mathrm{N}\Delta}(\boldsymbol{q}) = 0 .$$

If the same coupling constant is to be used for all processes, then $C_3 = \hat{1}[c^{[1]} C^{[1]}]^{[0]}$ and in the above matrix elements $y = 1$, and $(-)^y b_y [t|\tilde{C}^{[y]}|T] \rightarrow -2i$.

9.2.9. *The photon-delta interaction*

The photon-delta current interaction is of the form

$$G_{\gamma\Delta} = -ie \int \mathrm{d}^3 r : \big(\bar{\Phi} \gamma_\mu C \Theta + \bar{\Theta} \gamma_\mu C \Phi\big) A_\mu : ,$$

where the isospin operator projects on the charged states

$$C = \tfrac{1}{2} + C_3$$

and where C_3 is the usual $T = \frac{3}{2}$ isospin matrix (not to be confused with the isospin changing operators of the previous sections) which writes

$$C_3 = \begin{pmatrix} \frac{3}{2} & 0 & 0 & 0 \\ 0 & \frac{1}{2} & 0 & 0 \\ 0 & 0 & -\frac{1}{2} & 0 \\ 0 & 0 & 0 & -\frac{3}{2} \end{pmatrix} .$$

The structure of this interaction is similar to that of the $\pi\Delta\Delta$ pseudo-vector hamiltonian treated in sect. 6.11.2. It arises from that interaction by the replacement $\gamma_5 \partial_\mu \varphi \rightarrow A_\mu$. The evaluation is carried out term by term along the lines of that section, without any further complications. Since this is straightforward, the results will not be given here owing to the length of the expressions.

9.2.10. *The vector dominance interaction*

We consider the contribution to the electron scattering form factor of the process of the conversion of a virtual photon into a vector-meson ω or ρ_0, fig. 9.12. The

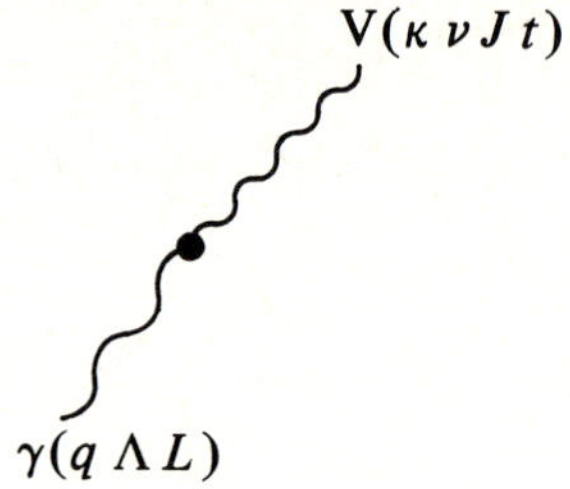

Fig. 9.12.

operator is of the form, according to sect. 2.5.3 and eq. (2.41)

$$\boldsymbol{G}_{\mathrm{VD}} = g_{\mathrm{VD}}(2M^2)\int \mathrm{d}^3r : P_0\boldsymbol{\phi}\cdot\boldsymbol{A} : = g_{\mathrm{VD}}(2M^2)\int \mathrm{d}^3r : P_0\boldsymbol{\phi}\cdot\boldsymbol{d}\mathrm{e}^{i\boldsymbol{q}\cdot\boldsymbol{r}} : ,$$

$$G^4_{\mathrm{VD}} = g_{\mathrm{VD}}(2M^2)\int \mathrm{d}^3r : P_0\phi_4 A_4 : = g_{\mathrm{VD}}(2M^2)\int \mathrm{d}^3r : P_0\phi_4\mathrm{e}^{i\boldsymbol{q}\cdot\boldsymbol{r}} : .$$

P_0 is unity for the ω and is the projection operator on charge 0 for the ρ. This projection operator is constructed with the isospin wave function $\eta^{[1]}$

$$P_0 = \hat{1}\left[c^{[1]}\eta^{[1]}\right]^{[0]}$$

with $c^{[1]}_m$ given by eq. (9.6).

We consider here the creation part of the vector meson field. The expressions for the annihilation part are obtained by hermitian conjugation. Using the notation of sect. 5.2.3 and inserting the plane wave expansion, the graph of fig. 9.13 for the

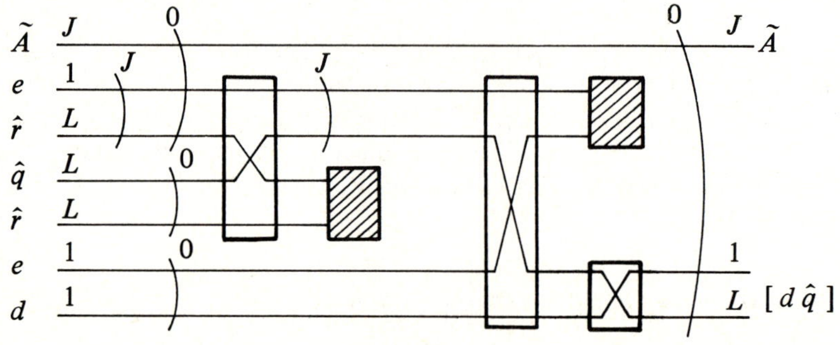

Fig. 9.13.

angular momentum recoupling yields

$$[\mathrm{V}|G_{\mathrm{VD}}|\gamma] = g_{\mathrm{VD}}(2M^2)4\pi\sum_n i^L\hat{J}C_\kappa(n,J)\int r^2\mathrm{d}r\, j_L(qr)\mathcal{W}_{\kappa\nu Jn}(r)\delta_{L,J+n},$$

$$[\mathrm{V}|G^4_{\mathrm{VD}}|\gamma] = g_{\mathrm{VD}}(2M^2)4\pi i^L\hat{J}\int r^2\mathrm{d}r\, j_L(qr)\mathcal{X}_{\nu J}(r)\delta_{LJ}.$$

The integrals over r can be readily carried out with the explicit expressions for the radial functions

$$\int r^2\mathrm{d}r\,\mathcal{W}_{\kappa\nu Jn}(r)\, j_L(qr)\delta_{L,J+n} = \sqrt{\frac{\pi}{2}}\,\frac{N_\kappa}{\sqrt{2E}}f_{\nu J}(q),$$

$$\int r^2\mathrm{d}r\,\mathcal{X}_{\nu J}(r)\, j_J(qr) = \sqrt{\frac{\pi}{2}}\sqrt{\frac{1}{2E}}\,\frac{q}{M}f_{\nu J}(q).$$

These results show the L- and n-independence of the radial integrals. The elementary invariant matrix elements are thus, now including isospin

$$[\mathrm{V}|G_{\mathrm{VD}}|\gamma] = g_{\mathrm{VD}}(2M^2)4\pi\sqrt{\frac{\pi}{2}}\, i^L\hat{J}\hat{t}\sum_n \frac{N_\kappa}{\sqrt{2E}}\, C_\kappa(n,J)f_{\nu J}(q)\delta_{L,J+n},$$

$$[\mathrm{V}|G^4_{\mathrm{VD}}|\gamma] = g_{\mathrm{VD}}(2M^2)4\pi\sqrt{\frac{\pi}{2}}\, i^L\hat{J}\hat{t}\sqrt{\frac{1}{2E}}\,\frac{q}{M}f_{\nu J}(q)\delta_{LJ}.$$

The final elementary vertex operators including the action of P_0 are ($t=0$ for ω, $t=1$ for ρ, and writing $c^{[0t]}$ with $c^{[00]}=1$ for the ω and $c^{[01]}=c^{[1]}=-i\delta_{m0}$ for the ρ)

$$G_{\mathrm{VD}}(\boldsymbol{q}) = \sum[\mathrm{V}|G_{\mathrm{VD}}|\gamma]:\Big[\big[\tilde{A}^{[Jt]}_{\kappa\nu}c^{[0t]}\big]^{[00]}\big[d^{[1]}\hat{q}^{\,[L]}\big]^{[J]}\Big]^{[0]}:,$$

$$G^4_{\mathrm{VD}}(\boldsymbol{q}) = \sum[\mathrm{V}|G^4_{\mathrm{VD}}|\gamma]:\Big[\big[\tilde{A}^{[Jt]}_{\mathfrak{L}\nu}c^{[0t]}\big]^{[00]}\hat{q}^{\,[J]}\Big]^{[0]}:.$$

9.3. STATIC MOMENTS

9.3.1. Quadrupole moment

For the static electric quadrupole interaction we only have from the Siegert theorem, eq. (2.35), the time-like contribution

$$Q = -i\int \mathrm{d}^3r:J_4\underset{\sim}{Q}:$$

with the wavy line denoting the quadrupole (rank-5) tensor which, written in invariant form, reads

$$\underset{\sim}{Q} = 2z^2 - x^2 - y^2 = \hat{2}\sqrt{\tfrac{16}{5}\pi}\,[w^{[2]}\hat{r}^{[2]}]^{[0]} r^2. \tag{9.7}$$

The operator amplitudes $w_m^{[2]}$ are introduced as usual to construct an invariant. They are given by

$$w_m^{[2]} = -\delta_{m0}.$$

Furthermore, the complete quadrupole operator must of course project on charged particles. This is achieved as before, using the isospin amplitudes (9.6).

The elementary vertex operators are obtained analogously to those of the electron scattering form factors of the previous sections, keeping the time-like parts and setting $L = 2$ with the expression for J_4 from sect. 2.5.1 and including isospin.

(i) Spin-0 field. The invariant matrix element for the interaction with the pion field is obtained from sect. 9.2.3

$$[\pi|Q_{\gamma\pi}|\pi']^P = -\tfrac{1}{4}ie\sqrt{6}\,[l|2|l']\int r^2\,\mathrm{d}r\sqrt{\tfrac{16}{5}\pi}\,r^2\Gamma^P_{\nu l\nu' l'}(r),$$

where P distinguishes between the different processes (scattering, pair production or annihilation) according to sect. 6.7. The full operator is

$$Q_{\gamma\pi} = \sum[\pi|Q_{\gamma\pi}|\pi']^P : [[\theta_\nu^{[l1]}\theta_{\nu'}^{[l'1]}c^{[01]}]^{[20]}w^{[2]}]^{[0]}:.$$

(ii) Spin-$\tfrac{1}{2}$ field. The invariant matrix element for the interaction with the nucleon, sect. 9.2.4, becomes

$$[\mathrm{N}|Q_{\gamma\mathrm{N}}(y)|\mathrm{N}'] = \tfrac{1}{4}e(-)^y[t|\tau^{[y]}|t]\sqrt{2}\sqrt{\tfrac{16}{5}\pi}$$

$$\times\left(\begin{bmatrix}\tfrac{1}{2} & l & j\\ \tfrac{1}{2} & l' & j'\\ 0 & 2 & 2\end{bmatrix}[l|2|l']\int r^2\,\mathrm{d}r\,u_{\nu l}(r)r^2u_{\nu' l'}(r)\right.$$

$$\left.+\begin{bmatrix}\tfrac{1}{2} & \lambda & j\\ \tfrac{1}{2} & \lambda' & j'\\ 0 & 2 & 2\end{bmatrix}[\lambda|2|\lambda']\int r^2\,\mathrm{d}r\,v_{\nu l\lambda}(r)r^2v_{\nu' l'\lambda'}(r)\right).$$

The quadrupole operator thus is

$$Q_{\gamma\mathrm{N}} = \sum[\mathrm{N}|Q_{\gamma\mathrm{N}}(y)|\mathrm{N}'] : [[\tilde{b}_{\nu l}^{[jt]}b_{\nu' l'}^{[j't]}c^{[0y]}]^{[20]}w^{[2]}]^{[0]}:.$$

(iii) Spin-1 field. Writing $\lambda = J + n$ the matrix element is for the scattering term

$$[\mathrm{V}|Q_{\gamma\mathrm{V}}|\mathrm{V}'] = \tfrac{1}{2}e\sqrt{6}\,\hat{1}\sum_{nn'}\begin{bmatrix} 1 & \lambda & J \\ 1 & \lambda' & J' \\ 0 & 2 & 2 \end{bmatrix}[\lambda|\lambda'|2]C_\kappa(n, J)C_{\kappa'}(n', J')$$

$$\times \int r^2 \mathrm{d}r \sqrt{\tfrac{16}{5}\pi}\, r^2 (\mathcal{W}_{\kappa\nu Jn}(r)\mathcal{Y}_{\kappa'\nu'J'n'}(r) + \mathcal{Y}_{\kappa\nu Jn}(r)\mathcal{W}_{\kappa'\nu'J'n'}(r)),$$

and the operator is

$$Q_{\gamma\mathrm{V}} = \sum [\mathrm{V}|Q_{\gamma\mathrm{V}}|\mathrm{V}']:[[\tilde{A}^{[J1]}_{\kappa\nu}A^{[J'1]}_{\kappa'\nu'}]^{[21]}c^{[01]}]^{[20]}w^{[2]}]^{[0]}:.$$

The operators for the production or annihilation of vector meson pairs are obtained by applying hermitian conjugation to one of the fields in these expressions.

(iv) Vector dominance. The evaluation of the contribution of the vector dominance term to the static quadrupole moment requires a special treatment. The terms above involve an interaction of the electromagnetic field with a conserved vector current. In contrast, here the interaction energy arises from the emission or absorption of a neutral particle.

Thus we shall start from the defining equation for the energy of a quadrupole moment denoted Q (the wavy line is for the quadrupole rank-5 tensor), interacting with an electric field gradient

$$H_Q = \int \mathrm{d}^3 r\, \underset{\sim}{Q} \otimes \underset{\sim}{\nabla \boldsymbol{E}}.$$

Here $\underset{\sim}{\nabla \boldsymbol{E}}$ is the rank-5 tensor constructed from ∇ and $\boldsymbol{E}$. The tensor product, indicated by $\otimes$, implies contraction of the tensor indices.

On the other hand, the vector dominance interaction energy is given by (2.41). Hence we shall recognize the vector dominance quadrupole moment Q by equating these energies,

$$g_{\mathrm{VD}}(2M^2)\int \mathrm{d}^3 r : \boldsymbol{\phi}\cdot\boldsymbol{A} := \int \mathrm{d}^3 r\, \hat{\underset{\sim}{Q}} \otimes \underset{\sim}{\nabla \boldsymbol{E}} \tag{9.8}$$

in the limit $q \to 0$.

For the photon vector potential $\boldsymbol{A}$ we use a polarized plane wave, the invariant form of which is at time $t = 0$

$$\boldsymbol{A} = \boldsymbol{d}\mathrm{e}^{i\boldsymbol{q}\cdot\boldsymbol{r}} = \hat{1}[d^{[1]}e^{[1]}]^{[0]}\mathrm{e}^{i\boldsymbol{q}\cdot\boldsymbol{r}}, \tag{9.9}$$

where the amplitudes $\boldsymbol{d}$ fulfill the transversality condition $\boldsymbol{d}\cdot\boldsymbol{q} = 0$. The electric field $\boldsymbol{E}$ thus is

$$\boldsymbol{E} = -\dot{\boldsymbol{A}} = iq\boldsymbol{d}\mathrm{e}^{i\boldsymbol{q}\cdot\boldsymbol{r}}.$$

With $\nabla = i\boldsymbol{q}$, we have for the rank-5 tensor (still indicated by the wavy line)

$$\underset{\sim}{\nabla \boldsymbol{E}} = -q\underset{\sim}{\boldsymbol{q}\boldsymbol{d}}\mathrm{e}^{i\boldsymbol{q}\cdot\boldsymbol{r}} = -q\hat{1}^2\underset{\sim}{\left[q^{[1]}e^{[1]}\right]^{[0]}\left[d^{[1]}e^{[1]}\right]^{[0]}}\mathrm{e}^{i\boldsymbol{q}\cdot\boldsymbol{r}},$$

Introducing the unit tensor $\hat{q}^{[1]}$ defined in sect. 4.2.4,

$$q_m^{[1]} = \sqrt{\tfrac{4}{3}\pi}\,\hat{q}_m^{[1]}q,$$

the invariant form of the tensor $\underset{\sim}{\nabla \boldsymbol{E}}$ is

$$\underset{\sim}{\nabla \boldsymbol{E}} = -\sqrt{\tfrac{4}{3}\pi}\,q^2\hat{2}\left[\left[\hat{q}^{[1]}d^{[1]}\right]^{[2]}\left[e^{[1]}e^{[1]}\right]^{[2]}\right]^{[0]}\mathrm{e}^{i\boldsymbol{q}\cdot\boldsymbol{r}}.$$

Now introducing the rank-5 unit tensor

$$\hat{Q}_m^{[2]} = \left[e^{[1]}e^{[1]}\right]_m^{[2]}, \qquad \text{and} \qquad \left[\hat{Q}^{[2]}|\hat{Q}^{[2]}\right] = \hat{2},$$

so that the invariant form of Q can be written as, eq. (9.7),

$$\underset{\sim}{Q} = \hat{2}Q(r)\left[w^{[2]}\hat{Q}^{[2]}\right]^{[0]},$$

we get for the right-hand side of eq. (9.8), in the limit $q \to 0$,

$$\begin{aligned} H_Q &= -\sqrt{\tfrac{4}{3}\pi}\,q^2(\hat{2})^2\int \mathrm{d}^3r\,Q(r)\left[w^{[2]}\hat{Q}^{[2]}\right]^{[0]} \otimes \left[\left[\hat{q}^{[1]}d^{[1]}\right]^{[2]}\hat{Q}^{[2]}\right]^{[0]} \\ &= -\sqrt{\tfrac{4}{3}\pi}\,q^2\hat{2}\left[w^{[2]}\hat{q}^{[1]}d^{[1]}\right]^{[0]}Q. \end{aligned} \tag{9.10}$$

The symbol $\otimes$ indicates that, after recoupling, only the term $L = 0$ is to be retained. In the last form of (9.10) the quantity Q is the contribution to the quadrupole moment from vector dominance.

On the other hand the left-side term of eq. (9.8) is calculated with the multipole $J = 2$ of the field ϕ, eq. (5.28). In the limit $q \to 0$, the dominant component is space-like with $n = -1$. Writing only the creation part of the quadrupole vector meson field, we have, omitting isospin

$$\begin{aligned} \boldsymbol{\phi}^{(+)}_{\lim p\to 0} = \sum_{\nu}\Big(&\left[\tilde{A}^{[2]}_{\mathcal{E}\nu}\left[e^{[1]}\hat{r}^{[1]}\right]^{[2]}\right]^{[0]}\sqrt{2+1}\,\mathcal{W}_{\mathcal{E}\nu 2(-1)} \\ &+ \left[\tilde{A}^{[2]}_{\mathcal{L}\nu}\left[e^{[1]}\hat{r}^{[1]}\right]^{[2]}\right]^{[0]}\sqrt{2}\,\mathcal{W}_{\mathcal{L}\nu 2(-1)}\Big). \end{aligned} \tag{9.11}$$

We substitute the plane wave expansion of eq. (9.9) for the vector potential $\boldsymbol{A}$ in eq. (9.8) together with the field expansion eq. (9.11). We note that the integration over r yields a delta function between the photon momentum q and the vector meson momentum p

$$\int r^2 \mathrm{d}r\, j_1(pr)\, j_1(qr) = \frac{\pi}{2} \frac{\delta(p-q)}{q^2}.$$

The recoupling geometry of fig. 9.13 yields for the left-hand side of eq. (9.8)

$$H_Q = ig_{\mathrm{VD}}(2M^2) \sum_\nu 4\pi\sqrt{\tfrac{1}{2}\pi}\, f_{\nu 2}(q) \left(\frac{\sqrt{3}}{\sqrt{2E}} \left[\tilde{A}^{[2]}_{\mathfrak{E}\nu} d^{[2]} \right]^{[0]} + \frac{\sqrt{E}}{M} \left[\tilde{A}^{[2]}_{\mathfrak{L}\nu} d^{[2]} \right]^{[0]} \right). \tag{9.12}$$

Here E and M are the energy and the mass of the vector meson and $E = M$ in the limit $q \to 0$. Furthermore, the quadrupole polarization tensor $d^{[2]}$ has been defined according to

$$d^{[2]} = \left[\hat{q}^{[1]} d^{[1]} \right]^{[2]}.$$

Here $d^{[1]}$ is the polarization vector of the real photon field; it is coupled with the momentum direction amplitude $\hat{q}^{[1]}$ of the real photon. For small values of q, the expression (3.8) yields

$$f_{\nu 2}(q)_{\lim q \to 0} = \alpha^{3/2} c_{\nu 2} (\alpha q)^2 L^{5/2}_{\nu - 1}(0).$$

Finally equating the two expressions for the energy, eqs. (9.12) and (9.10), we get the elementary quadrupole vector dominance vertex, including isospin t

$$[\mathrm{V}|Q_{\mathrm{VD}}|\gamma] = -ig_{\mathrm{VD}}(2M^2)\hat{t}\frac{\hat{1}}{2}\pi\left(\alpha^{7/2} c_{\nu 2} L^{5/2}_{\nu-1}(0)\right)\frac{1}{\sqrt{M}} n_\kappa,$$

with $n_\kappa = \sqrt{2}$ for $\kappa = \mathfrak{L}$ and $n_\kappa = \sqrt{3}$ for $\kappa = \mathfrak{E}$. The projection operator P_0 on charge zero has been taken into account and has contributed the factor $\hat{t}$.

The expression for the operator is thus,

$$Q_{\mathrm{VD}} = \sum [\mathrm{V}|Q_{\mathrm{VD}}|\gamma] : \left[\tilde{A}^{[2t]}_{\kappa\nu} d^{[20]} c^{[0t]} \right]^{[00]} : .$$

The sum of course is limited to $\kappa = \mathfrak{E}, \mathfrak{L}$. Annihilation of the vector meson is given by hermitian conjugation.

9.3.2. Magnetic moment

The magnetic dipole operator in the long wavelength limit has been given for the different fields in eqs. (2.36), (2.37) and (2.38). These expressions contain no anomalous part. In fact, the basic fields must be considered partly dressed and their particles have an anomalous (effective) magnetic intrinsic moment. Here we limit this correction to the important case of spin $\frac{1}{2}$ and we shall use the modified form (2.39) for the proton and (2.40) for the neutron.

Introducing the invariant form of the angular momentum operator

$$\boldsymbol{l} = \hat{1}[l^{[1]}d^{[1]}]^{[0]}, \tag{9.13}$$

the elementary vertex operators are given by the following expressions:

(i) Spin-0 field. The matrix element of (2.36) is for pion scattering and pion pair production or annihilation, including isospin

$$[\pi|\mu_{\gamma\pi}|\pi'] = -\frac{e}{2M}\sqrt{6}\,\delta_{\nu\nu'}\delta_{ll'}\hat{l}\sqrt{l(l+1)}\ .$$

The operator is

$$\mu_{\gamma\pi} = \sum [\pi|\mu_{\gamma\pi}|\pi']:\left[[\theta_{\nu}^{[l1]}\theta_{\nu}^{[l1]}]^{[11]}c^{[01]}d^{[10]}\right]^{[00]}:.$$

(ii) Spin-$\frac{1}{2}$ field. For the spin-$\frac{1}{2}$ field we treat separately the $\boldsymbol{j}$ and $\boldsymbol{s}$ parts, including anomalous moments as in eqs. (2.39) and (2.40). We obtain, as in sect. 9.2.4,

$$[\mathrm{N}|\mu_{\gamma\mathrm{N}}(y)|\mathrm{N}'] = -i\frac{e}{2M}A[t|\tau^{[y]}|t]$$

$$\times\left(B\hat{j}\sqrt{j(j+1)}\,\delta_{\nu\nu'}\delta_{ll'}\delta_{jj'}\right.$$

$$+C\frac{\sqrt{6}}{4}\left(\begin{bmatrix} s & l & j\\ s & l' & j'\\ 1 & 0 & 1\end{bmatrix}\delta_{ll'}\hat{l}\int r^2\mathrm{d}r\,u_{\nu l}(r)u_{\nu' l'}(r)\right.$$

$$\left.\left.+\begin{bmatrix} s & \lambda & j\\ s & \lambda' & j'\\ 1 & 0 & 1\end{bmatrix}\delta_{\lambda\lambda'}\hat{\lambda}\int r^2\mathrm{d}r\,v_{\nu l\lambda}(r)v_{\nu' l'\lambda'}(r)\right)\right).$$

For a charged particle, eq. (2.39), $A=(-)^y$, $B=1$, $C=g_p-1$. For a neutral particle, eq. (2.40), $A=1$, $B=0$, $C=g_n$. This way, the operator is

$$\mu_{\gamma\mathrm{N}} = \sum [\mathrm{N}|\mu_{\gamma\mathrm{N}}(y)|\mathrm{N}']:\left[[\tilde{b}_{\nu}^{[jt]}b_{\nu'}^{[j't]}]^{[1y]}c^{[0y]}d^{[10]}\right]^{[00]}:.$$

(iii) Spin-1 field. For the spin-1 field we neglect any anomalous magnetic moments. The total angular momentum operator (2.37) yields for the charged vector meson scattering term

$$[\mathrm{V}|\mu_{\gamma\mathrm{V}}|\mathrm{V}'] = -\frac{e}{2M}\sqrt{6}\,\hat{J}\sqrt{J(J+1)}\,\delta_{\kappa\kappa'}\delta_{\nu\nu'}\delta_{JJ'},$$

$$\mu_{\gamma\mathrm{V}} = \sum [\mathrm{V}|\mu_{\gamma\mathrm{V}}|\mathrm{V}']:\left[[\tilde{A}_{\mathfrak{M}\nu}^{[J1]}A_{\mathfrak{M}\nu'}^{[J1]}]^{[11]}c^{[01]}d^{[10]}\right]^{[00]}:.$$

(iv) Vector dominance. To evaluate the vector dominance contribution we follow a similar path as for the quadrupole moment.

On the one hand, the magnetic coupling energy H_μ is of the form

$$H_\mu = \boldsymbol{\mu} \cdot \boldsymbol{B}, \tag{9.14}$$

and the magnetic moment μ is defined by

$$\mu = H_\mu / |\boldsymbol{B}|\,.$$

With a plane wave potential $\boldsymbol{A}$, eq. (9.9), and formula (4.39) for the rotation we have

$$\begin{aligned}\boldsymbol{B} = \operatorname{rot} \boldsymbol{A} &= \hat{1}\sqrt{2}\,[e^{[1]}\,\nabla^{[1]} d^{[1]}]^{[0]} \mathrm{e}^{i\boldsymbol{q}\cdot\boldsymbol{r}} \\ &= i\sqrt{\tfrac{4}{3}\pi}\, q \hat{1}\sqrt{2}\,[e^{[1]}\hat{q}^{[1]} d^{[1]}]^{[0]} \mathrm{e}^{i\boldsymbol{q}\cdot\boldsymbol{r}},\end{aligned}$$

where again a unit tensor $\hat{q}^{[1]}$ has been introduced as in eq. (9.10). Hence with a simple recoupling, $q \to 0$,

$$H_\mu = i\sqrt{\tfrac{4}{3}\pi}\, q \hat{1}\sqrt{2}\, \mu\,[\hat{\mu}^{[1]}\hat{q}^{[1]} d^{[1]}]^{[0]}, \tag{9.15}$$

where μ is the magnetic moment and $\hat{\mu}^{[1]}$ a unit vector defined by

$$\boldsymbol{\mu} = \hat{1}\mu\,[\hat{\mu}^{[1]} e^{[1]}]^{[0]}.$$

On the other hand, we use the expression of the energy in terms of the interacting fields, eq. (2.41). We take the magnetic dipole component of the expansion (5.11)

$$H_\mu = g_{\mathrm{VD}}(2M^2)\int \mathrm{d}^3 r : \boldsymbol{A}\cdot\boldsymbol{\phi} :,$$

and go to the limit $q \to 0$. Thus

$$H_{\mu\,\lim q\to 0} = i g_{\mathrm{VD}}(2M^2) 2\pi^{3/2} \frac{\hat{1}}{\sqrt{M}} \sum_\nu \alpha^{5/2} c_{\nu 1} q L_{\nu-1}^{3/2}(0) : [\tilde{A}_{\mathfrak{N}\mathfrak{L}\nu}^{[1]} d^{[1]} \hat{q}^{[1]}]^{[0]} : . \tag{9.16}$$

Comparing (9.16) with the two expressions (9.14) and (9.15) after performing the expectation value of (9.16) with the appropriate state vector, the elementary matrix element is obtained

$$[\mathrm{V}|\mu_{\mathrm{VD}}|\gamma] = g_{\mathrm{VD}}(2M^2)\hat{\imath}\sqrt{\tfrac{3}{2}}\,\pi \frac{1}{\sqrt{M}} \alpha^{5/2} c_{\nu 1} L_{\nu-1}^{3/2}(0),$$

where the isospin has been taken into account. The operator for the vector dominance magnetic dipole moment is thus given by

$$\mu_{\mathrm{VD}} = \sum [\mathrm{V}|\mu_{\mathrm{VD}}|\gamma] : \left[\tilde{A}^{[1t]}_{\mathfrak{M}\nu} d^{[10]} \hat{q}^{[10]} c^{[0t]} \right]^{[00]} : .$$

9.4. ELECTROMAGNETIC MANY-BODY MATRIX ELEMENTS

9.4.1. *Structure of the matrix elements*

In carrying out the calculation of the many-body transition matrix elements we follow the notation of Chapter 7.

The transition matrix element between two states Ψ_m and Ψ_n is

$$K_{mn} = \langle \Psi_m | G | \Psi_n \rangle = \sum_{\alpha\alpha'} X^{m*}_{\alpha} X^{n}_{\alpha'} \langle \alpha IT | G | \alpha' I' T' \rangle .$$

Of course, with our phase definitions of Chapter 7 the amplitudes X are real.

The photon operator has been given in the previous sections in the form

$$G = \sum_{\zeta} G_{\zeta} = \sum \left[\mathsf{a} | G_{\zeta} | \mathsf{b} \right] \left[D^{[LZ]} : \Theta^{[LZ]}_{\zeta(\mathsf{ab})} : \right]^{[00]} .$$

Here the summations are over (i) the different processes, denoted by ζ, such as $\gamma\pi\pi'$, γNN′ etc..., (ii) the transferred total angular momentum and isospin of the photon L and Z respectively, (iii) the particle states entering the elementary vertices, denoted by the indices a, b. The quantity $D^{[LZ]}$ is an amplitude carrying the information about the orientation of the photon, viz. direction $\hat{\boldsymbol{q}}$ of the momentum transfer, polarization $\boldsymbol{d}$, isospin composition $\boldsymbol{c}$. The overall polarization tensor thus will be written as

$$D^{[LZ]} = \left[d^{[10]} \hat{q}^{[\Lambda 0]} c^{[0Z]} \right]^{[LZ]} .$$

They already have appeared in the results of sects. 9.2 and 9.3. The tensors $D^{[LZ]}$ represent the information about the angular correlation and the polarization dependence of the electromagnetic process. In those expressions Λ denotes the orbital angular momentum of the photon in the plane wave decomposition.

The configurational state vectors of eq. (7.1) now are written so as to exhibit isospin. For a system with three types, which was the example chosen in sec. 7.2.1 we have

$$|\alpha IT\rangle = (-)^{2I+2T} \hat{I}\hat{T} \left[W^{[IT]}_{\alpha} \left[\left[\tilde{\mathfrak{R}}^{[J_a T_a]} \tilde{\mathfrak{S}}^{[J_b T_b]} \right]^{[J_{ab} T_{ab}]} \tilde{\mathfrak{T}}^{[J_c T_c]} \right]^{[IT]}_{\alpha} \right]^{[00]} |0\rangle ,$$

where the indices a, b and c each refers to a different given type. In the same way we write for the operator

$$:\Theta^{[LZ]}_{\zeta}: = : \left[\left[\Theta^{[L_a Z_a]}_{\mathsf{R}} \Theta^{[L_b Z_b]}_{\mathsf{S}} \right]^{[L_{ab} Z_{ab}]} \Theta^{[L_c Z_c]}_{\mathsf{T}} \right]^{[LZ]} : .$$

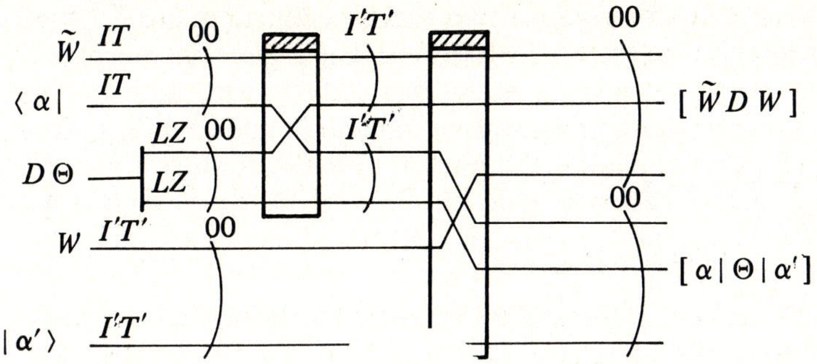

Fig. 9.14.

The transferred photon quantum numbers in each of the three types are denoted L_a, L_b, L_c for the angular momentum and Z_a, Z_b, Z_c for the isospin. For momentum transfers where one can neglect the effects of the boost, as discussed in sect. 9.2.2, $L_i = 0$, $Z_i = 0$ for the spectator types, i.e. their Θ_i is the unit operator.

Following the lines of sect. 7.2.2, the transition matrix elements for the individual configurations of the states Ψ_n and Ψ_m according to fig. 9.14 are, for a given process ζ

$$K_{\zeta\alpha\alpha'} = \langle \alpha IT | G_\zeta | \alpha' I'T' \rangle = (-)^{2I'+2T'} \left[\tilde{W}_m^{[IT]} D^{[LZ]} W_n^{[I'T']} \right]^{[00]}$$

$$\times \sum_{\mathsf{ab}} \eta_{\mathsf{ab}} \left[\mathsf{a} | G_\zeta | \mathsf{b} \right] \langle 0 | \left[[\mathfrak{R}\mathfrak{S}\mathfrak{T}]_\alpha^{[IT]} : \Theta_{\zeta(\mathsf{ab})}^{[LZ]} : [\tilde{\mathfrak{R}}\tilde{\mathfrak{S}}\tilde{\mathfrak{T}}]_{\alpha'}^{[I'T']} \right]^{[00]} | 0 \rangle . \quad (9.16)$$

In accordance with the discussion of sects. 4.4.1, 4.4.2 and 7.2.2 the phase η has been extracted from the configuration amplitudes as in eq. (7.2). This phase is still given by the rules developed in those sections, and only the interacting particles a, b, contribute to that phase. The values of η_{ab} for the different processes ζ are given in

TABLE 9.1
Phases η for electromagnetic processes with the conventions of table 7.1

$[\mathsf{a}\|G_\zeta\|\mathsf{b}]$	η_{ab}
$[\pi\|G_{\gamma\pi}\|\gamma\pi']$	$i^{l-l'}$
$[\pi\pi'\|G_{\gamma\pi}\|\gamma]$	$i^{l+l'+2}$
$[\,\|G_{\gamma\pi}\|\gamma\pi\pi']$	$i^{-l-l'-2}$
$[\mathrm{N}\|G_{\gamma\mathrm{N}}\|\gamma\mathrm{N}']$	$i^{l-l'}$
$[\rho\|G_{\gamma\mathrm{V}}\|\gamma\rho']$	$i^{J+\varepsilon_\kappa-J'-\varepsilon_{\kappa'}}$
$[\mathrm{N}\|G_{\gamma\mathrm{N}\Delta}\|\gamma\Delta]$	$i^{l_\mathrm{N}-l_\Delta}$
$[\Delta\|G_{\gamma\Delta}\|\gamma\Delta']$	$i^{l-l'}$
$[\rho\|G_{\mathrm{VD}}\|\gamma]$	$i^{J+\varepsilon_\kappa+1}$
$[\omega\|G_{\mathrm{VD}}\|\gamma]$	$i^{J+\varepsilon_\kappa}$

$\varepsilon_\kappa = 0$ for $\kappa = \mathfrak{M}$.
$\varepsilon_\kappa = 1$ for $\kappa = \mathfrak{E}, \mathfrak{L}$.

table 9.1 if one follows the phase conventions adopted in table 7.1. If one chooses instead the simpler convention $\xi = 1$ for all single particles except $\xi = (-i)^{1+t_h}$ for each vector meson, as explained in sect. 7.2.2, then $\eta_{ab} = \pm i$ for the vector dominance processes with creation (+) or annihilation (−) of a vector meson and $\eta_{ab} = 1$ for all other processes. The photon does not contribute to η since its phases are included in the transition operator G. Note also that the complete transition matrix elements $K_{\alpha\alpha'}$ are not real in contrast to the matrix elements $M_{\alpha\alpha'}$ of the hamiltonian, eq. (7.2).

We see that the many-body aspects of the matrix element (9.16) are contained in the evaluation of the vacuum expectation value of the particle operators, to which we now proceed.

9.4.2. *Separation of types*

The calculation of the vacuum expectation value in eq. (9.16) follows the three steps given in sect. 7.2.3: separation of types, separation of groups within a type, and analysis within a group. The only change is of course that the photon operator, unlike the scalar interactions, carries total angular momentum and isospin.

The evaluation of the type separation graph 9.15 yields

$$\left[[\mathfrak{R}\mathfrak{S}\mathfrak{T}]_{\alpha}^{[IT]}:[\Theta_{\mathrm{R}}\Theta_{\mathrm{S}}\Theta_{\mathrm{T}}]^{[LZ]}:[\tilde{\mathfrak{R}}\tilde{\mathfrak{S}}\tilde{\mathfrak{T}}]_{\alpha'}^{[I'T']}\right]^{[00]}$$

$$= P\left[\mathfrak{R}_{\alpha}^{[J_aT_a]}\tilde{\mathfrak{R}}_{\alpha'}^{[J'_aT'_a]}:\Theta_{\mathrm{R}}^{[L_aZ_a]}:\right]^{[00]}\left[\mathfrak{S}_{\alpha}^{[J_bT_b]}\tilde{\mathfrak{S}}_{\alpha'}^{[J'_bT'_b]}:\Theta_{\mathrm{S}}^{[L_bZ_b]}:\right]^{[00]}$$

$$\times\left[\mathfrak{T}_{\alpha}^{[J_cT_c]}\tilde{\mathfrak{T}}_{\alpha'}^{[J'_cT'_c]}:\Theta_{\mathrm{T}}^{[L_cZ_c]}:\right]^{[00]},$$

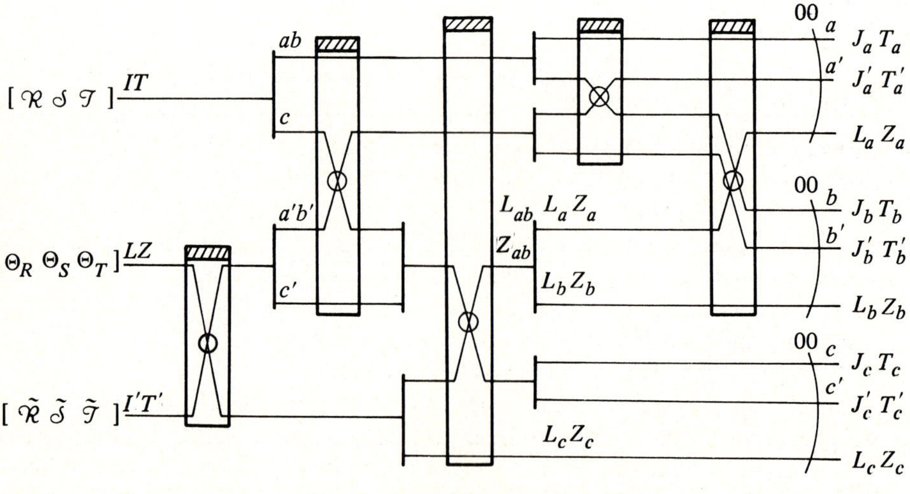

Fig. 9.15.

where the type separation box P is given by

$$P = \varepsilon(-)^{L+I'-I+Z+T'-T} \frac{\hat{L}\hat{Z}}{\hat{L}_a\hat{Z}_a\hat{L}_b\hat{Z}_b\hat{L}_c\hat{Z}_c} \begin{bmatrix} J_{ab} & J_c & I \\ J'_{ab} & J'_c & I' \\ L_{ab} & L_c & L \end{bmatrix}$$

$$\times \begin{bmatrix} T_{ab} & T_c & T \\ T'_{ab} & T'_c & T' \\ Z_{ab} & Z_c & Z \end{bmatrix} \begin{bmatrix} J_a & J_b & J_{ab} \\ J'_a & J'_b & J'_{ab} \\ L_a & L_b & L_{ab} \end{bmatrix} \begin{bmatrix} T_a & T_b & T_{ab} \\ T'_a & T'_b & T'_{ab} \\ Z_a & Z_b & Z_{ab} \end{bmatrix}.$$

All crossing signs associated with anticommuting operators are collected in the phase ε. As before in sect. 7.2.3 the interaction operators are brought to the bottom of the graph without performing any contractions, and again they will have to be recoupled to their proper position before evaluating the vacuum expectation values in each group. This procedure simplifies the evaluation of the many-body expectation values as it insures automatic book-keeping of the Fock-space phases in the following recoupling steps.

Next comes a simple operation Q where it is decided which of the types participate in the transition and which are spectators. It performs the vacuum expectation values for the spectator types and yields according to (4.48)

$$Q = \prod_i \langle 0 | \left[\mathfrak{R}_{\alpha_i}^{[J_i T_i]} \tilde{\mathfrak{R}}_{\alpha'_i}^{[J'_i T'_i]} 1^{[00]} \right]^{[00]} | 0 \rangle$$

$$= \prod_i (-)^{2J_i+2T_i} \hat{J}_i \hat{T}_i \delta_{J_i J'_i} \delta_{T_i T'_i} \delta_{\alpha_i \alpha'_i},$$

where i denotes the spectator types. For large momentum transfers, for $S > 0$ particles, the norms in Minkowski space are diminished from their values $\hat{J}_i$ and must be evaluated according to sect. 9.2.2.

9.4.3. *Evaluation within one type*

The evaluation of the remaining expectation values within one type is done along the lines of sects. 7.2.4 and 7.2.5. To be specific we shall consider as an example the case of the pion cloud, and the matrix elements for the form factor constructed with the elementary pion-photon operators $G_{\gamma\pi}$ of sect. 9.2.3 in the Hilbert spaces of sect. 5.4. These cases show all the features of the many-body problem. All other cases would be handled similarly. As stated in the introduction of this chapter we also present now a method which allows to generate automatically all matrix elements rather than presenting lists of specific matrix elements as in Chapter 7.

Within the type (here the pions) on which the photon operator acts, we have several groups of pions, each in a given orbital state π_i. A group is defined by the number of pions n_i in a state π_i and its couplings. Let the basis configurations have groups made of n_1, n_2, n_3 pions in states π_1, π_2, π_3 in the final state and groups of

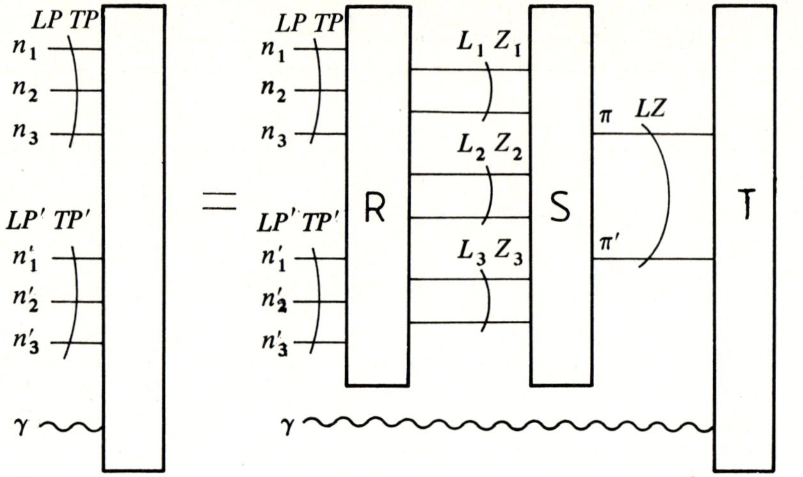

Fig. 9.16.

n'_1, n'_2, n'_3 pions in states π'_1, π'_2, π'_3 in the initial state. The general structure of the many-body matrix element within this type in terms of the contributions of the different groups is given by fig. 9.16. It involves three steps, viz., a recoupling box R, a separation and overlap box S, and the elementary pion-photon reaction box T.

The box R performs the recoupling of the pions from their initial and final configurational coupling scheme into a scheme where the initial and final groups are coupled. Box S separates the two pions which participate in the photon interaction and performs all the vacuum expectation values for the spectator pions. The box T is the vacuum expectation value of the interacting particle operators in the state vectors and the normal product operators arising from $:\Theta_\xi:$.

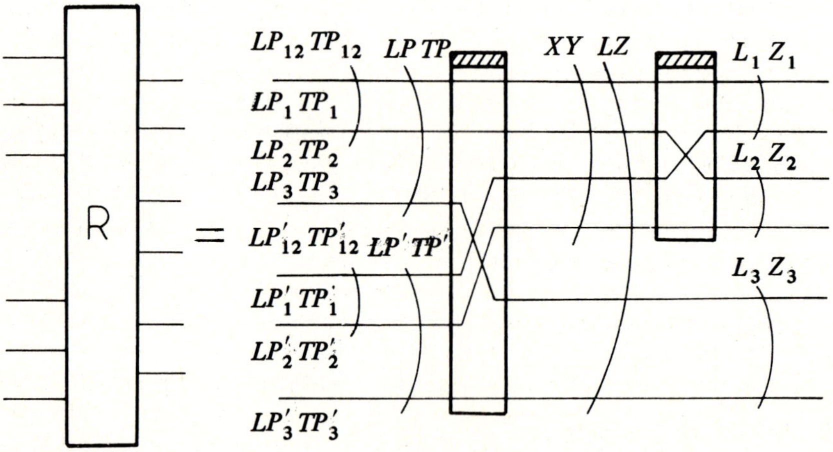

Fig. 9.17.

TABLE 9.2
Pion distributions among groups for the configurations of sect. 5.4.

Total number	n_1	n_2	n_3
1	1		
2	2		
	1	1	
3	3		
	2	1	
	1	1	1
4	4		
	3	1	
	2	2	
	2	1	1

The group recoupling box R represents the diagram of fig. 9.17, which defines the notation for the intermediate quantum numbers, with the couplings denoted as in eq. (5.38). Its value is

$$R=\sum_{XY}\begin{bmatrix} LP_1 & LP_2 & LP_{12} \\ LP_1' & LP_2' & LP_{12}' \\ L_1 & L_2 & X \end{bmatrix}\begin{bmatrix} LP_{12} & LP_3 & LP \\ LP_{12}' & LP_3' & LP' \\ X & L_3 & L \end{bmatrix}$$

$$\times\begin{bmatrix} TP_1 & TP_2 & TP_{12} \\ TP_1' & TP_2' & TP_{12}' \\ Z_1 & Z_2 & Y \end{bmatrix}\begin{bmatrix} TP_{12} & TP_3 & TP \\ TP_{12}' & TP_3' & TP' \\ Y & Z_3 & Z \end{bmatrix}.$$

We turn now to the structure of the separation box S. The photon-pion vertices $G_{\gamma\pi}$ connect only very few pion groups of the final state to a given pion group of the initial state. Let us consider the different distributions of the configurations of sect. 5.4 which we use for constructing our examples. They are given in table 9.2. Note that in this table, owing to the truncation of the model spaces of sect. 5.4, no distribution of the type $n_2 > n_1$ exists even though the groups π_1, π_2, π_3 are listed in the standard order. Furthermore for $n_i \geqslant 3$ only $1s^n$ groups exist, and owing to standard order π_2 and π_3 are never 1s states.

The separation and overlap box S may have three forms according to the photon process. These three cases are:

Case I, the initial distribution remains unchanged:

$$n_1 = n_1', \qquad n_2 = n_2', \qquad n_3 = n_3'.$$

Case II, the initial distribution changes but the number of pions remains the same:

$$n_i = n_i' - 1, \qquad n_j = n_j' + 1, \qquad n_k = n_k'.$$

Case III, two pions are created:

$$n_i = n_i' + 2, \qquad n_j = n_j', \qquad n_k = n_k'$$

or

$$n_i = n_i' + 1, \qquad n_j = n_j' + 1, \qquad n_k = n_k',$$

or two pions are annihilated:

$$n_i = n_i' - 2, \qquad n_j = n_j', \qquad n_k = n_k'$$

or

$$n_i = n_i' - 1, \qquad n_j = n_j' - 1, \qquad n_k = n_k'.$$

For case I, we get three terms, corresponding to the interaction of the photon with a pion in group 1, group 2 and group 3 respectively. These three contributions are given by fig. 9.18. The spectator groups terminate in vacuum expectation value boxes and contribute each the factor according to (4.48)

$$\widehat{LP_i}\widehat{TP_i}\delta_{L_i 0}\delta_{Z_i 0}.$$

The boxes B in group i separate out a particle from the initial and from the final state in group i and perform the overlap for the $n_i - 1$ other particles which do not interact with the photon, keeping the photon line at the bottom of the graph. In the

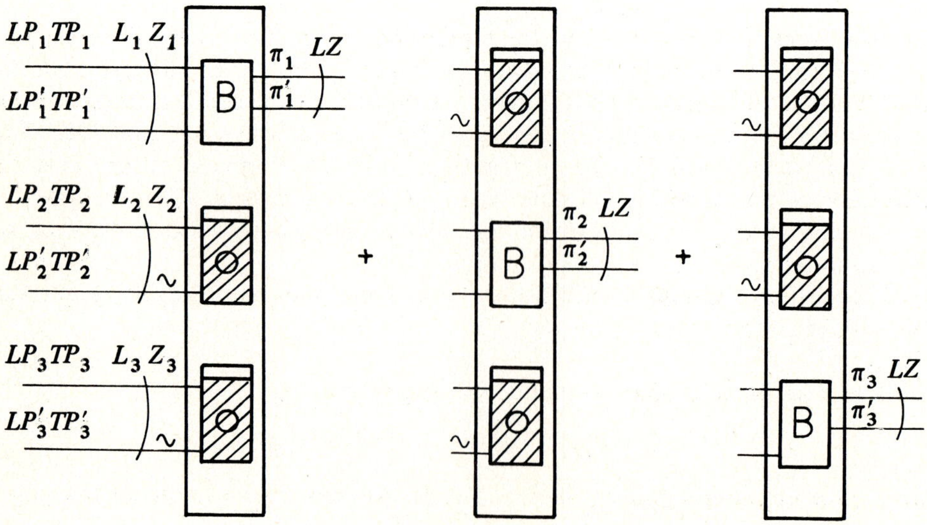

Fig. 9.18.

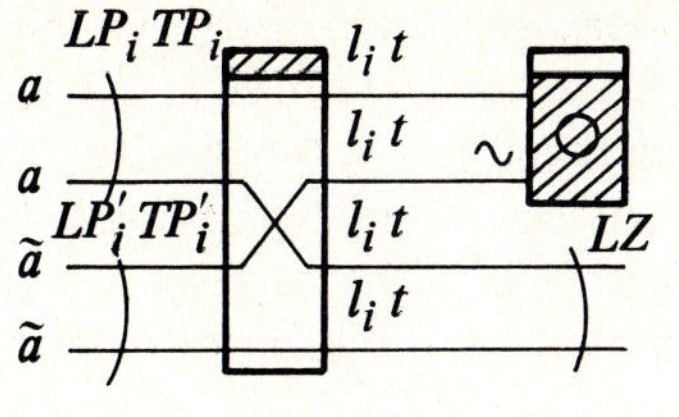

Fig. 9.19.

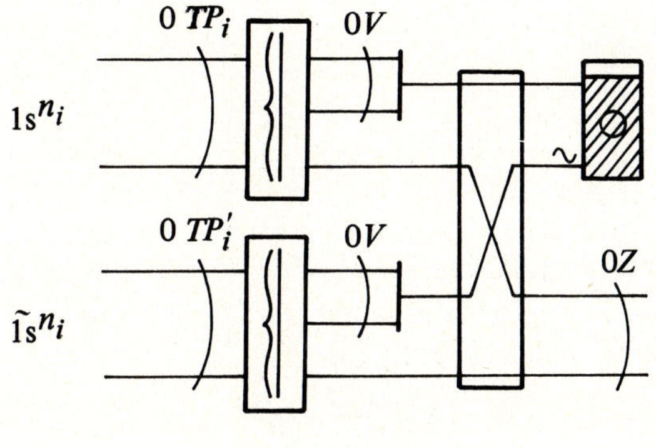

Fig. 9.20.

configuration spaces of Chapter 5 there are three cases according to the value of n_i:

$n_i = 1$

$$B(i) = 1.$$

$n_i = 2$, fig. 9.19

$$B(i) = \begin{bmatrix} l_i & l_i & LP_i \\ l_i & l_i & LP_i' \\ 0 & L & L \end{bmatrix} \begin{bmatrix} t & t & TP_i \\ t & t & TP_i' \\ 0 & Z & Z \end{bmatrix} \hat{l}_i \hat{t}$$

$n_i \geqslant 3$, $1s^n$ cloud, fig. 9.20

$$B(i) = \sum_V \mathrm{CFP}_1^{n_i}(TP_i, V)\mathrm{CFP}_1^{n_i}(TP_i', V) \begin{bmatrix} V & t & TP_i \\ V & t & TP_i' \\ 0 & Z & Z \end{bmatrix} \hat{V}\delta_{L0}.$$

For case II, the different changes of the distributions are shown symbolically in fig. 9.21. The departing arrow in group i means $n_i = n_i' - 1$ while the incoming arrow in group j means $n_j = n'_j + 1$. The structures of the box S are given in fig. 9.22 for the three processes of the top of fig. 9.21. The boxes C_- and C_+ separate a particle from the initial and final state respectively using the results of sect. 4.6.3, and perform the overlap for the spectator particles. The processes of the bottom of fig. 9.21 are obtained by interchanging in fig. 9.22 the boxes C_- and C_+. The values of the C boxes for the different distributions of our examples are:

$\underline{n_i = 1}$, $\underline{n'_i = 0}$ or $\underline{n_i = 0}$, $\underline{n'_i = 1}$

$$C_+(i) = C_-(i) = 1\,.$$

$\underline{n_i = 2}$, $\underline{n'_i = 1}$, fig. 9.23a

$$C_+(i) = \sqrt{2}\,\frac{\widehat{\mathrm{LP}_i}\widehat{\mathrm{TP}_i}}{\hat{l}_i\hat{t}}\,.$$

$\underline{n_i = 1}$, $\underline{n'_i = 2}$, fig. 9.23b

$$C_-(i) = \sqrt{2}\,\frac{\widehat{LP'_i}\widehat{TP'_i}}{\hat{l}_i\hat{t}}\,.$$

$n_i, n'_i \geqslant 3$, $1s^n$ clouds, $i = i' = 1$:
$\underline{n_1 = n'_1 + 1}$, fig. 9.24a

$$C_+(1) = \sqrt{n_1}\,\mathrm{CFP}_1^{n_1}(TP_1, TP'_1)\frac{\widehat{TP'_1}}{\hat{t}}\,.$$

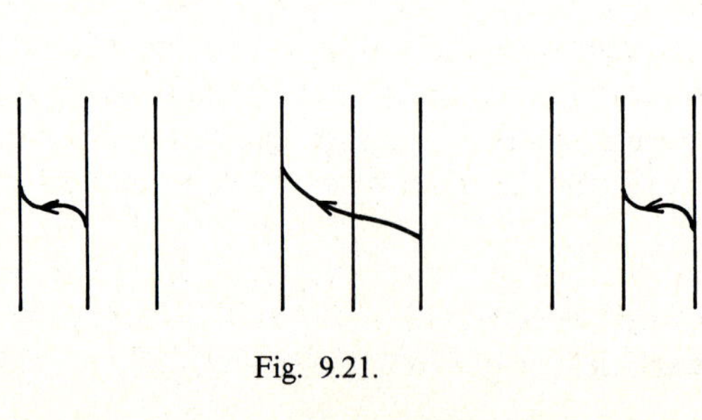

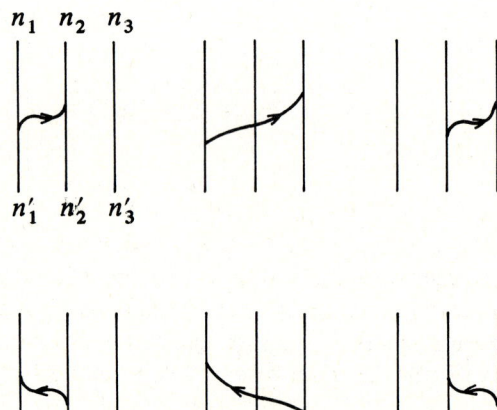

Fig. 9.21.

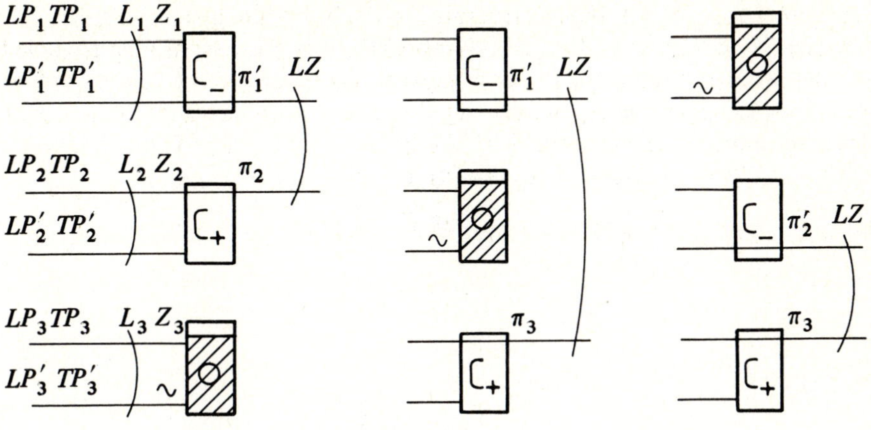

Fig. 9.22.

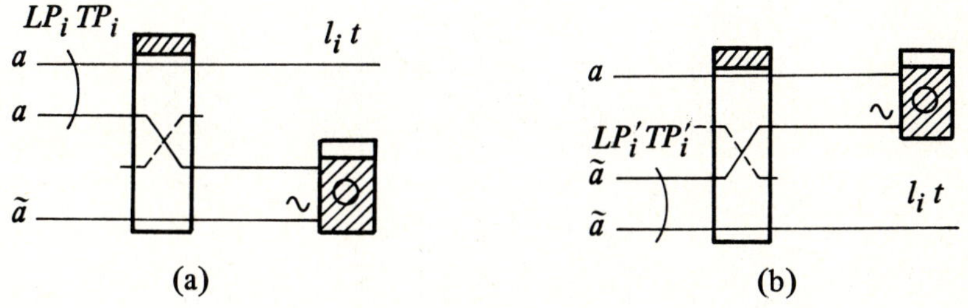

Fig. 9.23.

$n_1 = n'_1 - 1$, fig. 9.24b

$$C_-(1) = \sqrt{n'_1}\,\mathrm{CFP}_1^{n'_1}(TP'_1, TP_1)\frac{\widehat{TP_1}}{\hat{t}}\,.$$

The case III which leads to a change of distributions by $\Delta n = 2$ is shown in fig. 9.25. The top processes correspond to pair production, and the bottom to pair annihilation. The structures of the box S associated with these different processes are shown for pair production in fig. 9.26. Besides the vacuum expectation value boxes and the separation boxes C_+ defined above, a new operation, box D_+ appears. The operation D_+ performs the separation of two particles from a single group in the final state.

For pair annihilation, bottom graphs in fig. 9.25, the structures of the box S are the same, except for the replacement of C_+ by C_- and D_+ by D_-. The operation D_- separates off two particles in a given group in the initial state.

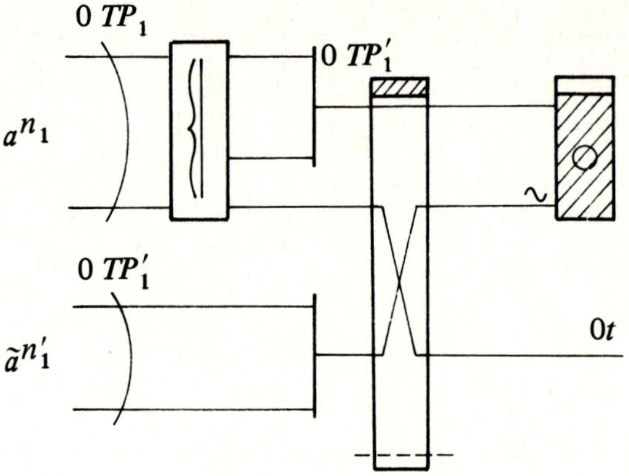

(a)

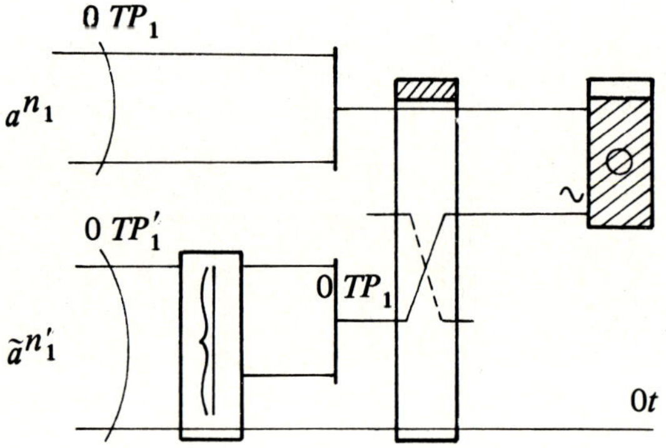

(b)

Fig. 9.24.

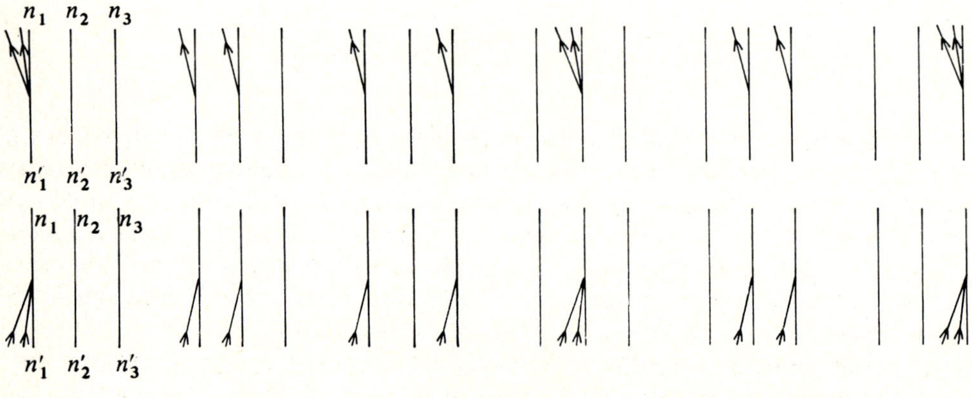

Fig. 9.25.

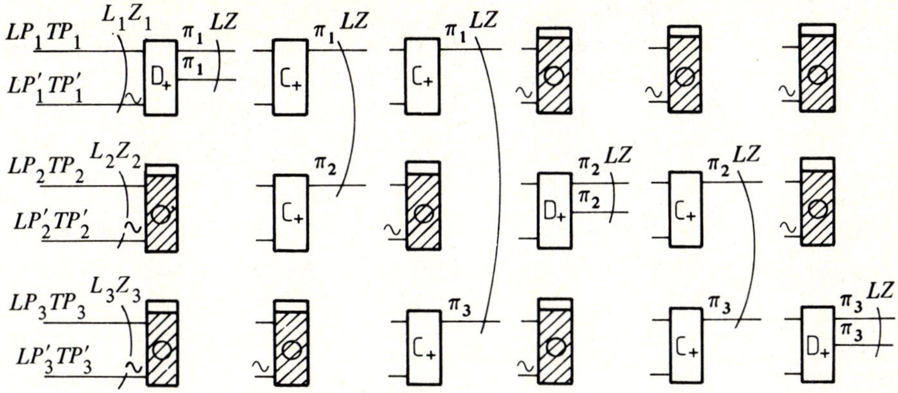

Fig. 9.26.

The values of D_+ and D_- for the cases of our examples are calculated with the results of sect. 4.6.3 and they are respectively:

$n_i = 2, n'_i = 0$ or $n_i = 0, n'_i = 2$

$$D_+(i) = D_-(i) = 1 .$$

$n_i \geqslant 3$, $1s^n$ cloud:

$n_1 = n'_1 + 2$, fig. 9.27a.

$$D_+(1) = \sqrt{n_1(n_1 - 1)}\, \mathrm{CFP}_2^{n_1}(TP_1, TP'_1, Z)\frac{TP_1}{\hat{Z}} .$$

$n_1 = n'_1 - 2$, fig. 9.27b.

$$D_-(1) = \sqrt{n'_1(n'_1 - 1)}\, \mathrm{CFP}_2^{n'_1}(TP'_1, TP_1, Z)\frac{TP'_1}{\hat{Z}} .$$

We now turn to the evaluation of the expectation value represented by the end box T, which terminates the many-body diagram of fig. 9.16. Denoting the pion lines entering the box T by π_a and π_b, its contribution is

$$\mathsf{m}T^P_{\mathrm{ab}} = \langle 0 | [[\theta_\mathrm{a}\theta_\mathrm{b}]^{[LZ]} {:} \Theta_P^{[LZ]} {:}]^{[00]} | 0 \rangle ,$$

where m is the operator multiplicity factor as discussed in sect. 7.2.6 and where P denotes the processes as defined in sect. 9.2.3. The vacuum expectation value is to be evaluated after recoupling the photon operator $:\Theta_P:$ into its proper position. One finds that in the present example the contributions of the end box T^P_{ab} for the three

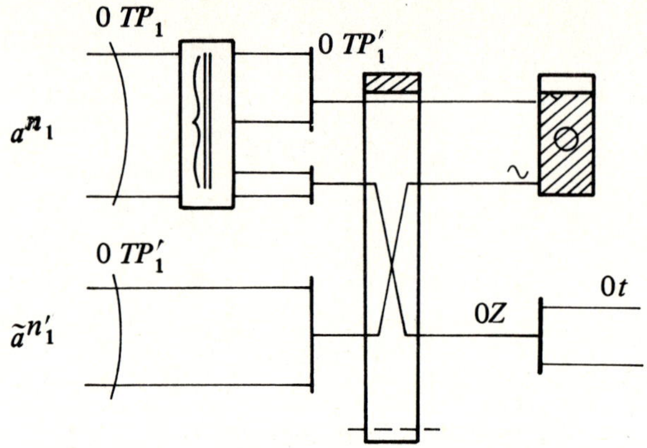

(a)

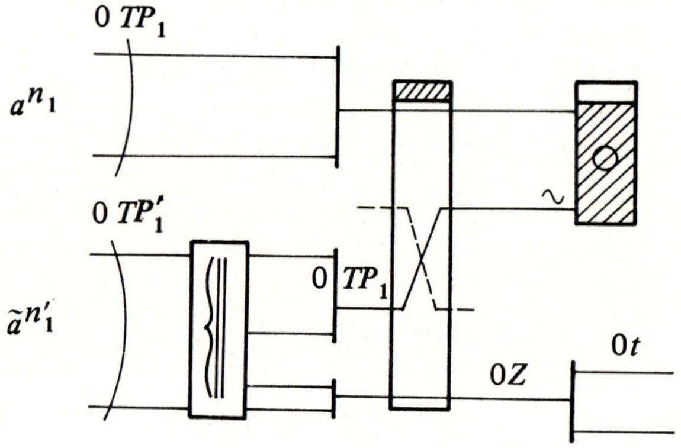

(b)

Fig. 9.27.

processes, pion pair annihilation, $P = 1$, pion scattering, $P = 2$, and pion pair production, $P = 3$, have the same form

$$T^P_{\mathrm{ab}} = \hat{L}\hat{Z}\,.$$

The calculation of the vacuum expectation value entering the matrix element (9.16) is then completed by multiplying the factors P, Q, R, S, T with the overall multiplicity M discussed in full detail in sect. 7.2.6, and given by eq. (7.3). Recall that it contains the operator multiplicity factor m. This terminates the computation of the transition matrix element (9.16).

The procedure just described systematizes the construction of the many-body matrix elements and allows their automatic generation by a computing program.

Appendix

A.1. THE COULOMB FORCE IN QED

The free fields for the vector mesons of non-vanishing mass of Chapters 2 and 5 form complete sets. This is shown by the commutation relations (5.23) they fulfill, together with the existence of three orthogonal polarization basis states. Hence any arbitrary massive vector field can be expanded in terms of these free fields.

For the case of vanishing mass, we have already noted that the longitudinal polarization mode is absent since from (2.20)

$$\operatorname{div} \boldsymbol{E} = 0 .$$

However in the presence of sources

$$\operatorname{div} \boldsymbol{E} = -4\pi e\rho(r, t) . \tag{A.1}$$

We note that this relation contains no time-derivatives, hence it is not a dynamical but a defining equation.

Thus for $M=0$ one requires a more general solution than that provided by transverse free fields. According to the theory of partial differential equations the most general solution is obtained by adding one particular solution of the inhomogeneous equations to the general solution of the homogeneous equations. In the present case the homogeneous equations are the free field equations of motion and the inhomogeneous parts are given by the interaction terms, i.e. the source terms. The solutions of the homogeneous equations here are the transverse free fields and we choose for the particular solution of the inhomogeneous source equations (A.1) a purely longitudinal field. To that end we write

$$\boldsymbol{E}^{(\ell)} = -\operatorname{grad} \lambda_0 , \tag{A.2}$$

so that

$$\operatorname{rot} \boldsymbol{E}^{(\ell)} = 0 .$$

Hence λ_0 is the solution of the Laplace equation

$$\operatorname{div} \boldsymbol{E}^{(\ell)} = -\nabla^2 \lambda_0 = -4\pi e\rho ,$$

and we have

$$\lambda_0(\boldsymbol{r}, t) = e\int \mathrm{d}^3r' \frac{\rho(\boldsymbol{r}', t)}{|\boldsymbol{r} - \boldsymbol{r}'|}\,. \tag{A.3}$$

One sees from eq. (2.18) that the function λ_0 is the time-like component of a four-vector. Owing to the gauge invariance of the lagrangian, one could choose as a particular solution any other four-vector field λ' connected to λ_0 by the gauge transformation

$$\lambda' = \lambda_0 + \gamma_\mu \partial_\mu \Lambda\,,$$

where Λ is an arbitrary solution of the $M = 0$ Klein-Gordon equation and where the γ_μ are the spin-1 γ-matrices. For example, Λ can be chosen such that $\partial_\mu \lambda'_\mu = 0$ (Lorentz gauge) or $\boldsymbol{\partial} \cdot \boldsymbol{\lambda}' = 0$ (Coulomb gauge).

The complete solution in the presence of sources for the electric field is thus

$$\boldsymbol{E} = \boldsymbol{E}^{(\mathrm{tr})} + \boldsymbol{E}^{(\ell)} = \boldsymbol{E}^{(\mathrm{tr})} - \operatorname{grad} \lambda_0\,.$$

Here the $\boldsymbol{E}^{(\mathrm{tr})}$ is the general solution of the homogeneous equations, i.e. the transverse fields $\bar{\pi}_{\mathfrak{M}}$ and $\bar{\pi}_{\mathfrak{E}}$ of eqs. (5.19) and (5.20). The fields $\boldsymbol{E}^{(\mathrm{tr})}$ and $\boldsymbol{E}^{(\ell)}$ are orthogonal in the following sense

$$\int \mathrm{d}^3r\, \boldsymbol{E}^{(\mathrm{tr})} \cdot \boldsymbol{E}^{(\ell)} = -\int \mathrm{d}^3r\, \boldsymbol{E}^{(\mathrm{tr})} \cdot \nabla \lambda_0 = \int \mathrm{d}^3r \left(\nabla \cdot \boldsymbol{E}^{(\mathrm{tr})}\right) \lambda_0 = 0\,.$$

Hence

$$\int \mathrm{d}^3r\, \boldsymbol{E}^2 = \int \mathrm{d}^3r \left((\boldsymbol{E}^{(\mathrm{tr})})^2 + (\boldsymbol{E}^{(\ell)})^2 \right).$$

The hamiltonian is written in terms of the canonical fields only. Thus the field λ_0 must be expressed in terms of the canonical fields of the charge carrying particles, i.e. the charge density $e\rho$,

$$\int \mathrm{d}^3r\, \tfrac{1}{2}(\boldsymbol{E}^{(\ell)})^2 = \tfrac{1}{2}\int \mathrm{d}^3r (\nabla\lambda_0)^2 = -\tfrac{1}{2}\int \mathrm{d}^3r\, \lambda_0 \nabla^2 \lambda_0$$

$$= -2\pi e^2 \int \mathrm{d}^3r \int \mathrm{d}^3r' \frac{\rho(\boldsymbol{r}, t)\rho(\boldsymbol{r}', t)}{|\boldsymbol{r} - \boldsymbol{r}'|}\,.$$

Together with the fourth component of the interaction

$$-4\pi \int \mathrm{d}^3r\, J_4 \phi_4 = 4\pi e \int \mathrm{d}^3r\, \rho \lambda_0 = 4\pi e^2 \int \mathrm{d}^3r \int \mathrm{d}^3r' \frac{\rho(\boldsymbol{r}, t)\rho(\boldsymbol{r}', t)}{|\boldsymbol{r} - \boldsymbol{r}'|}\,,$$

the complete hamiltonian for the massless vector field interacting with charge carrying fields is,

$$H = H_0 + \int \mathrm{d}^3r\left(\tfrac{1}{2}\left((\boldsymbol{E}^{(\mathrm{tr})})^2 + \boldsymbol{B}^2\right) - 4\pi \boldsymbol{J}\cdot\boldsymbol{\phi}\right) + 2\pi e^2 \int \mathrm{d}^3r \int \mathrm{d}^3r' \, \frac{\rho(\boldsymbol{r},t)\rho(\boldsymbol{r},t')}{|\boldsymbol{r}-\boldsymbol{r}'|}\,, \tag{A.4}$$

where H_0 is the free field hamiltonian for the charged fields and $\boldsymbol{J}\cdot\boldsymbol{\phi}$ the current interaction with the transverse modes of the vector field which may be treated along the lines of Chapter 9.

A.2. THE PARTICLE-HOLE REPRESENTATION

Consider a system of fermions. Let us denote by capital letters, $A, B, C\ldots$ the single-particle states above the Fermi level and by lower case letters, $a, b, c\ldots$ those below. The reference state, denoted by $|\ \rangle$, is a "closed shell". Its bra and ket are defined such that

$$\langle\ |\ \rangle = 1\,.$$

The hole creation operators are defined by

$$c_{m_a}^{(j_a)+} = (-)^{2j} b_{m_a}^{[j_a]}, \tag{A.5}$$

where $b_{m_a}^{[j_a]}$ are the annihilation operators relative to the vacuum $|0\rangle$, as given in sect. 4.6.1. This definition corresponds to the time-reversal operation

$$c_{m_a}^{(j_a)+} = \mathcal{T}\, b_{m_a}^{(j_a)} = (-)^{j_a+m_a} b_{-m_a}^{(j_a)}\,.$$

The hole annihilation operators are given by hermitian conjugation, table 4.1,

$$c_{m_a}^{(j_a)} = (-)^{j_a-m_a} \tilde{b}_{-m_a}^{[j_a]}\,.$$

The hole operators conform to our phase convention, i.e. they fulfill all the relations of table 4.1. In particular

$$c_{m_a}^{(j_a)+} = (-)^{2j_a} \tilde{c}_{m_a}^{[j_a]}, \tag{A.6}$$

$$c_{m_a}^{(j_a)} = (-)^{j_a+m_a} c_{-m_a}^{[j_a]}\,. \tag{A.7}$$

We also note

$$\tilde{c}_{m_a}^{[j_a]} = b_{m_a}^{[j_a]}, \tag{A.8}$$

$$c_{m_a}^{[j_a]} = (-)^{2j_a} \tilde{b}_{m_a}^{[j_a]}\,. \tag{A.9}$$

We have for the hole operators, relatively to the reference state,

$$c_{m_a}^{[j_a]}|\ \rangle = 0\,,$$

and consequently the anticommutator reads as usual

$$\langle\ |[c^{[j_a]}, \tilde{c}^{[j_a]}]^{[0]}_{+}|\ \rangle = (-)^{2j_a}\hat{j}_a\delta_{ab}. \tag{A.10}$$

Likewise for particle operators, for which of course $c^{[j_A]}_{m_A} \equiv b^{[j_A]}_{m_A}$ and $\tilde{c}^{[j_A]}_{m_A} \equiv \tilde{b}^{[j_A]}_{m_A}$, relatively to the reference state,

$$b^{[j_A]}_{m_A}|\ \rangle = 0,$$

and

$$\langle\ |[b^{[j_A]}, \tilde{b}^{[j_B]}]^{[0]}_{+}|\ \rangle = (-)^{2j_A}\hat{j}_A\delta_{AB}. \tag{A.11}$$

Comparing these expressions with those of sect. 4.6 one sees that all the rules of particle operators apply also to the hole operators. Thus, for example, the particle-hole ket writes

$$|IM\rangle = (-)^{2I}\hat{I}[W^{[I]}\tilde{b}^{[j_A]}\tilde{c}^{[j_a]}]^{[0]}|\ \rangle, \tag{A.12}$$

where again use is made of the fact that $(-)^{2j_A+2j_a} = (-)^{2I}$. The bra $\langle IM|$ is obtained following the rules of sect. 4.6.2 and we have

$$\langle IM|IM\rangle = 1.$$

Owing to the relations (A.8) and (A.9), one may work with the initial operators $b^{[\,]}, \tilde{b}^{[\,]}$ even for the hole states. In that case for states below the Fermi level the anticommutator (A.10) is replaced by

$$\langle\ |[\tilde{b}^{[j_a]}, b^{[j_b]}]^{[0]}_{+}|\ \rangle = \hat{j}_a\delta_{ab}, \tag{A.13}$$

and the ket (A.12) by

$$|IM\rangle = (-)^{2I}\hat{I}[W^{[I]}\tilde{b}^{[j_A]}b^{[j_a]}]^{[0]}|\ \rangle. \tag{A.14}$$

This way one may express the interaction operators of Chapter 6 in the particle-hole representation without any modification of the formulae for the invariant matrix elements given in that chapter. For example for the pion-nucleon interaction (pion emission part) we have

$$\begin{aligned} H_{\pi N} = \sum [\pi A|H_I|B]&:[\tilde{a}^{[l]}\tilde{b}^{[A]}b^{[B]}]^{[0]}: \\ + [\pi A|H_I|b]&:[\tilde{a}^{[l]}\tilde{b}^{[A]}b^{[b]}]^{[0]}: \\ + [\pi a|H_I|B]&:[\tilde{a}^{[l]}\tilde{b}^{[a]}b^{[B]}]^{[0]}: \\ + [\pi a|H_I|b]&:[\tilde{a}^{[l]}\tilde{b}^{[a]}b^{[b]}]^{[0]}:, \end{aligned} \tag{A.15}$$

where the different terms represent the emission of a pion by scattering of a particle, creation of a particle-hole pair, annihilation of a particle-hole pair, and scattering of a hole, respectively. The evaluation of the particle-hole matrix elements now follows the rules of Chapter 7, using the operators in the form (A.15), basis state vectors of the form (A.14), and contractions in the form (A.13) for hole states and (A.11) for particle states.

A.3. THE ONE-BOSON EXCHANGE INTERACTION

As an example of the perturbation treatment, we give the interaction between two nucleons via the exchange of a single meson, i.e. the second order term in the meson-nucleon coupling. The two nucleons are coupled to I, T. Their quantum numbers, ν, ℓ, j, t are denoted N′, M′ in the initial state and N, M in the final state. The summation is over all the different mesons and their quantum numbers, denoted by B, i.e. ν, J, τ, and, for vector mesons, in addition κ. The matrix elements of the propagator are diagonal in all quantum numbers except for the radial quantum numbers. This is indicated explicitly by using $\mathrm{B}, \bar{\mathrm{B}}, \mathrm{N}, \bar{\mathrm{N}}$ etc..., i.e., the barred and unbarred symbols denote the same quantum numbers except for the radial ones. Owing to time-ordering and exchange there are four terms:

$$\langle(\mathrm{M'N'})IT|H_I\frac{1}{E-H_0}H_I|(\mathrm{MN})IT\rangle$$

$$=\frac{1}{\sqrt{1+\delta_{\mathrm{MN}}}}\frac{1}{\sqrt{1+\delta_{\mathrm{M'N'}}}}\frac{1}{\hat{I}\hat{T}}\sum\frac{1}{\hat{J}\hat{\tau}}(-)^{J+\tau}\begin{bmatrix} t & t & \tau \\ t & t & \tau \\ T & T & 0 \end{bmatrix}\left(\begin{bmatrix} j_{\mathrm{M'}} & j_{\mathrm{M}} & J \\ j_{\mathrm{N'}} & j_{\mathrm{N}} & J \\ I & I & 0 \end{bmatrix}\right.$$

$$\times\left((-)^{j_{\mathrm{M'}}-j_{\mathrm{M}}}[\mathrm{M'}|H_I|\bar{\mathrm{B}}\bar{\mathrm{M}}]\left[\bar{\mathrm{B}}\bar{\mathrm{M}}\mathrm{N'}\left|\frac{1}{E-H_0}\right|\mathrm{BMN'}\right][\mathrm{BN'}|H_I|\mathrm{N}]\right.$$

$$\left.+(-)^{j_{\mathrm{N'}}-j_{\mathrm{N}}}[\mathrm{N'}|H_I|\bar{\mathrm{B}}\bar{\mathrm{N}}]\left[\bar{\mathrm{B}}\bar{\mathrm{N}}\mathrm{M'}\left|\frac{1}{E-H_0}\right|\mathrm{BNM'}\right][\mathrm{BM'}|H_I|\mathrm{M}]\right)$$

$$-(-)^{I+T}\begin{bmatrix} j_{\mathrm{N'}} & j_M & J \\ j_{\mathrm{M'}} & j_{\mathrm{N}} & J \\ I & I & 0 \end{bmatrix}\left((-)^{j_{\mathrm{N}}-j_{\mathrm{N'}}}[\mathrm{M'}|H_I|\bar{\mathrm{B}}\bar{\mathrm{N}}]\left[\bar{\mathrm{B}}\bar{\mathrm{N}}\mathrm{N'}\left|\frac{1}{E-H_0}\right|\mathrm{BNN'}\right]\right.$$

$$\left.\left.\times[\mathrm{BN'}|H_I|\mathrm{M}]+(-)^{j_{\mathrm{M}}-j_{\mathrm{M'}}}[\mathrm{N'}|H_I|\bar{\mathrm{B}}\bar{\mathrm{M}}]\left[\bar{\mathrm{B}}\bar{\mathrm{M}}\mathrm{M'}\left|\frac{1}{E-H_0}\right|\mathrm{BMM'}\right][\mathrm{BM'}|H_I|\mathrm{N}]\right)\right). \tag{A.16}$$

The matrix elements of the interaction operators H_I are given in Chapter 6. For the propagator the inversion of the invariant matrix

$$[\overline{\mathrm{B}}\overline{\mathrm{M}}\overline{\mathrm{N}}|E-H_0|\mathrm{BMN}]=\hat{I}\hat{T}\left(E-E^{\mathrm{B}}_{\bar{\nu}\nu}-E^{\mathrm{M}}_{\bar{\nu}\nu}-E^{\mathrm{N}}_{\bar{\nu}\nu}\right)$$

is required. Here the $E^{i}_{\bar{\nu}\nu}$ refer to the matrix elements of the free fields given in sect. 6.2.

Table index

Symbol index